HIPPOCRATES

VOLUME III

EDITED AND TRANSLATED BY

PAUL POTTER

HARVARD UNIVERSITY PRESS
CAMBRIDGE, MASSACHUSETTS
LONDON, ENGLAND
2025

First published 2025

Library of Congress Control Number 2025000795
CIP data available from the Library of Congress

ISBN 978-0-674-99768-4

*Composed in ZephGreek and ZephText by
Technologies 'N Typography, Merrimac, Massachusetts.
Printed on acid-free paper and bound by
Maple Press, York, Pennsylvania*

LOEB CLASSICAL LIBRARY
FOUNDED BY JAMES LOEB 1911

EDITED BY
JEFFREY HENDERSON

HIPPOCRATES
III

LCL 149

CONTENTS

LIST OF HIPPOCRATIC WORKS SHOWING THEIR DIVISION INTO VOLUMES IN THIS EDITION

Volume VI (LCL 473)

Diseases III. Internal Affections. Regimen in Acute Diseases (Appendix).

Volume VII (LCL 477)

Epidemics II. Epidemics IV. Epidemics V. Epidemics VI. Epidemics VII.

Volume VIII (LCL 482)

Places in Man. Glands. Fleshes. Prorrhetic I. Prorrhetic II. Physician. Use of Liquids. Ulcers. Hemorrhoids. Fistulas.

Volume IX (LCL 509)

Anatomy. Nature of Bones. Heart. Eight Months' Child. Coan Prenotions. Crises. Critical Days. Superfetation. Girls. Excision of the Fetus. Sight.

Volume X (LCL 520)

Generation. Nature of the Child. Diseases IV. Nature of Women. Barrenness.

Volume XI (LCL 538)

Diseases of Women I. Diseases of Women II.

GENERAL INTRODUCTION

E. T. Withington found it necessary to base the third Loeb *Hippocrates* volume (1928), which the present volume replaces, on the Greek text of J. E. Petrequin's *Chirurgie d' Hippocrate* (2 vols.; Paris, 1877–1878),[1] since he had only indirect access to volume 2 of H. Kühlewein's *Hippocratis Opera . . . omnia* (Leipzig, 1902), the edition which for the first time reported the readings of all three independent manuscripts, B, M, and V. In the intervening ninety-seven years, all of these texts have benefited from further study,[2] and although two of them, *Joints* and *Instruments of Reduction*, have not been edited in Greek since Kühlewein's edition, the other three have appeared in new critical editions based on a wider access to the indirect transmission and supplied with extensive philological and medical commentaries:

Hanson, Maury. *Hippocrates. On Head Wounds*. CMG I 4,1. Berlin, 1999.

Jouanna, J. et al. *Hippocrate. Sur les fractures*. Budé VII (1). Paris, 2022.

Jouanna, J. et al. *Hippocrate. L' Officine du médecin*. Budé VII (3). Paris, 2024.

1 See Withington, pp. vi and xxvi.

2 See the introductions to the individual treatises below for more details.

The five Hippocratic writings in this volume are all devoted to the branch of the medical art (τέχνη ἰατρική) that applies the hand (χειρουργίη = surgery) to treat disorders of the bones and their articulations.[3] In the first treatise, *Wounds in the Head*, four kinds of traumatic injuries to the skull are handled: fractures, contusions, depressions, and the *hedra* (= an imprint on the skull's surface). After an anatomical introduction to the skull and its sutures, various injuries are described and differentiated into specific categories, and many methods useful in diagnosis, prognosis, and management are expounded; bandaging and trephination play central roles in the therapeutic instruction.

The second work, *In the Surgery*, appears to be a collection of notes rather than a systematic treatise, describing the surgeon's office and its equipment, medical personnel, and several modes of treatment, including bandaging and massage. Various more specific statements pertaining to surgical practice make up a considerable portion of the work.

Fractures and *Joints*, which may have once been part of a single comprehensive treatise on disorders of the bones and joints, appear to have been excerpted and revised by the author of *Instruments of Reduction*, but how the two series of chapters *Joints* 17–29 and 81–87 are related to

[3] For surgical practices in other Hippocratic texts, see, e.g., *Hemorrhoids*, *Fistulas*, *Sight*, and *Excision of the Fetus*, as well as procedures described in textbooks of internal medicine, such as the removal of nasal polyps (*Diseases II* 33–37) and paracentesis (*Affections* 22), and in textbooks of gynecology, such as dilation of the os uteri (*Diseases of Women II* 24).

Instruments of Reduction 7–19 and 26–31 has been the subject of continuing scholarly discussion. In spite of the technical nature of their contents, *Fractures* and *Joints* provide evidence that their author was well versed in contemporary rhetorical practices,[4] and his frequent first-person appearance in the narrative give the texts both immediacy and focus.

Instruments of Reduction is a systematic account, arranged according to the anatomical principle of "from head to toe," of the bones, their injuries, and applicable treatments. The main topics are osteology, disorders of the head, disorders of the upper limb, disorders of the lower limb, abnormal spinal curvatures, and special topics in surgical practice.

These treatises are integral parts of the Hippocratic Collection and differ in no fundamental way from the rest of the writings in their scientific, professional, educational, or linguistic features; ἰατρός (literally, "healer") refers here as elsewhere equally to both the physician and the surgeon. Shared by practitioners in both fields are, for example, their forms of practice, as an itinerant craftsman or the resident of a populous town, their repeated reference to the personal qualities and habits necessary for clinical success, and their strict devotion to truth and honesty. And as much as the vocabulary and theories of the wide range of Hippocratic texts vary according to specific details of their scientific and practical requirements, their linguistic elements and patterns are indistinguishable.

These books received a great deal of attention and ad-

[4] Cf. Withington, pp. 85–89, for a detailed account of these relationships.

miration all through antiquity, and in the Byzantine age they were all included from the beginning in the two surviving collective Hippocratic manuscripts M and V. The illustrated surgical manuscript B, which also contains many texts by later Greek surgical writers, once contained all five of the treatises in the present volume.[5] Since the Renaissance, much scholarly and medical attention has been devoted to the texts, including the studies of Vidius (1544), Cocchi (1754), Petrequin (1877–1878), Gurlt (1898), and Majno (1975).

HISTORY AND CONSTITUTION OF THE TEXT

B	Laurentianus Graecus 74,7 (X c.)
M	Marcianus Venetus Graecus 269 (X c.)
V	Vaticanus Graecus 276 (XI/XII c.)
I	Parisinus Graecus 2140 (XII/XIII c.)
H	Parisinus Graecus 2142
Ha	older part (XII/XIII c.)
Hb	newer part (XIV c.)
Recentiores	approximately twenty manuscripts (XV/XVI c.), including:
N	Parisinus Graecus 2248 (XVI c.)
Holk	Holkhamensis Graecus 92 (XVI c.)

The *stemma codicum* appearing as Figure 1 provides an overview of the interdependencies among the manuscripts containing the five treatises in this volume. The particular

[5] See Jouanna *Frac.*, pp. lx–lxiv.

Figure 1. *Stemma Codicum*

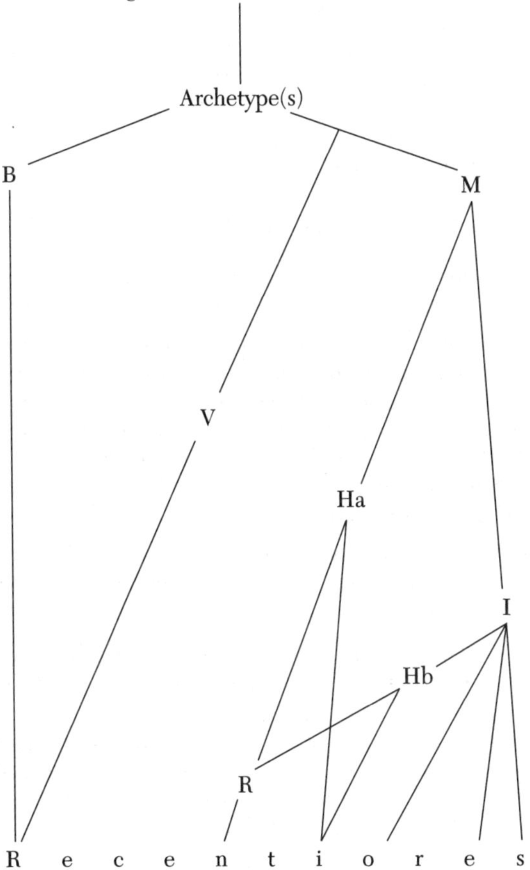

treatises are transmitted in the following independent witnesses:

Wounds in the Head	B V
In the Surgery	B M V
Fractures	B M V
Joints	B M V
Instruments of Reduction	M

NOTE ON TECHNICAL TERMS[6]

The following expressions require clarification as they are employed in this volume with particular significances.

(κατα-)τείνειν, κατάτασις/to extend, extension: either a) the natural opening of a folded joint such as the elbow or knee (the opposite of flexion); or b) the application of external force to pull two dislocated bones or two parts of a fractured bone apart from one another in order to return them to their normal relationship, as the first step of restoring them to health.

νάρθηξ/splint: a piece of wood or other material bound to a limb to hold its fractured or dislocated parts in place while they heal.

σωλήν/hollow splint, gutter, channel, "anciently applied to boxes or troughs in which a fractured limb, after being

[6] For the explanation of other technical terms, see also Loeb *Hippocrates* vol. 1 (LCL 147), pp. lxxii–lxxv; vol. 2 (LCL 148), pp. xii–xv; vol. 8 (LCL 482), pp. 10f.; vol. 10 (LCL 520), pp. x–xiii; and vol. 11 (LCL 538), pp. xvi f.

bound in splints was placed" (*Sydenham Lexicon*, s.v. Canales).

σκέλος/the whole lower limb, leg; πούς/the foot; κνήμη/the leg between the knee and the ankle, the calf; ἀντικνήμιον/the part of the leg in front of the κνήμη, the shin; μηρός/the thigh or thigh bone.

χείρ: this noun can signify either the whole upper limb or the hand alone; sometimes ἄκρη χείρ is used to distinguish the latter; πῆχυς/the forearm from the wrist to the elbow; βραχίων/the arm from the shoulder to the elbow, the upper arm.

This volume has benefited in many passages from corrections and suggestions made by my spouse, Beate Gundert. I am indebted to Dr. Vivian McAlister for his help in understanding surgical niceties on several occasions.

REFERENCES[7]

EDITIONS, TRANSLATIONS, AND COMMENTARIES

Hippocratic Collection

Early Works

Calvus	*Hippocratis Coi . . . octoginta volumina . . . per M. Fabium Calvum, Rhavennatem . . . latinitate donata* . . . Rome, 1525.
Aldina	*Omnia opera Hippocratis . . . in aedibus Aldi & Andreae Asulani soceri*. Venice, 1526.
Froben	*Hippocratis Coi . . . libri omnes . . .* [*per Ianum Cornarium*]. Basel, 1538.
Vidius	Vidius, V. *Chirurgia e graeco in latinum conversa.* Paris, 1544.
Cornarius	*Hippocratis Coi . . . opera . . . per Ianum Cornarium . . . Latina lingua conscripta.* Venice, 1546.
Zwinger	*Hippocratis Coi . . . viginti duo commentarii . . . Theod. Zvingeri studio & conatu.* Basel, 1579.

[7] Supplementary bibliographical information is given in the introductions to the individual treatises.

Foes	*Magni Hippocratis . . . opera quae extant . . . Latina interpretatione & annotationibus illustrata, Anutio Foesio . . . authore.* Geneva, 1657–1662.
Linden	*Magni Hippocratis Coi opera omnia graece & latine edita . . . industria & diligentia Joan. A. Vander Linden.* Leiden, 1665.
Mack	*Hippocratis opera omnia . . . studio et opera Stephani Mackii.* Vienna, 1743–1749.
Cocchi	Cocchi, A. *Graecorum chirurgici libri.* Florence, 1754.
Grimm	Grimm, J. F. K. *Hippokrates Werke aus dem griechischen. . . .* Altenburg, 1781–1792.

Post-Eighteenth Century

Adams	Adams, F. *The Genuine Works of Hippocrates.* London, 1849.
Daremberg	Daremberg, Ch. *Oeuvres choisies d' Hippocrate*[2]. Paris, 1855.
Diller *Schr.*	Diller, H. *Hippokrates. Schriften . . . übersetzt.* Hamburg, 1962.
Ermerins	Ermerins, F. Z. *Hippocratis . . . reliquiae.* Utrecht, 1859–1864.
Fuchs	Fuchs, R. *Hippokrates, sämmtliche Werke. Ins Deutsche übersetzt . . .* Munich, 1895–1900.
Gomperz	Gomperz, Th. *Die Apologie der Heilkunst*[2]. Leipzig, 1910.
Hanson	Hanson, Maury. *Hippocrates. On Head Wounds.* CMG I 4,1. Berlin, 1999.

Heiberg	Heiberg, J. *Hippocratis Opera*. CMG I 1. Leipzig and Berlin, 1927.
Jones *Loeb*	Jones, W. H. S. *Hippocrates with an English Translation.* London and New York, 1923–1931.
Jouanna *Frac.*	Jouanna, J. *Hippocrate. Sur les fractures*. Budé VII (1). Paris, 2022.
Jouanna *Off.*	———. Hippocrate. *Dans l'officine du médecin*. Budé VII (3). Paris, 2024.
Kühlewein	Kühlewein, H. *Hippocratis Opera . . . omnia*. Leipzig, 1894–1902.
Littré	Littré, E. *Oeuvres complètes d'Hippocrate.* Paris, 1839–1861.
Petrequin	Petrequin, J. E. *Chirurgie d' Hippocrate*. Paris, 1877–1878.
Reinhold	Reinhold, C. H. Th. *Ἱπποκράτης/Ψευδωνύμως Ἱπποκράτεια*. Athens, 1864–1867.
Withington	Withington, E. T. *Hippocrates with an English Translation.* London and New York, 1928.

Other Authors

Apoll. Cit	Kollesch, J., F. Kudlien, and D. Nickel. *Apollonios von Kition. Komementar zu Hippokrates über das Einrenken der Gelenke*. CMG X 1,1. Berlin, 1965.
Boudon	Boudon-Millot, V. *Galien . . . Sur ses propres livres*. Budé I. Paris, 2007.
Erotian	Nachmanson, E. *Erotiani Vocum Hippocraticarum collectio.* Gothenburg, 1918.

REFERENCES

Galen	Kühn, C. G. *Claudii Galeni opera omnia* . . . Leipzig, 1825–1833.
Herophilus	von Staden, H. *Herophilus. The Art of Medicine in Early Alexandria*. Cambridge, 1989.
Irmer	Irmer, D. *Palladius. Kommentar zu* Hippokrates *'De fracturis' und seine Parallelversion under dem Namen des* Stephanus von Alexandria. Hamburg, 1977.
Knutzen	Knutzen, G. H. *Technologie in den hippokratischen Schriften περὶ διαίτης ὀξέων, περὶ ἀγμῶν, περὶ ἄρθρων ἐμβολῆς*. Wiesbaden, 1964.
Kudlien	Kudlien, F. *Die handschriftliche Überlieferung des Galenkommentars zu Hippokrates De articulis*. Berlin, 1960.
Perilli	Perilli, L. *Galeni Vocum Hippocratis Glossarium.* CMG V 13,1. Berlin, 2017.
Pietrobelli	Pietrobelli, A. *Galien. Commentaire au Régime des madadies aiguës d' Hippocrate.* Budé IX (1). Paris, 2019.
Schöne	Schöne, H. *Apollonius von Kitium. Illustrierter Kommentar zu der hippokratischen Schrift* ΠΕΡΙ ΑΡΘΡΩΝ. Leipzig, 1896.
Schleiermach.	Schleiermacher, W. *Die Komposition der hippokratischen Schrift* ΠΕΡΙ ΑΓΜΩΝ-ΠΕΡΙ ΑΡΘΡΩΝ ΕΜΒΟΛΗΣ. Diss., Freiburg, 1929.
Vogt	Vogt, S. *De Galeni in libellum κατ᾽ ἰητρεῖον commentariis*. Diss., Marburg, 1910.

REFERENCES

GENERAL WORKS

Andorlini	Andorlini, I., ed. *Greek Medical Papyri I/II*. Florence, 2001/2009.
Budé	Collection des universités de France publiée sous le patronage de l'Association Guillaume Budé.
CMG	Corpus Medicorum Graecorum.
Dönt	Dönt, H. *Die Terminologie von Geschwür, Geschwulst und Anschwellung im Corpus Hippocraticum*. Diss., Vienna, 1968.
Ducatillon	Ducatillon, J. *Polémiques dans la Collection Hippocratique*. Paris, 1977.
Durling	Durling, R. J. *A Catalogue of Sixteenth Century Printed Books in the National Library of Medicine*. Bethesda, 1967.
Eijk	Eijk, Ph. J. van der. *Medicine and Philosophy in Classical Antiquity*. Cambridge, 2005.
Gurlt	Gurlt, E. *Geschichte der Chirurgie*. Berlin, 1898.
Index Hipp.	Kühn, J.-H., U. Fleischer et al. *Index Hippocraticus*. Göttingen, 1986–1989; Anastassiou, A., and D. Irmer, *Supplement*. 1999.
Jouanna	Jouanna, J. *Hippocrates*. Translated by M. B. DeBevoise. Baltimore, 1999.
Kibre	Kibre, P. *Hippocrates Latinus*. Rev. ed. New York, 1985.
LSJ	Liddell, H. G., R. Scott, and H. S. Jones, eds. *A Greek-English Lexicon*[9]. Oxford, 1940.

Majno	Majno, G. *The Healing Hand. Man and Wound in the Ancient World*. Cambridge MA, 1975.
Manetti/Ros.	Manetti, D., and A. Roselli. "Galeno commentatore di Ippocrate." In *Aufstieg und Niedergang der römischen Welt* (*ANRW*) II 37,2, edited by W. Haase and H. Temporini, pp. 1529–1635. Berlin, 1994.
Nachmanson	Nachmanson, E. *Erotianstudien*. Uppsala, 1917.
Pormann	Pormann, P., ed. *Hippocratic Commentaries in the Greek, Latin, Syriac and Arabic Commentaries Traditions*. Leiden, 2021.
Roselli	Roselli, A. "Galen's Surgical Commentaries on Hippocrates." In Pormann.
Sezgin	Sezgin, F. *Geschichte des arabischen Schriftums. Band III. Medizin—Pharmazie—Zoologie—Tierheilkunde bis ca 430 H*. Leiden, 1970.
Sydenham L.	Power, H., and L. W. Sedgwick, eds. *The New Sydenham Society's Lexicon of Medicine and the Allied Sciences*. London, 1881–1899.
Testimonien	Anastassiou, A., and D. Irmer. *Testimonien zum Corpus Hippocraticum*. Göttingen, (vol. I) 2006; (vol. II,1) 1997; (vol. II,2) 2001; (vol. III) 2012.
Ullmann	Ullmann, M. *Die Medizin im Islam*. Leiden, 1970.

GENERAL BIBLIOGRAPHY

EDITIONS, TRANSLATIONS, AND COMMENTARIES

Bourbon, F. *Hippocrate. Femmes stériles, Maladies des jeunes filles, Superfétation, Excision du fœtus.* Budé XII (4). Paris, 2017.

Ecca, G. *Die hippokratische Schrift Praecepta.* Wiesbaden, 2016.

García Gual, C. et al. *Tratados hipocráticos . . . introducciones, traducciones y notas.* Madrid, 1983–2003.

Giorgianni, F. *Hippokrates. Über die Natur des Kindes* (De genitura *und* De natura pueri). Wiesbaden, 2006.

Jouanna, J. *Hippocrate. Maladies II.* Budé X (2). Paris, 1983.

———. *Hippocrate. Des Vents. De l'Art.* Budé V (1). Paris, 1988.

———. *Hippocrate. De la maladie sacrée.* Budé II (3). Paris, 2003.

———. *Hippocrate. Pronostic.* Budé III (1). Paris, 2013.

Overwien, O. *Hippokrates. Über die Säfte.* CMG I 3,1. Berlin, 2014.

Pérez Cañizares, P. *Hippocrates. On Affections.* CMG I 2,6. Berlin, 2024.

Roselli, A. *Ippocrate. La malattia sacra.* Venice, 1996.

GENERAL BIBLIOGRAPHY

GENERAL WORKS

Bliquez, L. *The Tools of Asclepius.* Leiden, 2014.

Bruno Celli, B. *Bibliografia Hipocrática.* Caracas, 1984.

Byl, S. "Les dix dernières années (1982–1992) de la recherche hippocratique." In *Université Jean Monnet-Saint Etienne. Centre Jean Palerne. Lettre d'Informations* 22 (1993): 1–39.

Craik, E. M. *The 'Hippocratic' Corpus: Content and Context.* London, 2015.

Dean-Jones, L. A., and R. M. Rosen, eds. *Ancient Concepts of the Hippocratic. Papers Presented at the XIIIth International Hippocratic Colloquium, Austin, Texas 2008.* Leiden, 2015.

Eijk, Ph. J. van der, ed. *Hippocrates in Context. Papers read at the XIth International Hippocratic Colloquium (Newcastle-upon-Tyne 2002).* Leiden, 2005.

Eijk, Ph. J. van der, H. F. Horstmanshoff, and P. H. Schrijvers, eds. *Ancient Medicine in its Socio-cultural Context.* Amsterdam, 1995.

Fichtner, G. *Corpus Hippocraticum. Verzeichnis der hippokratischen und pseudohippokratischen Schriften.* Berlin, 2015.

Flashar, H. *Hippokrates. Meister der Heilkunst.* Munich, 2016.

Harris, W. V., ed. *Mental Disorders in the Classical World.* Leiden, 2013.

Horstmanshoff, H. F., ed. *Hippocrates and Medical Education: Selected Papers Read at the XII International Hippocrates Colloquium.* Leiden, 2010.

Jouanna, J. *Greek Medicine from Hippocrates to Galen.* Leiden, 2012.

Leven, K.-H., ed. *Antike Medizin. Ein Lexikon.* Munich, 2005.

Maloney, G. *Index Inverses du vocabulaire Hippocratique.* Québec, 1987.

Maloney, G., and W. Frohn. *Concordance des oeuvres Hippocratiques*. Québec, 1984, and Hildesheim, 1986.

Maloney, G., and R. Savoie. *Cinq cents ans de bibliographie Hippocratique*. Québec, 1982.

Nutton, V. *Ancient Medicine.* London, 2004.

Oser-Grote, C. M. *Aristoteles und das Corpus Hippocraticum. Die Anatomie und Physiologie des Menschen.* Stuttgart, 2004.

Perilli, L., Ch. Brockmann, K.-D. Fischer, and A. Roselli, eds. *Officina Hippocratica*. Berlin, 2011.

Pormann, P., ed. *Cambridge Companion to Hippocrates*. Cambridge, 2018.

Thivel, A., and A. Zucker, eds. *Le normal et le pathologique dans la Collection hippocratique. Actes du Xème colloque international hippocratique (1999).* Nice, 2001.

WOUNDS IN THE HEAD

INTRODUCTION

Erotian includes eight words from *Wounds in the Head* in his glossary, for one of which he refers to Bacchius of Tanagra's (ca. 275–200 BC) opinion on its spelling, which places the treatise among the earliest group of medical writings that came later to form the Hippocratic Collection.[1] In another article, Erotian cites a passage from chapter 11 of the work, naming it with its present title.[2]

Galen, too, is familiar with the writing. He includes five terms from the text in his Hippocratic *Glossary*, as well as including two other glosses that he refers to a now lost *Appendix to Wounds in the Head*.[3] In his *Commentary on Hippocrates'* In the Surgery, he also discusses passages from the treatise's chapters 4 to 7,[4] and in his *On My Own Books* he refers to a commentary he wrote on *Wounds in the Head*, which is now lost in Greek.[5]

[1] Cf. Nachmanson, pp. 361–63; Erotian B8, p. 28; Loeb *Hippocrates* vol. 1 (LCL 147), pp. xlvii–xlix.

[2] Cf. Nachmanson, p. 344; Erotian E30, p. 37.

[3] Cf. Perilli: α158, *ἀρτίστομα* (ch. 11); μ16, *μέλανι φαρμάκῳ* (ch. 14); σ47, *στεῖλαι* (ch. 14); *ὑπόπνον* (ch. 15); χ3, *χαράκτῳ* (ch. 21).

[4] Cf. Hanson, pp. 37f.

[5] Cf. Boudon, p. 160, n. 6.

Otherwise, the treatise receives only infrequent mention in other ancient and medieval writers.[6]

Wounds in the Head deals mainly with injuries of the skull vault, only occasionally referring to the brain or blood vessels of the head, or to symptoms resulting from their damage. The author distinguishes four basic injuries of the skull:

1. fracture (*ῥηγνύναι/ῥωγμή*): a crack-like break in the skull, possibly with fragmentation or detachment of bone pieces
2. contuse/contusion (*φλᾶν/φλάσις*): a local crushing of bone tissue without a fracture or separation of bone from the skull
3. depress/depression inward (*ἐσφλᾶν/ἔσφλασις ἔσω*): an inward displacement of a detached fragment of skull bone below the surface of the skull
4. *hedra* (*ἕδρη*) = impression/imprint left in the sutures on the outer surface of the skull by a weapon striking it, and possibly remaining fixed there

The chapters of the treatise are organized as follows:

1 Morphology of the skull: the sutures and layers of the bone.
2 Thickness and strength of the skull in different parts of the head.
3 The *hedra* and its significance in various locations.

[6] See *Testimonien*, vol. I, pp. 452–55; vol. II,1, pp. 452–55; vol. II,2, pp. 354f.; vol. III, pp. 385f.; Sezgin, p. 45; Ullmann, p. 32; Kibre, p. 234.

4–8 Seven basic types (τρόποι) of head wounds, their varieties, and their clinical features: ch. 4, fracture and contusion (type 1); ch. 5, contusion (type 2); ch. 6, depression (type 3); ch. 7, *hedra* with fracture and contusion (type 4), *hedra* with contusion (type 5), *hedra* alone (type 6); ch. 8, injury elsewhere than at the site of the injury (type 7).

9 Indications for trephining.

10–12 Clinical evaluation and examination of the wound (ch. 10); history of the injury (ch. 11); involvement of the sutures (ch. 12).

13–20 Treatment: use of lint dressings, poultices, and bandages; extension of a scalp wound by incising; use of black dye and rasping; incipient suppuration and need to keep the wound clean and dry; separation of necrotic bone fragments; growth of granulation material; the skulls of children; prognosis of head wounds; facial edema.

21 Trephination: timing, technique, management; use of the perforator.

The treatise shares a number of practical and terminological similarities with other works of the Collection, which suggests a common medical milieu. The closest of these are with *Ulcers* 1, 3, 5, 6, 8, and 11, *Epidemics* V 27 and 28, and *Use of Liquids* 1 and 5.[7]

[7] Cf. Hanson, pp. 53–55.

Beginning in the Renaissance with publication of the text in Latin and Greek, and in particular with *Chirurgia e Graeco in Latinum conversa, V. Vidio Florentino interprete, cum nonnullis eiusdem Vidii commentariis*, Paris, 1544 (pp. 61–125), *Wounds in the Head* became a standard work of the surgical literature, appearing in many editions and translations, often with commentaries.[8] Besides finding a place in all the collected Hippocratic editions and translations, including F. Adams' *Genuine Works of Hippocrates* (vol. 1, pp. 421–66 and plates 1–3), and J. E. Petrequin's *Chirurgie d' Hippocrate* (vol. 1, pp. 411–558), it was more recently the subject of the following monographic study:

Hanson, Maury. *Hippocrates. On Head Wounds*. CMG I 4,1. Berlin, 1999.

The present edition is heavily indebted to Hanson's work, as well as to E. T. Withington's Loeb *Hippocrates* vol. 3 (1928).

[8] See Littré, vol. 3, pp. 178–80; Petrequin, vol. 1, pp. 432f.

ΠΕΡΙ ΤΩΝ ΕΝ ΚΕΦΑΛΗΙ ΤΡΩΜΑΤΩΝ

III 182 Littré 1. Τῶν ἀνθρώπων αἱ κεφαλαὶ οὐδὲν ὁμοίως σφίσιν
αὐταῖς, οὐδὲ αἱ ῥαφαὶ τῆς κεφαλῆς πάντων κατὰ
ταὐτὰ πεφύκασιν. ἀλλ᾽ ὅστις μὲν ἔχει ἐκ τοῦ ἔμ-
προσθεν τῆς κεφαλῆς προβολὴν—ἡ δὲ προβολή ἐστι
τὸ τοῦ ὀστέου ἔξεχον στρογγύλον παρὰ τὸ ἄλλο—
τούτου εἰσὶν αἱ ῥαφαὶ πεφυκυῖαι ἐν τῇ κεφαλῇ ὥσπερ
γράμμα τὸ ταῦ γράφεται, τὴν μὲν γὰρ βραχυτέρην
γραμμὴν ἔχει πρὸ τῆς προβολῆς ἐπικαρσίην πεφυ-
κυῖαν· τὴν δὲ ἑτέρην μακροτέρην[1] γραμμὴν ἔχει διὰ
μέσης τῆς κεφαλῆς κατὰ μῆκος πεφυκυῖαν ἐς τὸν
τράχηλον αἰεί. ὅστις δ᾽ ὄπισθεν τῆς κεφαλῆς τὴν
προβολὴν ἔχει, αἱ ῥαφαὶ τούτῳ πεφύκασι τὰ ἐναντία
ἢ τῷ προτέρῳ· ἡ μὲν γὰρ βραχυτέρη γραμμὴ πρὸ
τῆς προβολῆς πέφυκεν ἐπικαρσίη· ἡ δὲ μακροτέρη
γραμμὴ διὰ μέσης τῆς κεφαλῆς πέφυκε κατὰ μῆκος
ἐς τὸ μέτωπον αἰεί. ὅστις δὲ καὶ ἀμφοτέρωθεν τῆς
184 κεφαλῆς προβολὴν ἔχει, ἔκ τε τοῦ ἔμπροσθεν καὶ | ἐκ
τοῦ ὄπισθεν, τούτῳ αἱ ῥαφαί εἰσιν ὁμοίως πεφυκυῖαι
ὡς γράμμα τὸ ἦτα γράφεται· πέφυκασι δὲ τῶν γραμ-
μέων αἱ μὲν μακραὶ πρὸ τῆς προβολῆς ἑκατέρης ἐπι-

WOUNDS IN THE HEAD

1. Human heads are not at all like one another, nor are the sutures of all heads arranged in the same way. For example, in a person who has a protuberance at the front of the head—a protuberance is a rounded projection of the bone extending beyond the rest—the sutures in the head will be arranged the way the letter tau is written (T), with the shorter stroke set transversely in front of the protuberance, and the other longer stroke running through the middle of the head lengthwise right to the neck. But in a person with a protuberance at the back of the head, the sutures will be turned the opposite way from this: the shorter stroke in front of the (sc. posterior) protuberance will run transversely, while the longer stroke will run through the middle of the head lengthwise right to the forehead. Then, in a person who has two protuberances in the head, one at the front and one at the back, the sutures will be arranged the way the letter eta is written (H), with the long strokes of the letter crosswise in front of each of

---

[1] μάκ. B: om. V

κάρσιαι πεφυκυῖαι· ἡ δὲ βραχείη διὰ μέσης τῆς κεφαλῆς κατὰ μῆκος πρὸς ἑκατέρην τελευτῶσα τῇσι μακρῇσι γραμμῇσι. ὅστις δὲ μηδὲ ἑτέρωθι μηδεμίην προβολὴν ἔχει, οὗτος ἔχει τὰς ῥαφὰς τῆς κεφαλῆς ὡς γράμμα τὸ χεῖ γράφεται· πεφύκασι δὲ αἱ γραμμαὶ ἡ μὲν ἑτέρη ἐπικαρσίη πρὸς τὸν κρόταφον ἀφήκουσα· ἡ δὲ ἑτέρη κατὰ μῆκος διὰ μέσης τῆς κεφαλῆς.

Διπλόον δ' ἐστὶ τὸ ὀστέον κατὰ μέσην τὴν κεφαλήν· σκληρότατον δὲ καὶ πυκνότατον αὐτοῦ πέφυκεν τό τε ἀνώτατον ἣ[2] *ὁμόχροια τοῦ ὀστέου ἡ ὑπὸ τῇ*
186 *σαρκὶ | καὶ τὸ κατώτατον τὸ πρὸς τῇ μήνιγγι ἣ*[3] *ὁμόχροια τοῦ ὀστέου ἡ κάτω. ἀποχωρέον δὲ ἀπὸ τοῦ ἀνωτάτου ὀστέου καὶ τοῦ κατωτάτου, ἀπὸ τῶν σκληροτάτων καὶ πυκνοτάτων ἐπὶ τὸ μαλθακώτερον καὶ ἧσσον πυκνὸν καὶ ἐπικοιλότερον ἐς τὴν διπλόην αἰεί.*
188 *ἡ δὲ διπλόη κοιλότατον καὶ μαλθακώτατον καὶ | μάλιστα σηραγγῶδές ἐστιν. ἔστι δὲ καὶ πᾶν τὸ ὀστέον τῆς κεφαλῆς, πλὴν κάρτα ὀλίγου τοῦ τε ἀνωτάτου καὶ τοῦ κατωτάτου σπόγγῳ ὅμοιον· καὶ ἔχει τὸ ὀστέον ἐν ἑωυτῷ ὅμοια σαρκία πολλὰ καὶ ὑγρά, καὶ εἴ τις αὐτὰ διατρίβοι τοῖσι δακτύλοισιν αἷμα ἂν διαγίνοιτο ἐξ αὐτῶν· ἔνεστι δ' ἐν τῷ ὀστέῳ καὶ φλέβια λεπτότερα καὶ κοιλότερα αἵματος πλέα. σκληρότητος μέν νυν καὶ μαλθακότητος καὶ κοιλότητος ὧδε ἔχει.*

2. *Παχύτητι δὲ καὶ λεπτότητι ξυμπάσης τῆς κεφαλῆς τὸ ὀστέον λεπτότατόν ἐστι καὶ ἀσθενέστατον τὸ κατὰ βρέγμα, καὶ σάρκα ὀλιγίστην καὶ λεπτοτάτην ἔχει ἐπὶ ἑωυτῷ ταύτῃ τῆς κεφαλῆς τὸ ὀστέον,*

the protuberances, and the short stroke running lengthwise through the middle of the head and ending where it meets each long stroke. Finally, in a person with no protuberance at either end of the head, the sutures will be arranged the way the letter chi is written (X): one of the two strokes will run crosswise to the temple (sc. on each side), with the other stroke lengthwise through the middle of the head.

The bone present in the central part of the head is double: hardest and densest are its smooth surface just under the skin at the very top, and its smooth surface lowest down next to the membrane (i.e., the dura mater). Moving away from the upper and the lower extremes of the bone into the diploe, there is a change from what is hardest and densest to what is softer, less dense, and hollower; the diploe is very hollow and soft, and particularly porous. In fact, all the bone of the head, except for its very thin (sc. tables) on the top and bottom, is like a sponge: this part is uniformly fleshy and moist, and if someone were to press it with the fingers, it would give off blood; within the bone there are fine, hollow little vessels full of blood. These are the respective areas of hardness, softness and hollowness.

2. As far as thickness and thinness are concerned, the thinnest and weakest bone of the whole head is to be found at the bregma, where it is covered by the least and thinnest flesh of any place in the head, and under which

---

2 ἥ V: ἦν B
3 ἥ V: ᾗ ἥ B

καὶ ὁ ἐγκέφαλος κατὰ τοῦτο τῆς κεφαλῆς πλεῖστος
ὕπεστιν. καὶ δὴ ὅτι οὕτω ταῦτα ἔχει, τῶν τε τρωσίων
190 | καὶ τῶν βελέων ἴσων τε ἐόντων κατὰ μέγεθος καὶ
ἐλασσόνων, καὶ ὁμοίως τε τρωθεὶς καὶ ἧσσον, τὸ
ὀστέον ταύτῃ τῆς κεφαλῆς φλᾶταί τε μᾶλλον καὶ ῥή-
γνυται καὶ ἔσω ἐσφλᾶται, καὶ θανασιμώτερά ἐστι καὶ
χαλεπώτερα ἰητρεύεσθαί τε καὶ ἐκφυγγάνειν τὸν θά-
νατον ταύτῃ ἤ που ἄλλοθι τῆς κεφαλῆς. ἐξίσων τε
ἐόντων τῶν τρωμάτων καὶ ὁμοίως τε τρωθεὶς καὶ ἧσ-
σον, ἀποθνήσκει ὁ ἄνθρωπος, ὁκόταν καὶ ἄλλως
μέλλῃ ἀποθανεῖσθαι ἐκ τοῦ τρώματος, ἐν ἐλάσσονι
χρόνῳ ὁ ταύτῃ ἔχων τὸ τρῶμα τῆς κεφαλῆς ἤ που
ἄλλοθι. ὁ γὰρ ἐγκέφαλος τάχιστά τε καὶ μάλιστα
κατὰ τὸ βρέγμα αἰσθάνεται τῶν κακῶν τῶν γινο-
μένων ἔν τε τῇ σαρκὶ καὶ τῷ ὀστέῳ· ὑπὸ λεπτοτάτῳ
γὰρ ὀστέῳ ἐστὶ ταύτῃ ὁ ἐγκέφαλος καὶ ὀλιγίστῃ
σαρκί, καὶ ὁ πλεῖστος ἐγκέφαλος ὑπὸ τῷ βρέγματι
κεῖται. τῶν δὲ ἄλλων τὸ κατὰ τοὺς κροτάφους ἀσθενέ-
στατόν ἐστιν· συμβολή τε γὰρ τῆς κάτω γνάθου
πρὸς τὸ κρανίον, καὶ κίνησις ἔνεστιν ἐν τῷ κροτάφῳ
ἄνω καὶ κάτω ὥσπερ ἄρθρον· καὶ ἡ ἀκοὴ πλησίον
γίνεται αὐτοῦ, καὶ φλὲψ διὰ τοῦ κροτάφου τέταται
κοίλη τε καὶ ἰσχυρή.

Ἰσχυρότερον δ᾽ ἐστι τῆς κεφαλῆς τὸ ὀστέον ἅπαν
192 τὸ ὄπισθεν τῆς κορυφῆς καὶ τῶν | οὐάτων ἢ ἅπαν
τὸ πρόσθεν, καὶ σάρκα πλείονα καὶ βαθυτέρην ἐπὶ
ἑωυτῷ ἔχει τοῦτο τὸ ὀστέον. καὶ δὴ τούτων οὕτως
ἐχόντων, ὑπό τε τῶν τρωσίων καὶ τῶν βελέων ἴσων

the greater part of the brain is located. Here is proof that this is so: with wounds from weapons of equal size or less, inflicted with equal or less violence, the bone in that part of the head is more liable to be contused, broken and depressed: furthermore, injuries there are more deadly and difficult to treat, and death is more difficult to escape than injury in any other part of the head. And when comparable wounds are inflicted with equal or less violence, a person will die—especially if he is going to die from the injury—in a shorter time, when he has a head wound there, than in any other place in the head. For the brain is affected more quickly and forcefully through the bregma by the injuries to the flesh and the bone, since at the bregma the brain is lying under very thin bone and very sparse flesh, and most of the brain lies under the bregma. Otherwise, the skull bone is weakest at the temples, since the lower jaw is attached to the cranium at the temple, providing up and down movement as if in a joint. The ear is close to this, and a major hollow vessel passes through the temple.

The skull everywhere behind the crown and the ears is stronger than in the whole anterior part, and is covered by more and thicker flesh. This being so, in a person suffering from equal and greater wounds and weapons, inflicted *ἐόντων, καὶ ὁμοίων καὶ μεζόνων καὶ [μειόνων][4] ὁμοίως τιτρωσκόμενος καὶ μᾶλλον, ταύτῃ τῆς κεφαλῆς τὸ ὀστέον ἧσσον ῥήγνυται καὶ φλᾶται ἔσω, κὴν μέλλῃ ὤνθρωπος ἀποθνῄσκειν καὶ ἄλλως ἐκ τοῦ τρώματος, ἐν τῷ ὄπισθεν τῆς κεφαλῆς ἔχων τὸ τρῶμα ἐν πλείονι χρόνῳ ἀποθανεῖται· ἐν πλείονι γὰρ χρόνῳ τὸ ὀστέον ἐμπυΐσκεταί τε καὶ διαπυΐσκεται κάτω ἐπὶ τὸν ἐγκέφαλον διὰ τὴν παχύτητα τοῦ ὀστέου· καὶ ἐλάσσων ταύτῃ τῆς κεφαλῆς ὁ εγκέφαλος ὕπεστι. καὶ πλέονες ἐκφυγγάνουσι τὸν θάνατον τῶν ὄπισθεν τιτρωσκομένων τῆς κεφαλῆς ὡς ἐπὶ τὸ πολὺ ἢ τῶν ἔμπροσθεν. καὶ ἐν χειμῶνι πλείονα χρόνον ζῇ ὤνθρωπος ἢ ἐν θέρει, ὅστις ἂν καὶ ἄλλως μέλλῃ ἀποθανεῖσθαι ἐκ τοῦ τρώματος ὅπου ἂν τῆς κεφαλῆς ἔχων τὸ τρῶμα.*

3. *Αἱ δὲ ἕδραι τῶν βελέων τῶν ὀξέων καὶ κουφοτέ-*
194 *ρων, αὐταὶ | ἐπὶ σφῶν αὐτῶν γινόμεναι ἐν τῷ ὀστέῳ*
ἄνευ ῥωγμῆς τε καὶ φλάσιος καὶ ἔσω ἐσφλάσιος— αὗται δὲ γίνονται ὁμοίως ἔν τε τῷ ἔμπροσθεν τῆς κεφαλῆς καὶ ἐν τῷ ὄπισθεν—ἐκ τούτων ὁ θάνατος οὐ γίνεται κατά γε δίκην, οὐδ᾽ ἢν γένηται. ῥαφὴ δὲ ἐν ἕλκει φανεῖσα, ὀστέου ψιλωθέντος, πανταχοῦ τῆς κεφαλῆς τοῦ ἕλκεος γενομένου, ἀσθενέστατον γίνεται τῇ τρώσει καὶ τῷ βέλει ἀντέχειν, εἰ τύχοι τὸ βέλος ἐς αὐτὴν τὴν ῥαφὴν στηριχθέν. πάντων δὲ μάλιστα, ἢν τὸ βέλος ἐν τῷ βρέγματι γενόμενον κατὰ τὸ ἀσθενέστατον τῆς κεφαλῆς, καὶ αἱ ῥαφαὶ εἰ τύχοιεν ἐοῦσαι [ἢ][5] περὶ τὸ ἕλκος καὶ τὸ βέλος αὐτῶν τύχοι τῶν ῥαφέων.

with equal or greater violence, the bone there is broken and depressed less: also, if the person is going to die, especially from the wound, he will die after a longer time if he is wounded in the posterior part of the head; this is because, due to the bone's thickness there, it suppurates and sends pus down on to the brain only after a longer time; furthermore, less of the brain is in that part of the head. Usually patients escape death more often when they are wounded in the back of the head than in the front. A person also generally survives longer in winter than in summer, in particular when he is going to die from a wound in any part of the head.

3. *Hedrae* made in the bone by sharp, light projectiles—which occur equally at the front of the head and at the back—by themselves in the absence of any fracture, contusion, or depression are never the true cause of death, even if such a patient does die. If a suture becomes visible in a lesion when the bone has been stripped bare—anywhere this happens—the bone resists the wound and the weapon least successfully if the weapon has become fixed in the suture itself: in particular if the weapon is fixed in the weakest part of the head, the bregma, and there happen to be sutures around the lesion, and the weapon happens to strike the sutures themselves.

---

4 *μειόνων* B[3]V: del. Littré
5 *ἢ* BV: del. Aldina

4. Τιτρώσκεται δὲ ὀστέον τὸ ἐν τῇ κεφαλῇ τοσ-
ούσδε τρόπους· τῶν δὲ τρόπων ἑκάστου πλείονες ἰδέαι
196 γίνονται τοῦ κατήγματος | ἐν τῇ τρώσει. ὀστέον ῥή-
γνυται τιτρωσκόμενον καὶ τῇ ῥωγμῇ ἐν τῷ περιέχοντι
ὀστέῳ τὴν ῥωγμήν, ἀνάγκη φλάσιν προσγενέσθαι,
ἥνπερ ῥαγῇ. τῶν γὰρ βελέων ὅ τι περ ῥήγνυσι τὸ
ὀστέον, τὸ αὐτὸ τοῦτο καὶ φλᾷ τὸ ὀστέον ἢ μᾶλλον ἢ
ἧσσον, αὐτό τε, ἐν ᾧπερ καὶ ῥήγνυσι τὴν ῥωγμὴν καὶ
τὰ περιέχοντα ὀστέα τὴν ῥωγμήν· εἷς οὗτος τρόπος.
ἰδέαι δὲ ῥωγμέων παντοῖαι γίνονται· καὶ γὰρ λεπτότε-
ραί τε καὶ λεπταὶ πάνυ, ὥστε οὐ καταφανέες γίνονται,
ἔστιν αἳ τῶν ῥωγμέων, οὔτε αὐτίκα μετὰ τὴν τρῶσιν,
οὐδὲ ἐν τῇσιν ἡμέρῃσιν, ἐν ᾗσιν ἂν καὶ πόνων ὄφελος
200 γένοιτο <καὶ>[6] | τοῦ θανάτου τῷ ἀνθρώπῳ. αἱ δ᾽ αὖ
παχύτεραί τε καὶ εὐρύτεραι ῥήγνυνται τῶν ῥωγμέων,
ἔνιαι δὲ καὶ πάνυ εὐρέαι. ἔστι δὲ αὐτῶν καὶ αἳ μὲν ἐπὶ
μακρότερον ῥήγνυνται, αἳ δὲ ἐπὶ βραχύτερον· καὶ αἱ
μὲν ἰθύτεραι, αἱ δὲ ἰθεῖαι πάνυ, αἱ δὲ καμπυλώτεραί
τε καὶ <πάνυ>[7] καμπύλαι· καὶ <αἱ μὲν>[8] βαθύτεραί, αἱ
δὲ ἐς τὸ[9] κάτω καὶ διὰ παντὸς τοῦ ὀστέου.

5. Φλασθείη δ᾽ ἂν τὸ ὀστέον μένον ἐν τῇ ἑωυτοῦ φύσει, καὶ ῥωγμὴ τῇ φλάσει οὐκ ἂν προσγένοιτο ἐν τῷ ὀστέῳ οὐδεμία· δεύτερος οὗτος τρόπος. ἰδέαι δὲ τῆς φλάσιος πλείους γίνονται· καὶ γὰρ μᾶλλόν τε καὶ ἧσσον φλᾶται καὶ ἐς βαθύτερόν τε καὶ διὰ παντὸς τοῦ ὀστέου, καὶ ἧσσον ἐς βαθὺ καὶ οὐ διὰ παντὸς τοῦ ὀστέου, καὶ ἐπὶ πλέον τε καὶ ἔλασσον μήκεός τε καὶ πλατύτητος. ἀλλὰ οὐ τούτων τῶν ἰδεῶν οὐδεμίαν

4. Injuries of the bone in the head are of the following types, and of each type of injury there are several varieties. The bone, on being wounded, fractures, and in the bone around the fracture a contusion necessarily follows, since the same weapon that caused the fracture must also cause a contusion, either more or less severe, in the same place and in the bone around the fracture. *This is type one*. Of fractures there are many different varieties: some fractures are quite thin and very fine, so that they are not apparent either immediately after the break takes place, nor in the days when the person could be helped for their pains or saved from death; other fractures are thicker and wider, and others very wide; some fractures extend for a greater length, others for a shorter length. Some fractures are fairly straight, others very straight, others somewhat curved, and others very curved; some fractures are moderately deep, while others go all the way through the bone to the bottom.

5. A bone can be contused, but remain in its natural position and without any fracture accompanying the contusion in the bone. *This is the second type*. Of contusions there are many varieties, some being more serious and others less serious, some very deep and through the whole of the bone and others less deep and not through the whole bone, and yet others of greater or less length and depth. But none of these varieties can be identified by

---

[6] Add. Littré (vol. 10, p. xxiii)

[7] Add. Ermerins

[8] Add. Martin

[9] *αἱ δὲ ἐς τὸ* V: *τε ἐκ τοῦ* B

ἐστὶν ἰδόντα τοῖσιν ὀφθαλμοῖσι γνῶναι ὁποίη τίς
202 ἐστιν τὴν ἰδέην καὶ ὁπόση | τις τὸ μέγεθος· οὐδὲ γὰρ
εἰ πέφλασται ἢ μὴ πέφλασται[10] ἐόντων πεφλασμένων
καὶ τοῦ κακοῦ γεγενημένου γίνεται τοῖσιν ὀφθαλ-
μοῖσι καταφανὲς ἰδεῖν αὐτίκα μετὰ τὴν τρῶσιν,
ὥσπερ οὐδὲ τῶν ῥωγμέων ἔνιαι ἑκὰς ἐοῦσαί τε καὶ
ἐρρωγότος τοῦ ὀστέου.

204 6. Ἐσφλᾶται τὸ ὀστέον ἐκ τῆς | φύσιος τῆς ἑωυτοῦ
ἔσω σὺν ῥωγμῇσιν· ἄλλως γὰρ οὐκ ἂν ἐσφλασθείη·
τὸ γὰρ ἐσφλώμενον εἴσω,[11] ἀπορρηγνύμενόν τε καὶ
καταγνύμενον, ἐσφλᾶται ἔσω ἀπὸ τοῦ ἄλλου ὀστέου
μένοντος ἐν τῇ φύσει τῇ ἑωυτοῦ· καὶ δὴ οὕτω ῥωγμὴ
ἂν προσείη τῇ ἐσφλάσει· τρίτος οὗτος τρόπος. ἐσφλᾶ-
ται δὲ τὸ ὀστέον πολλὰς ἰδέας· καὶ γὰρ ἐπὶ πλέον τοῦ
ὀστέου καὶ ἐπ᾽ ἔλασσον, καὶ μᾶλλόν τε καὶ ἐς βα-
θύτερον κάτω, καὶ ἧσσόν τε καὶ ἐπιπολαιότερον.

7. Καὶ ἕδρης ἐγγενομένης ἐν τῷ ὀστέῳ βέλεος
προσγένοιτ᾽ ἂν ῥωγμὴ τῇ ἕδρῃ. τῇ τε ῥωγμῇ[12] καὶ
206 φλάσιν προσγενέσθαι ἀναγκαῖόν ἐστι ἢ | μᾶλλον ἢ
ἧσσον, ἤνπερ καὶ ῥωγμὴ προσγένηται, ἔνθαπερ καὶ
ἕδρη ἐγένετο καὶ ἡ ῥωγμή <καὶ>[13] ἐν τῷ ὀστέῳ τῷ
περιέχοντι τήν τε ἕδρην καὶ τὴν ῥωγμήν· τέταρτος
οὗτος τρόπος. καὶ ἕδρη μὲν ἂν γένοιτο φλάσιν ἔχουσα
τοῦ ὀστέου περὶ αὐτήν, ῥωγμὴ δὲ οὐκ ἂν προσγένοιτο
208 | τῇ ἕδρῃ καὶ τῇ φλάσει ὑπὸ τοῦ βέλεος· πέμπτος
οὗτος τρόπος. καὶ ἕδρη δὲ τοῦ βέλεος γίνεται ἐν τῷ
ὀστέῳ· ἕδρη δὲ καλέεται, ὅταν μένον τὸ ὀστέον ἐν τῇ
ἑωυτοῦ φύσει τὸ βέλος στηρίξαν ἐς τὸ ὀστέον δῆλον

seeing them with the eyes, with regard to their precise form or their size; for not even whether a contusion has actually occurred or not—when in fact a contusion has occurred and caused damage—can be judged by seeing it with the eyes immediately after the injury, just as some fractures cannot be recognized when they are located far away (sc. from the wound), although the bone has been broken.

6. Depression of a bone from its natural position requires a fracture, since otherwise this movement could not occur. For the piece of bone that is depressed inward, being broken off and shattered, is depressed inward away from the rest of the bone that remains in its natural position, i.e., there is a fracture besides the depression. *This is the third type*. The bone can suffer depressed fractures of many kinds; for the fracture may involve more or less of the skull, and the bone can be displaced further and deeper down, or only depressed a smaller distance and remain nearer the surface.

7. If a *hedra* is made in the skull bone by a weapon, there may also be a fracture, and with the fracture there must then also be a contusion—either greater or less, when a fracture is added to a *hedra*—in the place where the *hedra* and the fracture occur, and also in the bone surrounding the *hedra* and the fracture. *This is the fourth type*. A *hedra* can also occur together with a contusion of the bone around it, but a fracture must not necessarily be added to a *hedra* and contusion caused by a weapon. *This is the fifth type.* A weapon *hedra* is made in the bone: it is called a *hedra* when, with the bone remaining in its natu-

---

10 ἢ μὴ πέφλασται B: om. V 11 εἴσω B: om. V
12 τῇ τε ῥωγμῇ V: om. B 13 Add. καὶ Littré

ποιήσῃ ὅπου ἐστήριξεν. ἐν δὲ τῷ τρόπῳ ἑκάστῳ πλείονες ἰδέαι γίνονται· καὶ περὶ μὲν φλάσιός τε καὶ ῥωγμῆς, κἢν ἄμφω ταῦτα προσγένηται τῇ ἕδρῃ, καὶ ἢν φλάσις μούνη γένηται. ἤδη πέφρασται ὅτι πολλαὶ ἰδέαι γίνονται καὶ τῆς φλάσιος καὶ τῆς ῥωγμῆς.

Ἡ δὲ ἕδρη αὐτὴ ἐφ᾽ ἑωυτῆς γίνεται, μακροτέρη καὶ βραχυτέρη ἐοῦσα, καὶ καμπυλωτέρη, καὶ ἰθυτέρη, καὶ κυκλοτερής· καὶ πολλαὶ ἄλλαι ἰδέαι τοιούτου τρόπου, ὁποῖον ἄν τι καὶ τὸ στόμα[14] τοῦ βέλεος ᾖ· αἱ δὲ αὐταὶ καὶ βαθύτεραι τὸ κάτω καὶ μᾶλλον καὶ ἧσσον, καὶ στενότεραί τε καὶ ἧσσον στεναὶ καὶ εὐρύτεραί τε καὶ πάνυ εὐρεῖαι, ᾗ διακεκόφαται. διακοπὴ δὲ ὁποσητισοῦν γινομένη μήκεός τε καὶ εὐρύτητος ἐν τῷ ὀστέῳ, ἕδρη ἐστίν, ἢν τὰ ἄλλα | ὀστέα τὰ περιέχοντα τὴν διακοπὴν μένῃ ἐν τῇ φύσει τῇ ἑωυτῶν, καὶ μὴ συνεσφλᾶται τῇ διακοπῇ ἔσω ἐκ τῆς φύσιος τῆς ἑωυτῶν· οὕτω δὲ ἔσφλασις ἂν εἴη καὶ οὐκέτι ἕδρη.

8. Ὀστέον τιτρώσκεται ἄλλῃ τῆς κεφαλῆς ἢ ᾗ[15] τὸ ἕλκος ἔχει ὤνθρωπος καὶ τὸ ὀστέον ἐψιλώθη τῆς σαρκός· ἕβδομος οὗτος τρόπος. καὶ ταύτην τὴν ξυμφορήν, ὁκόταν γένηται, οὐκ ἂν ἔχοις ὠφελῆσαι οὐδέν. οὐδὲ γὰρ εἰ πέπονθε τὸ κακὸν τοῦτο, οὐκ ἔστιν ὅκως χρὴ αὐτὸν ἐξελέγξαντα εἰδέναι, εἰ πέπονθε τὸ κακὸν τοῦτο ὤνθρωπος, οὐδὲ ὅπῃ τῆς κεφαλῆς.

9. Τούτων τῶν τρόπων τῆς κατήξιος ἐς πρῖσιν ἀφήκει ἥ τε φλάσις ἡ ἀφανὴς ἰδεῖν καὶ ἤν πως τύχῃ

[14] στόμα B: σχῆμα V [15] ἢ ᾗ Martin: ἣν B: ᾗ V

ral position, a weapon, by being fixed in it, makes a distinct mark where it was fixed. In each of these types of contusions and of fractures there are many forms, both when the two of them are added to a *hedra*, or if a contusion alone accompanies it; as I have often indicated, there are many forms of contusions and fractures.

A *hedra* can also arise by itself, being longer or shorter, more curved or straighter, or also more circular; many other varieties are determined by the shape of the weapon's edge.[1] These forms may be more or less deep, and at their base more or less narrow, or wider, or very wide, where there is a cleft through the bone; for a cleft, of any length or breadth, is a *hedra*, if the other bones around it stay in their natural place and are not crushed inward away from their natural location along with the cleft—if that happened, it would be a depressed fracture and no longer a *hedra*.

8. The skull may also be wounded some place in the head other than where there is a lesion and the bone is stripped of flesh. *This is the seventh type.* If this sort of injury occurs, you will have no means of providing aid, since whether a patient has suffered the injury cannot be determined, and if he has suffered an injury, its location is impossible to know.

9. Of these types of skull injury, both an invisible contusion, and even one that happens to become visible, will

[1] See Hanson, pp. 103f., for a discussion of this numbering system, in which he agrees with Ermerins in making a *hedra* without any complication of the sixth type.

φανερὴ γενομένη καὶ ἡ ῥωγμὴ ᾗ[16] ἀφανὴς ἰδεῖν καὶ ἢν φανερὴ ᾖ. καὶ ἤν, ἕδρης γενομένης τοῦ βέλεος ἐν τῷ ὀστέῳ, προσγένηται ῥωγμὴ καὶ φλάσις τῇ ἕδρῃ, καὶ ἢν φλάσις μούνη προσγένηται ἄνευ ῥωγμῆς τῇ ἕδρῃ, καὶ αὕτη ἐς πρῖσιν ἀφήκει. τὸ δὲ ἔσω ἐσφλώμενον ὀστέον ἐκ τῆς φύσιος τῆς ἑωυτοῦ ὀλίγα τῶν
212 πολλῶν πρίσιος δεῖται·[17] | καὶ τὰ μάλιστα ἐσφλασθέντα[18] καὶ μάλιστα καταρραγέντα, ταῦτα πρίσιος ἥκιστα κέχρηται· οὐδὲ ἕδρη αὐτὴ ἐφ᾽ ἑωυτῆς γενομένη ἄτερ ῥωγμῆς καὶ φλάσιος, οὐδὲ αὕτη[19] πρίσιος δεῖται· οὐδ᾽ ἢν διακοπὴ μεγάλη καὶ εὐρεῖα, οὐδ᾽ αὕτη· διακοπὴ γὰρ καὶ ἕδρη τωὐτόν ἐστιν.

10. Πρῶτον δὲ χρὴ τὸν τρωματίην σκοπεῖσθαι, ὅπῃ ἔχει τὸ τρῶμα τῆς κεφαλῆς, εἴτ᾽ ἐν τοῖσιν ἰσχυροτέροισιν εἴτ᾽ ἐν τοῖσιν ἀσθενεστέροισι, καὶ τὰς τρίχας καταμανθάνειν τὰς περὶ τὸ ἕλκος, εἰ διακεκόφαται ὑπὸ τοῦ βέλεος, καὶ εἰ ἔσω ἤϊσαν ἐς τὸ τρῶμα· καὶ ἢν τοῦτο ᾖ, φάναι κινδυνεύειν τὸ ὀστέον ψιλὸν εἶναι τῆς σαρκὸς ‹καὶ›[20] ἔχειν τι σίνος τὸ ὀστέον ὑπὸ τοῦ βέλεος. ταῦτα μὲν οὖν χρὴ ἀπόπροσθεν σκεψάμενον λέξαι, μὴ ἁπτόμενον τοῦ ἀνθρώπου· ἁπτόμενον δὲ τοῦ ἀνθρώπου[21] ἤδη πειρᾶσθαι εἰδέναι σάφα εἰ ἔστι ψιλὸν τὸ ὀστέον τῆς σαρκὸς ἢ οὔ· καὶ ἢν μὲν καταφανὲς ᾖ τὸ ὀστέον τοῖσι ὀφθαλμοῖσι, ψιλόν ‹ἐόν›.[22]
214 εἰ δὲ μή, τῇ μήλῃ σκέψασθαι. καὶ ἢν μὲν εὕρῃς | ψιλὸν ἐὸν τὸ ὀστέον τῆς σαρκὸς καὶ μὴ ὑγιὲς ἀπὸ τοῦ τρώματος, χρὴ τοῦ ἐν τῷ ὀστέῳ ἐόντος τὴν διάγνωσιν

come to need trephination, as does an invisible fracture or one that is visible. Also, where a weapon *hedra* in the bone is accompanied by a fracture and a contusion, or by a contusion alone without any fracture, this too will arrive at sawing (trephination); but a bone that is depressed below its natural position only rarely requires trephination; and the more serious a fracture or depression a bone suffers, the less often it needs trephination. Nor does a *hedra* by itself without a fracture or contusion need trephination; nor does a large wide cleft require it, since such a cleft is the same thing as a *hedra*.

10. First, you must examine such an injury to determine where in the head it is located, i.e., whether in the stronger parts or where the bone is weaker, and investigate carefully the hair in the region of the wound, to see whether it has been cut through by the weapon, and whether it has fallen into the wound. If it has, announce that there is a danger that the bone there is stripped of flesh, and that the bone itself has been damaged by the weapon. This should be stated on your first inspection, when you have not yet touched the patient. Then, when you palpate the wound, try to ascertain clearly whether or not the bone is in fact denuded of flesh. If it is clearly visible that the bone is stripped of flesh, fine; but if it is not clear, examine the wound with a probe. If you discover that the bone has been denuded, and rendered unsound by the wound, you must first determine the situation in

---

16 ᾗ Littré: ἧν V: om. B   17 δεῖται B: προσδεῖται V

18 καὶ τ. μ. ἐσφλασθέντα V: om. B

19 αὕτη Withington: αυτη B: αυτῇ V

20 Add. καὶ Littré   21 τοῦ ἀνθρώπου B: om. V

22 ψιλὸν ἐόν Hanson: ἐὸν ψιλόν Reinhold: ψιλόν BV

πρῶτα ποιέεσθαι, ὁρῶντα ὁκόσον τέ ἐστι τὸ κακὸν καὶ τίνος δεῖται ἔργου. χρὴ δὲ καὶ ἐρωτᾶν τὸν τετρωμένον ὅκως ἔπαθε καὶ τίνα τρόπον.

Ἢν δὲ μὴ καταφανὲς ᾖ τὸ ὀστέον, εἰ ἔχει τι κακὸν ἢ μὴ ἔχει, πολλῷ ἔτι χρὴ μᾶλλον τὴν ἐρώτησιν ποιέεσθαι <ἢ>[23] ψιλοῦ τε ἐόντος τοῦ ὀστέου, τὸ τρῶμα ὅκως ἐγένετο καὶ ὅντινα τρόπον. τὰς γὰρ φλάσιας καὶ τὰς ῥωγμὰς τὰς οὐ φαινομένας ἐν τῷ ὀστέῳ, ἐνεούσας δέ, ἐκ τῆς ἀποκρίσιος τοῦ τετρωμένου πρῶτον διαγινώσκειν πειρᾶσθαι, εἴ τι πέπονθε τούτων τὸ ὀστέον ἢ οὐ πέπονθεν. ἔπειτα δὲ καὶ λόγῳ καὶ ἔργῳ ἐξελέγχειν, πλὴν μηλώσιος. μήλωσις γὰρ οὐκ ἐξελέγχει, εἰ πέπονθέ τι τούτων τῶν κακῶν τὸ ὀστέον, καὶ εἴ τι ἐνέχει ἐν ἑωυτῷ, ἢ οὐ πέπονθεν. ἀλλ᾽ ἕδρην τε τοῦ βέλεος ἐξελέγχει μήλωσις, καὶ ἢν ἐμφλασθῇ τὸ ὀστέον εἴσω ἐκ τῆς φύσιος τῆς ἑωυτοῦ, καὶ ἢν ἰσχυρῶς ῥαγῇ τὸ ὀστέον, ἅπερ καὶ τοῖσιν ὀφθαλμοῖσι καταφανέα ἐστὶν ὁρῶντα γινώσκειν.

11. Ῥήγνυται δὲ τὸ ὀστέον τάς τε ἀφανέας ῥωγμὰς καὶ τὰς φανεράς, καὶ φλᾶται τὰς ἀφανέας φλάσιας, καὶ ἐσφλᾶται ἔσω ἐκ τῆς φύσιος τῆς ἑωυτοῦ, μάλιστα
216 ὁκόταν ἕτερος ὑφ᾽ ἑτέρου τιτρωσκόμενος | ἐπίτηδες τρῶσαι βουλόμενος ἢ ὁκόταν ἀέκων[24]—καὶ ὁκόταν ἐξ ὑψηλοτέρου γίνηται ἡ βολὴ ἢ ἡ πληγή, ὁκοτέρη ἂν ᾖ, μᾶλλον ἢ ὁκόταν ἐξ ἰσοπέδου τοῦ χωρίου, καὶ ἢν περικρατῇ τῇ χειρὶ τὸ βέλος, ἤν τε βάλλῃ ἤν τε τύπτῃ, καὶ ἰσχυρότερος ἐὼν ἀσθενέστερον τιτρώσκῃ.

the bone, finding out how serious the injury is, and what action is required; you must also ask the wounded person what injury he suffered, and how it took place.

If you cannot see whether or not the bone is damaged, you must question the patient more thoroughly than when the bone is denuded, about how the wound happened and in what manner. For regarding contusions and fractures of the bone that are real, but invisible, try first to clarify by questioning the patient whether the skull suffered any of these things or not. Then confirm your hypothesis by questioning and examining, except with the probe—since the probe will not reveal whether or not the bone has suffered any of these injuries in it. The probe can, however, expose a *hedra* made by a weapon, or the depression of a piece of bone inward out of its natural position, or that a bone is severely fractured, just as these things may be clearly known if they are visible to the eyes.

11. The skull is more likely to suffer a visible or invisible fracture, an invisible contusion, and to be depressed inward out of its natural position when a person is wounded by someone wishing to inflict harm, than if there is no intention; also when a missile or blow—whichever the case happens to be—comes from above rather than from level ground, and if the assailant has a firm hold of the weapon with his hand, whether he is throwing it or striking with it, and if it is a stronger man wounding a weaker one.

---

23 Add. ἢ Ermerins

24 *μάλιστα . . . ἀέκων*: *μάλιστα* BV: *μᾶλλον* Hanson *τιτρωσκόμενος* BV: *τιτρώσκεται* Hanson *βουλόμενος* BV: *βουλομένου* Hanson (following Scaliger) *ἀέκων* B: *ἀέκοντος* Hanson (following Littré's note): *ἀέκων—ὁκόταν* om. V

ὁκόσοι δὲ πίπτοντες τιτρώσκονται πρός τε τὸ ὀστέον καὶ αὐτὸ τὸ ὀστέον, ὁ ἀπὸ ὑψηλοτάτου πίπτων καὶ ἐπὶ σκληρότατον καὶ ἀμβλύτατον, τούτῳ κίνδυνος τὸ |
218 ὀστέον ῥαγῆναι καὶ φλασθῆναι καὶ εἴσω ἐσφλασθῆναι ἐκ τῆς φύσιος τῆς ἑωυτοῦ. τῷ δ' ἐξ ἰσοπέδου μᾶλλον τοῦ χωρίου πίπτοντι καὶ ἐπὶ μαλθακώτερον, ἧσσον ταῦτα πάσχει τὸ ὀστέον ἢ οὐκ ἂν πάθοι. ὁκόσα δ' ἐσπίπτοντα ἐς τὴν κεφαλὴν βέλεα τιτρώσκει πρὸς τὸ ὀστέον καὶ αὐτὸ τὸ ὀστέον, τὸ ἀπὸ ὑψηλοτάτου ἐμπεσὸν καὶ ἥκιστα ἐξ ἰσοπέδου, καὶ σκληρότατόν τε ἅμα καὶ ἀμβλύτατον καὶ βαρύτατον, καὶ ἥκιστα κοῦφον καὶ ἥκιστα ὀξὺ καὶ μαλθακόν, τοῦτο ἂν ῥήξειε τὸ ὀστέον καὶ φλάσειεν.

Καὶ μάλιστά γε κίνδυνος ταῦτα πάσχειν τὸ ὀστέον, ὁκόταν ταῦτά τε γίνηται καὶ ἐς ἰθὺ τρωθῇ καὶ κατ' ἀντίον γένηται τὸ ὀστέον τοῦ βέλεος, ἤν τε πληγῇ ἐκ χειρὸς ἤν τε βληθῇ ἤν τέ τι ἐμπέσῃ αὐτῷ, καὶ ἢν αὐτὸς καταπεσὼν τρωθῇ, καὶ ὁπωσοῦν τρωθεὶς κατ' ἀντίον γενομένου τοῦ ὀστέου τῷ βέλει. τὰ δ' ἐς πλάγιον τοῦ ὀστέου παρασύραντα βέλεα ἧσσον καὶ ῥήγνυσι τὸ ὀστέον καὶ φλᾷ καὶ εἴσω ἐσφλᾷ, κἢν ψιλωθῇ τὸ ὀστέον τῆς σαρκός. ἔνια δὲ[25] τῶν τρωμάτων τῶν οὕτω τρωθέντων οὐδὲ ψιλοῦται τὸ ὀστέον τῆς σαρκός.

Τῶν δὲ βελέων ῥήγνυσι μάλιστα τὸ ὀστέον τάς τε
220 φανερὰς ῥωγμὰς καὶ | τὰς ἀφανέας καὶ φλᾷ τε καὶ ἐσφλᾷ εἴσω ἐκ τῆς φύσιος τῆς ἑωυτοῦ τὸ ὀστέον τὰ στρογγύλα τε καὶ περιφερέα καὶ ἀρτίστομα, ἀμβλέα

Those wounded around the skull or in the skull itself by falling are in danger, if they fall from a great height and on to something very hard and blunt, that their skull will be fractured, contused, and depressed inward out its natural position; in a person who falls from a more level height and the fall is on to something softer, the bone suffers such things less severely, or not at all. Projectiles falling on to the head can injure both the area around the skull and the skull itself: a projectile falling from a great height as opposed to a more even height, and being both hard and at the same time very blunt and heavy, as opposed to light, sharp and soft, has a greater tendency to fracture and contuse the bone.

There is in particular a danger that the bone will suffer these injuries when the blow happens to be straight and the bone is facing the weapon directly, whether the blow was struck with the hand or a projectile was thrown, or something fell on the person, or if he was injured by falling down himself, or if he was wounded in any way whatever with the bone directly facing a weapon. Projectiles that graze the bone at an angle are less likely to fracture it, contuse it, or depress it inward, even if the bone is stripped of its flesh: and some wounds that happen like this do not even strip the bone of flesh.

The weapons most likely to afflict the bone with fractures, both visible and invisible, to contuse it, and to depress it inward out of its natural position, are those which have a rounded spherical shape and smooth edges, and are

---

25 δὲ B: γὰρ V

τε ἐόντα καὶ βαρέα καὶ σκληρά. καὶ τὴν σάρκα ταῦτα φλᾷ τε καὶ πέπειραν ποιέει καὶ κόπτει. καὶ τὰ ἕλκεα γίνεται ὑπὸ τῶν τοιούτων βελέων, ἔς τε πλάγιον καὶ ἐν κύκλῳ ὑπόκοιλα, καὶ διάπυά τε μᾶλλον γίνεται καὶ ὑγρά ἐστι καὶ ἐπὶ πλέονα χρόνον καθαίρεται. ἀνάγκη γὰρ τὰς σάρκας τὰς φλασθείσας καὶ κοπείσας πυός γενομένας ἐκτακῆναι. τὰ δὲ βέλεα τὰ προμήκεα ἐπὶ πολὺ λεπτὰ ἐόντα καὶ ὀξέα καὶ κοῦφα, τήν τε σάρκα διατάμνει μᾶλλον ἢ φλᾷ καὶ τὸ ὀστέον ὡσαύτως. καὶ ἕδρην μὲν ἐμποιέει αὐτὸ καὶ διακόψαν—διακοπὴ γὰρ καὶ ἕδρη τωὐτόν ἐστι—φλᾷ δὲ οὐ μάλα τὸ ὀστέον τὰ τοιαῦτα βέλεα οὐδὲ ῥήγνυσιν οὐδ᾿ ἐκ τῆς φύσιος εἴσω ἐσφλᾷ.

Ἀλλὰ χρὴ πρὸς τῇ ὄψει τῇ ἑωυτοῦ, ὅ τι ἄν σοι φαίνηται ἐν τῷ ὀστέῳ, καὶ ἐρώτησιν ποιέεσθαι πάντων τούτων· τοῦ γὰρ μᾶλλόν τε καὶ ἧσσον τρωθέντος ταῦτά ἐστι σημεῖα, καὶ ἢν ὁ τρωθεὶς καρωθῇ καὶ σκότος περιχυθῇ καὶ ἢν δῖνος ἔχῃ καὶ πέσῃ. |

222 12. Ὁκόταν δὲ τύχῃ ψιλωθὲν τὸ ὀστέον τῆς σαρκὸς ὑπὸ τοῦ βέλεος, καὶ τύχῃ κατ᾿ αὐτὰς τὰς ῥαφὰς γενόμενον τὸ ἕλκος, χαλεπὸν γίνεται καὶ τὴν ἕδρην τοῦ βέλεος φράσασθαι τὴν ἐν τῷ ἄλλῳ ὀστέῳ φανερὴν γενομένην, εἴτ᾿ ἔνεστιν ἐν τῷ ὀστέῳ εἴτε μὴ ἔνεστιν, καὶ ἢν τύχῃ γενομένη ἡ ἕδρη ἐν αὐτῇσι τῇσι ῥαφῇσιν. συγκλέπτει γὰρ αὐτὴ ἡ ῥαφὴ τρηχυτέρη ἐοῦσα
224 τοῦ | ἄλλου ὀστέου, καὶ οὐ διάδηλον ὅ τι τε αὐτοῦ ῥαφή ἐστι καὶ ὅ τι τοῦ βέλεος ἕδρη, ἢν μὴ κάρτα μεγάλη γένηται ἡ ἕδρη. προσγίνεται δὲ καὶ ῥῆξις τῇ

blunt, heavy and hard; these also contuse the flesh, crush and macerate it. The lesions that result from such weapons become hollowed out, to one side or all around, and these have a greater tendency to suppurate, to be moist, and to be delayed in becoming clean, since tissues that are contused and macerated have to melt away into pus. Weapons that are elongated, being generally thin, sharp and light, cut through the flesh rather than contusing it, and the same with bone; such a weapon is also likely to produce a *hedra* and to cleave the bone—actually, cleaving the bone and making a *hedra* is the same thing—but such a weapon does not usually contuse the bone, or fracture it, or depress it inward out of its natural position.

But besides inspecting the wound with your own eyes, you must make a thorough inquiry about all the following points too—for they are signs that indicate how serious such a wound is: whether after being injured the person became drowsy, was cloaked in darkness, had vertigo, or fell down.

12. When the bone of the skull happens to be stripped of flesh by a weapon, and the wound happens to be located in the sutures themselves, it is difficult to make out whether a *hedra* from the weapon, that elsewhere would be obvious, is present or not, and especially if the *hedra* happens to be in the sutures. For a suture, being itself rougher than the rest of the bone, is deceptive, making it unclear what is in fact suture and what is a *hedra* caused by the weapon, unless the *hedra* is very large. Further-

ἕδρῃ ὡς ἐπὶ τὸ πολὺ τῇ ἐν τῇσι ῥαφῇσι γινομένῃ, καὶ
γίνεται καὶ αὐτὴ ἡ ῥῆξις χαλεπωτέρη φράσασθαι,
ἐρρωγότος τοῦ ὀστέου, διὰ τοῦτο ὅτι κατ᾽ αὐτὴν τὴν
ῥαφὴν ἡ ῥῆξις γίνεται, ἣν ῥηγνύηται, ὡς ἐπὶ τὸ πολύ·
ἕτοιμον γὰρ ταύτῃ ῥήγνυσθαι τὸ ὀστέον καὶ δια-
226 χαλᾶν διὰ τὴν ἀσθενείην | τῆς φύσιος τοῦ ὀστέου
ταύτῃ καὶ διὰ τὴν ἀραιότητα, καὶ δὴ ἅτε τῆς ῥαφῆς
ἑτοίμης ἐούσης ῥήγνυσθαι καὶ διαχαλᾶν. τὰ δ᾽ ἄλλα
ὀστέα τὰ περιέχοντα τὴν ῥαφὴν μένει ἀρραγέα, ὅτι
ἰσχυρότερά ἐστι τῆς ῥαφῆς. ἡ δὲ ῥῆξις ἡ κατὰ τὴν
ῥαφὴν γινομένη καὶ διαχάλασίς ἐστι τῆς ῥαφῆς, καὶ
228 φράσασθαι οὐκ εὐμαρής, | οὔτε[26] ἀπὸ ἕδρης τοῦ
βέλεος γενομένης ἐν τῇ ῥαφῇ, ἐπειδὰν ῥαγῇ καὶ δια-
χαλάσῃ, οὔτε ἢν φλασθέντος τοῦ ὀστέου κατὰ τὰς
ῥαφὰς[27] ῥαγῇ καὶ διαχαλάσῃ.[28] ἀλλ᾽ ἔστι χαλεπώτε-
ρον φράσασθαι τὴν ἀπὸ τῆς φλάσιος ῥωγμήν. συγ-
κλέπτουσι γὰρ τὴν γνώμην καὶ τὴν ὄψιν τοῦ ἰητροῦ
αὐταὶ αἱ ῥαφαὶ ῥωγμοειδέες φαινόμεναι καὶ τρηχύτε-
ραι ἐοῦσαι τοῦ ἄλλου ὀστέου, ὅτι μὴ ἰσχυρῶς δι-
εκόπη καὶ διεχάλασεν—διακοπὴ δὲ καὶ ἕδρη τωὐτόν
ἐστιν. ἀλλὰ χρή, εἰ κατὰ τὰς ῥαφὰς τὸ τρῶμα γένοιτο
καὶ πρός γε τὸ ὀστέον καὶ ἐς τὸ ὀστέον στηρίξειε τὸ
βέλος, προσέχοντα τὸν νόον ἀνευρίσκειν ὅ τι ἂν πε-
πόνθῃ τὸ ὀστέον. ἀπὸ γὰρ ἴσων τε βελέων τὸ μέγεθος
καὶ ὁμοίων καὶ πολλὸν [τε][29] ἐλασσόνων, καὶ ὁμοίως
τε τρωθεὶς καὶ πολὺ ἧσσον, πολλῷ μέζον ἐκτήσατο
τὸ κακὸν ἐν τῷ ὀστέῳ ὁ ἐς τὰς ῥαφὰς δεξάμενος τὸ

more, any *hedra* present in the sutures will generally be accompanied by a fracture, and such a fracture is more difficult to detect even though the bone is broken, because it is generally in the suture itself where the break has occurred. The bone there is likely to be broken and to come apart due to the natural weakness of the bone at that point, and its looseness, and particularly because any suture is prone to break and come apart. The rest of the bone around the suture, though, will remain unbroken, since it is stronger than the suture. A fracture that takes place in a suture will include an opening of the suture, although this is not easy to detect; it is also not easy to distinguish whether the bone has fractured and broken apart due to a *hedra* caused by a weapon in a suture, or fractured and come apart due to a contusion of the bone at the sutures. In fact, it is even more difficult to recognize a fracture resulting from a contusion, since the sutures interfere with the physician's vision and ability to make a judgment, due to their broken appearance and their being rougher than the rest of the bone, unless the bone is markedly cleft and opened—in fact, a cleft and a *hedra* are practically the same thing. But you must, if a fracture occurs in the sutures, and the weapon strikes the bone and becomes fixed in it, consider carefully in order to understand exactly what the bone has suffered. For from comparable weapons of the same or even much smaller size, striking the same way or even much less forcefully, a person will suffer much more damage in the bone when a blow hits the

---

26 *οὔτε* Hanson: *οὔτε η* B: *ἦ οὔτε* V
27 *ῥαφὰς* Littré: *σάρκας* B: om. V 28 *οὔτε ἦν φλασθέντος . . . διαχαλάσῃ* B: om. V 29 *τε* BV: del. Littré

βέλος ᾖ ὁ μὴ ἐς τὰς ῥαφὰς δεξάμενος. καὶ τούτων τὰ πολλὰ πρίεσθαι δεῖ. ἀλλ' οὐ χρὴ αὐτὰς τὰς ῥαφὰς πρίειν, ἀλλ' ἀποχωρήσαντα ἐν τῷ πλησίον ὀστέῳ τὴν πρῖσιν ποιέεσθαι, ἢν πρίῃς.

13. Περὶ δὲ ἰήσιος τρωσίων τῶν ἐν τῇ κεφαλῇ καὶ ὅπως χρὴ ἐλέγχειν[30] τὰς πάθας τὰς ἐν τῷ ὀστέῳ γινομένας τὰς μὴ φανεράς, ὧδέ μοι δοκέει. ἕλκος ἐν τῇ κεφαλῇ οὐ χρὴ τέγγειν οὐδενί, οὐδέ οἴνῳ, ἄλλως ἥκι-
230 στα· οὐδὲ καταπλάσσειν, οὐδὲ μοτῷ τὴν | ἴησιν ποιέεσθαι, οὐδ' ἐπιδεῖν χρὴ ἕλκος ἐν τῇ κεφαλῇ, ἢν μὴ ἐν τῷ μετώπῳ ᾖ τὸ ἕλκος ἐν τῷ ψιλῷ τῶν τριχῶν, ἢ περὶ τὴν ὀφρὺν καὶ τὸν ὀφθαλμόν. ἐνταῦθα δὲ γινόμενα τὰ ἕλκεα καταπλάσιος καὶ ἐπιδέσιος μᾶλλον κέχρηται ἤ που ἄλλοθι τῆς κεφαλῆς τῆς ἄλλης· περιέχει γὰρ ἡ κεφαλὴ ἡ ἄλλη τὸ μέτωπον πᾶν· ἐκ δὲ τῶν περιεχόντων τὰ ἕλκεα, καὶ ἐν ὅτῳ ἂν ᾖ τὰ ἕλκεα, φλεγμαίνει καὶ ἐπανοιδίσκεται δι' αἵματος ἐπιρροήν. χρὴ δὲ οὐδὲ τὰ ἐν τῷ μετώπῳ διὰ παντὸς τοῦ χρόνου καταπλάσσειν καὶ ἐπιδεῖν, ἀλλ' ἐπειδὰν παύσηται φλεγμαίνοντα, καὶ τὸ οἴδημα καταστῇ παύσασθαι καταπλάσσοντα καὶ ἐπιδέοντα. ἐν δὲ τῇ ἄλλῃ κεφαλῇ ἕλκος οὔτε μοτοῦν χρὴ οὔτε καταπλάσσειν οὔτ' ἐπιδεῖν, εἰ μὴ καὶ τομῆς δέοιτο.

Τάμνειν δὲ χρὴ τῶν ἑλκέων τῶν ἐν κεφαλῇ γινο-
232 μένων, | καὶ ἐν τῷ μετώπῳ, ὅπου ἂν τὸ μὲν ὀστέον ψιλὸν ᾖ τῆς σαρκός, καὶ δοκέῃ τι σίνος ἔχειν ὑπὸ τοῦ

[30] ἐλέγχειν B: ἐξελέγχειν V

sutures than when it does not hit them. Most of such cases require trephination: but you must not trephine the sutures themselves, but instead keep away from them as you saw the bone next to them, if you do (sc. decide to) trephine.

13. The following is my opinion on the treatment of wounds in the head, and how to identify injuries of the skull that are not obvious. A lesion in the head should not be moistened with any (sc. liquid), not even wine—or if so, with very little.[2] Nor should you apply plasters, nor carry out treatment with a lint dressing, nor bind anything on to a lesion in the head, unless the wound is in the forehead, in the part without hair or around the eyebrow and the eye. Lesions occurring there, however, do benefit more from plastering and a bandaging than those anywhere else in the rest of the head. The rest of the head surrounds the whole forehead, and from these surrounding parts, lesions, no matter where they happen to develop, become inflamed and swell up, due to an afflux of blood. You should not treat even wounds of the forehead continuously with plasters and bandages, but once their inflammation ends and their swelling goes down, stop applying plasters and bandages altogether. In the rest of the head, there is no need to use a lint dressing for a lesion, nor to apply plasters and bandages, unless an incision is also necessary.

You must incise wounds in the head and the forehead wherever the bone is stripped of flesh and there seems to be a wound from a weapon, but the lesion is not suffi-

[2] See *Ulcers* and *Use of Liquids* 1 and 5.

βέλεος, τὰ δὲ ἕλκεα μὴ ἱκανὰ τὸ μέγεθος τοῦ μήκεος καὶ τῆς εὐρύτητος ἐς τὴν σκέψιν τοῦ ὀστέου, εἴ τι πέπονθεν ὑπὸ τοῦ βέλεος κακὸν καὶ ὁποῖόν τι πέπονθε, καὶ ὁκόσον μὲν ἡ σὰρξ πέφλασται καὶ τὸ ὀστέον ἔχει τι σίνος, καὶ δ' αὖτε εἰ ἀσινές τέ ἐστι τὸ ὀστέον ὑπὸ τοῦ βέλεος καὶ μηδὲν πέπονθε κακόν, καὶ ἐς τὴν ἴησιν, ὁκοίης τινὸς δεῖται τὸ ἕλκος, ἥ τε σὰρξ καὶ ἡ πάθη τοῦ ὀστέου· τὰ δὴ τοιαῦτα τῶν ἑλκέων[31] τομῆς δεῖται. καὶ ὅταν τὸ μὲν ὀστέον ψιλωθῇ τῆς σαρκός, ὑπόκοιλα δὲ ᾖ ἐς πλάγιον ἐπὶ πολύ, ἐπανα-
234 τάμνειν τὸ κοῖλον, ὅκου μὴ | εὐμαρές ἐστι τῷ φαρμάκῳ ἀφικέσθαι, ὁκοίῳ ἄν τινι χρῇ·[32] καὶ τὰ κυκλοτερέα τῶν ἑλκέων καὶ ὑπόκοιλα ἐπὶ πολὺ καὶ τὰ τοιαῦτα ἐπανατάμνων τὸν κύκλον διχῇ κατὰ μῆκος, ὡς πέφυκεν ὤνθρωπος, μακρὸν ποιέειν τὸ ἕλκος.

Τάμνοντι δὲ κεφαλήν, τὰ μὲν ἄλλα τῆς κεφαλῆς ἀσφάλειαν ἔχει ταμνόμενα· ὁ δὲ κρόταφος, καὶ ἄνωθεν ἔτι τοῦ κροτάφου, κατὰ τὴν φλέβα τὴν διὰ τοῦ κροτάφου φερομένην, τοῦτο δὲ τὸ χωρίον μὴ τάμνειν. σπασμὸς γὰρ ἐπιλαμβάνει τὸν τμηθέντα. καὶ ἢν μὲν ἐπ' ἀριστερὰ τμηθῇ κροτάφῳ, τὰ ἐπὶ δεξιὰ ὁ σπασμὸς ἐπιλαμβάνει, ἢν δ' ἐπὶ τὰ δεξιὰ τμηθῇ κροτάφῳ, τὰ ἐπ' ἀριστερὰ ὁ σπασμὸς ἐπιλαμβάνει.[33]

14. Ὅταν οὖν τάμνῃς ἕλκος ἐν κεφαλῇ τοῦ ὀστέου εἵνεκα τῆς σαρκὸς ἐψιλωμένου, θέλων εἰδέναι εἴ τι ἔχει τὸ ὀστέον κακὸν ὑπὸ τοῦ βέλεος ἢ καὶ οὐκ ἔχει, τάμνειν χρὴ τὸ μέγεθος τῆς τομῆς, ὅση ἂν[34] δοκέῃ ἀποχρῆναι. τάμνοντα δὲ χρὴ ἀναστεῖλαι τὴν σάρκα

ciently long and wide to make inspection of the bone possible, to determine whether it has suffered any injury from the weapon, and what in particular it has suffered, how much of the flesh is contused and the bone injured, or, on the other hand, that the bone is uninjured by the weapon and has not been hurt; also regarding treatment, what attention the lesion requires, both the flesh and the damaged bone. Such wounds must be incised: when the bone is stripped of flesh and is deeply undermined on one side, open up the undermined area by incising at the place where it is not easy for the particular medication you are employing to penetrate. More circular lesions that are markedly undermined, you should open up by making two incisions at the circumference of the circle in the line of the person's body, so that the lesion will run lengthwise.

For a person wishing to make an incision in the head, the rest of the head may be incised with safety, but not the temple and the region above it containing the particular vessel, that must be spared, since incising there would provoke convulsions. If the temple is incised on the patient's left side, the spasm will be on the right of his body, whereas if the right temple is incised, the spasm will affect the left side.

14. When you incise a lesion in the head because flesh has been stripped from the bone, in order to discover whether or not the bone has been damaged by the weapon, make the incision sufficiently large. In making the incision, draw the flesh back from the bone where it adheres

---

31 ἑλκέων V: ὀστέων B 32 χρῃή Littré: χρή BV
33 ἣν δ' ἐπὶ τὰ δεξιὰ . . . ἐπιλαμβάνει V: om. B
34 ὅση ἂν V: ὁκοσηι ην B

ἀπὸ τοῦ ὀστέου ᾖ πρὸς τῇ μήνιγγι καὶ πρὸς τῷ ὀστέῳ
236 | πέφυκεν. ἔπειτα διαμοτῶσαι πᾶν τὸ ἕλκος μοτῷ, ὅστις εὐρύτατον τὸ ἕλκος παρέξει ἐς τὴν ὑστεραίην σὺν ἐλαχίστῳ πόνῳ· μοτώσαντα δὲ χρὴ καταπλάσματι χρῆσθαι ὁκόσον ἄν περ χρόνον καὶ τῷ μοτῷ μάζης ἐκ λεπτῶν ἀλφίτων· ἐν ὄξει δὲ μάσσειν, ἕψειν δὲ καὶ γλίσχρην ποιέειν ὡς μάλιστα.

Τῇ δὲ ὑστεραίῃ ἡμέρῃ, ἐπειδὰν ἐξέλῃς τὸν μοτόν, κατιδὼν τὸ ὀστέον ὅ τι πέπονθεν, ἐὰν μή σοι καταφανὴς ᾖ ἡ τρῶσις, ὁκοίη τίς ἐστιν ἐν τῷ ὀστέῳ, μηδὲ διαγινώσκῃς εἴ τέ τι ἔχει τὸ ὀστέον κακὸν ἐν ἑωυτῷ, ἢ καὶ οὐκ ἔχει, τὸ δὲ βέλος δοκέῃ ἀφικέσθαι ἐς τὸ ὀστέον καὶ σίνασθαι, ἐπιξύειν χρὴ τῷ ξυστῆρι κατὰ βάθος καὶ κατὰ μῆκος τοῦ ἀνθρώπου, ὡς πέφυκε, καὶ αὖθις ἐπικάρσιον τὸ ὀστέον τῶν ῥηξίων εἵνεκα τῶν ἀφανέων ἰδεῖν καὶ τῆς φλάσιος εἵνεκα τῆς ἀφανέος τῆς οὐκ ἐσφλωμένης εἴσω ἐκ τῆς φύσιος τῆς κεφαλῆς τοῦ ἄλλου ὀστέου. ἐξελέγχει γὰρ ἡ ξύσις μάλα τὸ κακόν, ἢν μὴ καὶ ἄλλως καταφανέες ἔωσιν αὗται αἱ
238 πάθαι ἐοῦσαι ἐν | τῷ ὀστέῳ.[35] καὶ ἢν ἕδρην ἴδῃς ἐν τῷ ὀστέῳ τοῦ βέλεος, ἐπιξύειν χρὴ αὐτήν τε τὴν ἕδρην καὶ τὰ περιέχοντα αὐτὴν ὀστέα, μὴ πολλάκις τῇ ἕδρῃ προσγένηται ῥῆξις καὶ φλάσις, ἢ μόνη φλάσις, ἔπειτα λανθάνῃ οὐ καταφανέα ἐόντα.

Ἐπειδὰν δὲ ξύσῃς τὸ ὀστέον τῷ ξυστῆρι, ἢν μὲν δοκέῃ ἐς πρῖσιν ἀφήκειν ἡ τρῶσις τοῦ ὀστέου, πρίειν χρή, καὶ τὰς τρεῖς ἡμέρας μὴ ὑπερβάλλειν ἀπρίωτον, to the membrane (i.e., pericranium) and the bone. Then pack the whole wound with a lint dressing, in order to hold the lesion as wide open as possible with the least pain until the following day. When you pack (sc. the wound) apply a poultice at the same time containing a plaster of dough made from fine barley meal: knead this with vinegar, and boil it to render it as viscid as possible.

On the next day when you remove the packing, investigate the wound to determine its damage, if you are unable to see directly what kind of wound the bone has received, or if it has or has not been damaged; if the weapon seems to have reached the bone and harmed it, scrape the bone with a rasp both lengthwise in line with the person's body and also crosswise, in order to find any potentially invisible fractures or contusion in the bone when the bone in the rest of the head has not been crushed inward out of its natural position: the rasping will expose such damage clearly, if these injuries present in the bone are not otherwise apparent. If you see the *hedra* of a weapon in the bone, rasp the bone at the *hedra* and around it, in case there is, as often happens, in addition to the *hedra*, a fracture and a contusion, or just a contusion, lying invisible and unseen.

Then, when you rasp the bone with the scraper, if the bone's wound seems to require trephination, trephine it before three days pass, i.e., within three days, especially if

---

35 Add. *τοῦ βέλεος* V

ἀλλ᾽ ἐν ταύτῃσι πρίειν, ἄλλως τε καὶ τῆς θερμῆς ὥρης, ἢν ἐξ ἀρχῆς λαμβάνῃς τὸ ἴημα.

Ἢν δὲ ὑποπτεύῃς μὲν τὸ ὀστέον ἐρρωγέναι ἢ πεφλάσθαι, ἢ ἀμφότερα ταῦτα, τεκμαιρόμενος ὅτι ἰσχυρῶς τέτρωται ἐκ τῶν λόγων τοῦ τρωματίου, καὶ ὅτι ὑπὸ ἰσχυροτέρου τοῦ τρώσαντος, ἢν ἕτερος ὑφ᾽ ἑτέρου τρωθῇ, καὶ τὸ βέλος ὅτῳ ἐτρώθη, ὅτι τῶν κακούργων βελέων ἦν, ἔπειτα τὸν ἄνθρωπον ὅτι δῖνός τε ἔλαβε καὶ σκότος, καὶ ἐκαρώθη καὶ κατέπεσεν· τούτων δὲ οὕτω γινομένων, ἢν μὴ διαγινώσκῃς εἰ ἔρρωγε τὸ ὀστέον ἢ πέφλασται, ἢ καὶ ἀμφότερα ταῦτα, μήτε
240 ἄλλως ὁρέων, δεῖ δὴ | ἐπὶ τὸ ὀστέον τὸ τηκτὸν τὸ μελάντατον δεύσαντα, τῷ μέλανι φαρμάκῳ τῷ τηκομένῳ στεῖλαι τὸ ἕλκος· ὑποτείνας ὀθόνιον ἐλαίῳ τέγξας, εἶτα καταπλάσας τῇ μάζῃ ἐπιδῆσαι. τῇ δὲ ὑστεραίῃ ἀπολύσας, ἐκκαθήρας τὸ ἕλκος ἐπιξῦσαι. καὶ ἢν μὴ ὑγιὲς ᾖ, ἀλλ᾽ ἐρρώγῃ καὶ πεφλασμένον ᾖ, τὸ μὲν ἄλλο ἔσται ὀστέον λευκὸν ἐπιξυόμενον· ἡ δὲ ῥωγμὴ
242 καὶ ἡ φλάσις, κατατακέντος τοῦ | φαρμάκου, δεξαμένη τὸ φάρμακον ἐς ἑωυτὴν μέλαν ἐόν, ἔσται μέλαινα ἐν λευκῷ τῷ ὀστέῳ τῷ ἄλλῳ. ἀλλὰ χρὴ αὖθις τὴν ῥωγμὴν ταύτην φανεῖσαν μέλαιναν ἐπιξέειν κατὰ βάθος· καὶ ἢν μὲν ἐπιξύων τὴν ῥωγμὴν ἐξέλῃς καὶ ἀφανέα ποιήσῃς, φλάσις μὲν γεγένηται τοῦ ὀστέου ἢ μᾶλλον ἢ ἧσσον, ἥτις περιέρρηξε καὶ τὴν ῥωγμὴν τὴν ἀφανισθεῖσαν ὑπὸ τοῦ ξυστῆρος· ἧσσον δὲ φοβερὸν καὶ ἧσσον ἂν πρῆγμα ἀπ᾽ αὐτῆς γίνοιτο ἀφανισθείσης τῆς ῥωγμῆς. ἢν δὲ κατὰ βάθος ᾖ καὶ μὴ ἐθέλῃ ἐξιέναι

this happens during the hot season, and you are taking on the therapy from the beginning.

If you suspect that the bone has in fact been fractured or contused, or both, on taking into account evidence that the person was wounded by someone stronger—the two combatants being of different strength—and that the weapon was of a pernicious kind, and that the wounded person has been seized by vertigo and darkness, and lost consciousness and fell down, if this is what happened, but you cannot be sure by inspecting the wound in any other way whether the bone has been fractured or contused, or both, then you should wipe a very dark solution on to the bone, applying the dark agent to the lesion, and then after applying a linen underdressing moistened with olive oil and plastering the wound with barley dough, bind it up. On the following day remove this, clean the lesion out, and rasp it. If it is not sound, but is broken and contused, the rest of the bone on being rasped will be white, while the fracture or the contusion, on imbibing the melted dark agent into itself, will show up black against the white bone. Then scrape the fracture, which appears dark, down to its base: if, by rasping the fracture, you remove it and cause it to disappear, a contusion of the bone is present, more or less severe, and this caused the fracture that has now been removed by being rasped; such a contusion is less dire and significant once the fracture has been obliterated. But if the fracture is deep and refuses to go away on being ἐπιξυομένη, ἐς πρῖσιν ἀφήκει ἤδη τοιαύτη συμφορή. ἀλλὰ χρὴ πρίσαντα τὰ λοιπὰ ἰητρεύειν τὸ ἕλκος.

15. Φυλάσσεσθαι δὲ χρὴ ὅπως μή τι κακὸν ἀπολαύσῃ τὸ ὀστέον ἀπὸ τῆς σαρκός, ἢν κακῶς ἰητρεύηται. ὀστέῳ γὰρ καὶ πεπρισμένῳ καὶ ἄλλως ἀπρίστῳ ἐψιλωμένῳ δέ, καὶ ὑγιεῖ δὲ ἐόντι καὶ ἔχοντί τι σίνος ὑπὸ τοῦ βέλεος, δοκέοντι δὲ ὑγιεῖ εἶναι, κίνδυνός ἐστι μᾶλλον ὑπόπυον γενέσθαι, ἢν μὴ καὶ ἄλλως μὴ μέλλῃ, ἢν καὶ ἡ σὰρξ ἡ περιέχουσα τὸ ὀστέον κακῶς θεραπεύηται, καὶ φλεγμαίνῃ τε καὶ περισφίγγηται. πυρετῶδες γὰρ γίνεται καὶ πολλοῦ φλογμοῦ πλέον· καὶ δὴ τὸ ὀστέον ἐκ τῶν περιεχουσῶν σαρκῶν ἐς ἑωυτὸ θέρμην τε καὶ φλογμὸν καὶ ἄραδον ἐμποιέει καὶ σφυγμόν, καὶ ὁκόσα περ ἡ σὰρξ ἴσχει κακὰ ἐν ἑωυτῇ, καὶ ἐκ τούτων ὧδε ὑπόπυον γίνεται. κακὸν δὲ καὶ ὑγρήν τε εἶναι τὴν σάρκα ἐν τῷ ἕλκει καὶ μυδῶσαν καὶ ἐπὶ πολλὸν χρόνον καθαίρεσθαι. ἀλλὰ χρὴ διάπυον μὲν ποιῆσαι τὸ ἕλκος ὡς τάχιστα· οὕτω γὰρ ἂν
244 ἥκιστα φλεγμαίνοι τὰ περιέχοντα | τὸ ἕλκος καὶ τά-
χιστ' ἂν καθαρὸν εἴη. ἀνάγκη γὰρ ἔχει τὰς σάρκας τὰς κοπείσας καὶ φλασθείσας ὑπὸ τοῦ βέλεος, ὑποπύους γενομένας, ἐκτακῆναι. ἐπειδὰν δὲ καθαρθῇ, ξηρότερον χρὴ γίνεσθαι τὸ ἕλκος· οὕτως γὰρ ἂν τάχιστα ὑγιὲς γένοιτο, ξηρῆς σαρκὸς βλαστούσης καὶ μὴ ὑγρῆς, καὶ οὕτως οὐκ ἂν ὑπερσαρκήσειε τὸ ἕλκος. ὁ δὲ αὐτὸς λόγος καὶ ὑπὲρ τῆς μήνιγγος τῆς περὶ τὸν ἐγκέφαλον· ἢν γὰρ αὐτίκα ἐκπρίσας τὸ ὀστέον καὶ ἀφελὼν ἀπὸ τῆς μήνιγγος ψιλώσῃς, αὐτὴν καθαρὴν

rasped, the case requires trephination; after you trephine, apply the other treatments required to heal such a lesion.

15. Care must also be taken to prevent the bone from being harmed by the flesh as the consequence of some incorrect treatment; for if the bone has been trephined, or not trephined but denuded of flesh—both if it remains healthy itself, and even being damaged in some way by the weapon it retains the appearance of soundness—there is a danger that it will suppurate, in spite of not being disposed itself to suppuration, as the result of the contiguous flesh being poorly managed, so that it becomes inflamed and forms sinuses, which generate fever and great heat. When this happens, the bone attracts to itself from the surrounding flesh heat, inflammation, irritation, throbbing, and all the evils contained in the flesh, and from these suppuration follows. It is also bad for the lesion in the flesh to be moist and clammy, and to remain uncleaned for a long time. In fact, you should bring the lesion to suppuration as soon as possible, in order that the area around it will suppurate as little as possible and (sc. the wound) will be quickly cleaned. Tissues that have been struck and contused by a weapon must necessarily suppurate and then be sloughed off; on being cleaned, the lesion dries up, with dry tissue growing in, free of moisture, which leads most promptly to recovery and the lesion not being overgrown (sc. with proud flesh). It is the same with the membrane around the brain: if you trephine through the bone at once and remove it from the membrane, thereby exposing the membrane, you should clean and dry it as

χρὴ ποιῆσαι ὡς τάχιστα καὶ ξηρήν, ὡς μὴ ἐπὶ πολὺν χρόνον ὑγρὴ ἐοῦσα μυδᾷ τε καὶ ξαίνηται·[36] τούτων γὰρ οὕτως γινομένων σαπῆναι αὐτὴν κίνδυνος. |

246 16. Ὀστέον δὲ ὅ τι δεῖ ἀποστῆναι ἀπὸ τοῦ ἄλλου ὀστέου, ἕλκεος ἐν κεφαλῇ γενομένου, ἕδρης τε ἐούσης τοῦ βέλεος ἐν τῷ ὀστέῳ,[37] ἢ ἄλλως ἐπὶ πολὺ ψιλωθέντος τοῦ ὀστέου, ἀφίσταται ἐπὶ πολὺ ἔξαιμον γενόμενον. ἀναξηραίνεται γὰρ τὸ αἷμα ἐκ τοῦ ὀστέου ὑπό τε τοῦ χρόνου καὶ ὑπὸ φαρμάκων τῶν πλείστων. τάχιστα δ' ἂν ἀποσταίη, εἴ τις τὸ ἕλκος ὡς τάχιστα καθήρας ξηραίνοι τὸ λοιπὸν τό τε ἕλκος καὶ τὸ ὀστέον, καὶ τὸ μέζον καὶ τὸ ἧσσον. τὸ γὰρ τάχιστα ἀποξηρανθὲν καὶ ἀποστρακωθὲν τούτῳ μάλιστα ἀφίσταται ἀπὸ τοῦ ἄλλου ὀστέου τοῦ ἐναίμου τε καὶ ζῶντος. αὐτὸ ‹δ'› ἔξαιμόν τε γενόμενον καὶ ξηρὸν ‹ἐν›[38] τῷ ἐναίμῳ καὶ ζῶντι μάλα ἀφίσταται. |

248 17. Ὅσα δὲ τῶν ὀστέων ἐσφλᾶται ἔσω ἐκ τῆς φύσιος τῆς ἑωυτῶν, καταρραγέντα ἢ καὶ διακοπέντα πάνυ εὐρέα, ἀκινδυνότερα τὰ τοιαῦτα γίνεται, ἐπὴν ἡ μῆνιγξ ὑγιὴς ᾖ· καὶ τὰ πλέοσι ῥωγμῇσιν ἐσκαταρραγέντα καὶ εὐρυτέρῃσιν ἔτι ἀκινδυνότερα καὶ εὐμαρέστερα ἐς τὴν ἀφαίρεσιν γίνεται. καὶ οὐ χρὴ πρίειν τῶν τοιούτων οὐδέν, οὐδὲ κινδυνεύειν τὰ ὀστέα πειρώμενον ἀφαιρέειν πρὶν ἢ αὐτόματα ἐπανίῃ· †εἰκὸς πρῶτον χαλάσαντος.†[39] ἐπανέρχεται δὲ τῆς σαρκὸς |

36 ξαίνηται B: ἐξαίρηται V

37 The text in B ends at the bottom of a leaf with ὀ[στέῳ].

soon as you can, lest, by remaining moist, clammy and irritated over a long period, it runs the risk of becoming septic.

16. Any piece of bone bound to separate from the rest of the skull as the result of a wound in the head, when there is a *hedra* in the bone made by a weapon, or the bone is extensively denuded (sc. of flesh) in some other way, usually separates after becoming bloodless; for the blood dries up out of the bone with the passage of time and the application of most medications. Separation will proceed most rapidly if someone cleans the wound as soon as possible, and dries up the rest of the lesion and the bone, whether the affected area happens to be larger or smaller. For if the (sc. necrotic) bone is dried very soon and takes on the appearance of a potsherd, it will separate readily from the remaining bone that retains blood and life. An area of bone situated in the midst of bone that contains blood and life, which spontaneously becomes bloodless and dry, will also separate readily.

17. Any bones depressed inward out of their natural position, if they are comminuted or become widely separated, are less in danger of developing complications if their membrane remains intact; depressed, comminuted fractures with many widely separated fragments are even less dangerous and easier to remove. None of these require trephination, nor do you need to run the risk of trying to remove the bony fragments before they separate spontaneously † it is likely to come apart first † since they

[38] Add. δ' and ἐν Kollesch/Nickel (in Hanson).
[39] Obel. Hanson as corrupt.

ὑποφυομένης· ὑποφύεται δ᾽ ἐκ τῆς διπλόης τοῦ ὀστέου καὶ ἐκ τοῦ ὑγιέος, ἢν ἡ ἄνωθεν μοῖρα τοῦ ὀστέου μούνη σφακελίσῃ· οὕτω δ᾽ ἂν τάχιστα ἥ τε σὰρξ ὑποφύοιτο καὶ βλαστάνοι καὶ τὰ ὀστέα ἐπανίοι, εἴ τις τὸ ἕλκος ὡς τάχιστα διάπυον ποιήσας καθαρὸν ποιήσηται. καὶ ἢν διὰ παντὸς τοῦ ὀστέου ἄμφω αἱ μοῖραι ἐσφλασθῶσιν ἔσω ἐς τὴν μήνιγγα, ἥ τε ἄνω μοῖρα τοῦ ὀστέου καὶ ἡ κάτω, ἰητρεύοντι ὡσαύτως τὸ ἕλκος ὑγιὲς τάχιστα ἔσται, καὶ τὰ ὀστέα τάχιστα ἐπάνεισι τὰ ἐσφλασθέντα ἔσω.

18. Τῶν δὲ παιδίων τὰ ὀστέα καὶ λεπτότερά ἐστι καὶ μαλθακώτερα διὰ τοῦτο, ὅτι ἐναιμότερά ἐστι, καὶ κοῖλα καὶ σηραγγώδεα καὶ οὔτε πυκνὰ οὔτε στερεά. καὶ ὑπὸ τῶν βελέων ἴσων τε ἐόντων καὶ ἀσθενεστέρων, καὶ τρωθέντων ὁμοίως τε καὶ ἧσσον, τὸ τοῦ νεωτέρου παιδίου καὶ μᾶλλον καὶ θᾶσσον ὑποπυΐσκεται ἢ τὸ τοῦ πρεσβυτέρου, καὶ ἐν ἐλάσσονι χρόνῳ· καὶ ὅσα ἂν ἄλλως μέλλῃ ἀποθανεῖσθαι ἐκ τοῦ τρώματος, ὁ νεώτερος τοῦ πρεσβυτέρου θᾶσσον ἀπόλλυται.

Ἀλλὰ χρή, ἢν ψιλωθῇ τῆς σαρκὸς τὸ ὀστέον, προσέχοντα τὸν νόον, πειρᾶσθαι διαγινώσκειν ὅ τι μὴ ἔστι τοῖσιν ὀφθαλμοῖσιν ἰδεῖν, καὶ γνῶναι εἰ ἔρρωγε τὸ ὀστέον καὶ πέφλασται, ἢ εἰ[40] μοῦνον πέφλασται, καὶ εἰ, ἕδρης γενομένης τοῦ βέλεος, πρόσεστι φλάσις ἢ ῥωγμὴ ἢ ἄμφω ταῦτα. καὶ ἤν τι τούτων πεπόνθῃ τὸ ὀστέον, ἀφεῖναι τοῦ αἵματος τρυπῶντα τὸ ὀστέον σμικρῷ τρυπάνῳ, φυλασσόμενον ἐπ᾽ ὀλίγον·

come up as the flesh grows beneath them: this flesh grows out of the diploe of the bone from what is healthy when only the upper table of the bone is necrotic. This flesh grows and burgeons beneath the bone, making it move upward most quickly, if someone induces suppuration and cleaning as rapidly as possible. If the entire bone, including both tables on the top and the bottom, is depressed inward against the membrane, by application of the treatment I have indicated the lesion will be most readily healed, and the pieces of bone that were depressed inward will most quickly come (sc. back) up.

18. The skull bones of children are quite delicate and soft because they contain more blood, and are hollow and porous rather than dense and solid. If wounded with weapons that are similar or even less dangerous, inflicted with equal or less force, the bone of a younger child suppurates more and quicker than the bone of an older child, and for a shorter time. Among children destined to die from a wound, a younger one dies sooner than an older one.

You must, if the bone is denuded of flesh, apply your mind to infer things that cannot be seen or known with the eyes alone, *viz.* whether the bone is fractured and contused, or only contused, and when there is a *hedra* from a weapon, whether there is also a contusion, a fracture, or both of these. If the skull bone has suffered any of these (sc. injuries), perforate it with a small perforator in order to let the blood escape, taking care to check your progress

---

40 πέφλ., ἢ εἰ Ermerins: εἰ πέφλ. ἢ V

λεπτότερον γὰρ τὸ ὀστέον καὶ ἐπιπολαιότερον τῶν νέων ἢ τῶν πρεσβυτέρων.

19. Ὅστις δὲ μέλλει ἐκ τρωμάτων ἐν κεφαλῇ ἀπο-
252 θνῄσκειν, | καὶ μὴ δυνατὸν αὐτὸν ὑγιᾶ γενέσθαι μηδὲ σωθῆναι, ἐκ τῶνδε τῶν σημείων χρὴ τὴν διάγνωσιν ποιέεσθαι τοῦ μέλλοντος ἀποθνῄσκειν, καὶ προλέγειν τὸ μέλλον ἔσεσθαι. πάσχει γὰρ τάδε· ὁπόταν τις ὀστέον κατεηγὸς ἢ ἐρρωγὸς ἢ πεφλασμένον, ἢ ὅτῳ γοῦν τρόπῳ κατεηγὸς ἐννοήσας ἁμάρτῃ, καὶ μήτε ξύσῃ μήτε πρίσῃ †μήτε δεόμενον, μήτε δὲ†[41] ὡς ὑγιέος ὄντος τοῦ ὀστέου, πρὸ τῶν τεσσερεσκαίδεκα ἡμερέων πυρετὸς ἐπιλήψεται, ὡς ἐπὶ πολὺ ἐν χειμῶνι, ἐν δὲ τῷ θέρει μετὰ τὰς ἑπτὰ ἡμέρας ὁ πυρετὸς ἐπιλαμβάνει. καὶ ἐπειδὰν τοῦτο γένηται, τὸ ἕλκος ἄχροον γίνεται καὶ ἐξ αὐτοῦ ἰχὼρ ῥεῖ σμικρός, καὶ τὸ φλεγμαῖνον
254 ἐκτέθνηκεν ἐξ αὐτοῦ, καὶ βλιχῶδες γίνεται | καὶ φαίνεται ὥσπερ τάριχος, χροιὴν πυρρόν, ὑποπέλιον. καὶ τὸ ὀστέον σφακελίζειν τηνικαῦτα ἄρχεται, καὶ γίνεται περκνόν, λεῖον ὄν· τελευταῖον δὲ ἅπαν ὠχρὸν γενόμενον ἢ ἔκλευκον. ὅταν δ᾽ ἤδη ὑπόπυον ᾖ, ἐπὶ τῇ γλώσσῃ φλύκταιναι γίνονται, καὶ παραφρονέων τελευτᾷ. καὶ σπασμὸς ἐπιλαμβάνει τοὺς πλείστους τὰ ἐπὶ θάτερα τοῦ σώματος· ἢν μὲν ἐν τῷ ἐπ᾽ ἀριστερὰ τῆς κεφαλῆς ἔχῃ τὸ ἕλκος, τὰ ἐπὶ δεξιὰ τοῦ σώματος ὁ σπασμὸς λαμβάνει· ἢν δ᾽ ἐν τῷ ἐπὶ δεξιὰ τῆς κεφαλῆς ἔχῃ τὸ ἕλκος, τὰ ἐπ᾽ ἀριστερὰ τοῦ σώματος ὁ σπασμὸς ἐπιλαμβάνει. εἰσὶ δ᾽ οἳ καὶ ἀπόπληκτοι γίνονται, καὶ οὕτως ἀπόλλυνται πρὸ ἑπτὰ ἡμέρων ἐν

at short intervals, since the bone in young people is thinner and shallower than in older ones.

19. In anyone likely to succumb as the result of a wound in the head, because unable either to recover (sc. by himself) or to be saved (sc. by medical intervention), you must make your evaluation from the following signs with regard to impending death, and predict the probable disease course. The symptoms are as follows. In the case where someone recognizes that the skull has been injured by sustaining a fracture or contusion, or in some other way, but fails either to rasp it or to trephine it, in the belief that this is unnecessary in spite of the bone's injury, fever will as a rule set in before fourteen days in the winter, and just after seven days in the summer. As this happens, the lesion will become discolored, a little serous discharge with flow out of it, and what is inflamed will die away out of it; the lesion becomes clammy and takes on the appearance of salt fish, becoming purple in color and somewhat livid. The bone begins to necrose, while remaining smooth, becomes dusky in color, and finally turns pale yellow and whitish. When it has already become somewhat purulent, sores will form on the tongue, and the patient will become delirious and die. In most patients there is a spasm on the opposite side of the body: if the wound is on the left side of the head, the spasm will be on the right side of the body, whereas if the lesion is on the right side, the spasm will be on the left side. Other patients become paralyzed, in which case they will die before the seventh day in summer

---

41 Obel. Hanson as corrupt.

θέρει ἢ τεσσάρων καὶ δέκα ἐν χειμῶνι. ὁμοίως δὲ τὰ σημεῖα ταῦτα σημαίνει, καὶ ἐν πρεσβυτέρῳ ἐόντι τῷ τρωματίῃ καὶ ἐν νεωτέρῳ.

Ἀλλὰ χρή, εἰ ἐννοοίης τὸν πυρετὸν ἐπιλαμβάνοντα καὶ τῶν ἄλλων τι σημεῖον τούτῳ προσγενόμενον, μὴ διατρίβειν, ἀλλὰ πρίσαντα τὸ ὀστέον πρὸς τὴν μήνιγγα ἢ καταξύσαντα τῷ ξυστῆρι—καπυρὸν δὲ γίνεται καὶ εὔξυστον—ἔπειτα τὰ λοιπὰ οὕτως ἰητρεύειν ὅπως ἂν δοκέῃ ξυμφέρειν, πρὸς τὸ γινόμενον ὁρῶν.

20. Ὅταν δ' ἐπὶ τρώματι ἐν κεφαλῇ ἀνθρώπου ἢ πεπριωμένου ἢ ἀπριώτου, ἐψιλωμένου δὲ τοῦ ὀστέου, οἴδημα ἐπιγένηται ἐρυθρὸν καὶ ἐρυσιπελατῶδες ἐν τῷ
256 προσώπῳ καὶ ἐν τοῖσιν ὀφθαλμοῖσιν | ἀμφοτέροισιν ἢ τῷ ἑτέρῳ, καὶ εἴ τις ἅπτοιτο τοῦ οἰδήματος, ὀδυνῷτο, καὶ πυρετὸς ἐπιλαμβάνοι καὶ ῥῖγος, τὸ δὲ ἕλκος αὐτό τε ἀπὸ τῆς σαρκὸς καλῶς ἔχοι ἰδέσθαι καὶ τἀπὸ τοῦ ὀστέου, καὶ τὰ περιέχοντα τὸ ἕλκος ἔχοι καλῶς, πλὴν τοῦ οἰδήματος τοῦ ἐν προσώπῳ, καὶ ἄλλην ἁμαρτάδα μηδεμίαν ἔχοι τὸ οἴδημα τῆς ἄλλης διαίτης, τούτου χρὴ τὴν κάτω κοιλίην ὑποκαθῆραι φαρμάκῳ ὅ τι χόλην ἄγει· καὶ οὕτω καταρθέντος, ὅ τε πυρετὸς ἀφίησι καὶ τὸ οἴδημα καθίσταται καὶ ὑγιὴς γίνεται. τὸ δὲ φάρμακον χρὴ διδόναι πρὸς τὴν δύναμιν τοῦ ἀνθρώπου ὁρῶν, ὡς ἂν ἔχῃ ἰσχύος.

21. Περὶ δὲ πρίσιος, ὅταν καταλάβῃ ἀνάγκη πρῖσαι ἄνθρωπον, ὧδε γινώσκειν. ἢν ἐξ ἀρχῆς λαβὼν τὸ ἴημα πρίῃς, οὐ χρὴ ἐκπρίειν τὸ ὀστέον πρὸς τὴν μήνιγγα αὐτίκα· οὐ γὰρ συμφέρει τὴν μήνιγγα ψιλὴν

and the fourteenth day in winter. These signs have the same significance in both older and younger patients.

But if you see that fever is developing together with some of the other signs, you should not hesitate, but trephine the skull right down to the membrane, or scrape it down with a rasp—the bone will be brittle and easy to abrade—and manage your further treatment in the way that most promises to bring benefit, paying attention to how the case develops.

20. When, in a person with a wound in the head—whether it has been trephined or not—the bone is denuded (sc. of flesh), a reddish edema and erysipelas symptoms develop in the face and either one or both eyes—when anyone touches the edema it is painful—and fever and chills come on, but the lesion in the flesh has a favorable appearance, as does the bone, the area around the lesion is healthy except for the edema in the face, and the edema has not been complicated by any mistake in the rest of the regimen, in this case the person's cavity must be cleaned downward with a purgative that moves bile: when the evacuation has taken place, the fever will subside, the swelling will go down, and the person will recover. The purge must be adapted to the person's strength, whatever force he happens to command.

21. Concerning trephination: when it becomes necessary to trephine someone, consider the following. If you are trephining in a case you have managed since its onset, you do not need to saw all the way to the membrane immediately, since it is not good for the membrane to be bare

εἶναι τοῦ ὀστέου ἐπὶ πολὺν χρόνον κακοπαθοῦσαν, ἀλλὰ τελευτῶσά πῃ καὶ διεμύδησεν. ἔστι δὲ καὶ ἕτερος κίνδυνος, ἢν αὐτίκα ἀφαιρῇς πρὸς τὴν μήνιγγα ἐκπρίσας τὸ ὀστέον, τρῶσαι ἐν τῷ ἔργῳ τῷ πρίονι
258 τὴν μήνιγγα. ἀλλὰ χρὴ πρίοντα, | ἐπειδὰν ὀλίγον πάνυ δέῃ διαπεπρίσθαι, καὶ ἤδη κινῆται τὸ ὀστέον, παύσασθαι πρίοντα, καὶ ἐᾶν ἐπὶ τὸ αὐτόματον ἀποστῆναι τὸ ὀστέον. ἐν γὰρ τῷ διαπριωτῷ ὀστέῳ καὶ ἐπιλελειμμένῳ τῆς πρίσιος οὐκ ἂν ἐπιγένοιτο κακὸν οὐδέν, λεπτὸν γὰρ τὸ λειπόμενον ἤδη γίνεται. τὰ δὲ λοιπὰ ἰᾶσθαι χρή, ὡς ἂν δοκέῃ ξυμφέρειν τῷ ἕλκει.

Πρίοντα δὲ χρὴ πυκινὰ ἐξαιρέειν τὸν πρίονα τῆς θερμασίης εἵνεκα τοῦ ὀστέου, καὶ ὕδατι ψυχρῷ ἐναποβάπτειν. θερμαινόμενος γὰρ ὑπὸ τῆς περιόδου ὁ πρίων καὶ τὸ ὀστέον ἐκθερμαίνων καὶ ἀναξηραίνων κατακαίει, καὶ μέζον ποιέει ἀφίστασθαι τὸ ὀστέον τὸ περιέχον τὴν πρῖσιν ἢ ὅσον μέλλει ἀφίστασθαι. καὶ ἢν αὐτίκα βούλῃ ἐκπρῖσαι τὸ πρὸς τὴν μήνιγγα, ἔπειτα ἀφελέειν τὸ ὀστέον, ὡσαύτως χρὴ πυκινά τε ἐξαιρέειν τὸν πρίονα καὶ ἐναποβάπτειν τῷ ὕδατι τῷ ψυχρῷ.

Ἢν δὲ μὴ ἐξ ἀρχῆς λαμβάνῃς τὸ ἴημα, ἀλλὰ παρ᾽ ἄλλου παραδέχῃ ὑστερίζων τῆς ἰήσιος, πρίονι χρὴ χαρακτῷ ἐκπρίειν μὲν αὐτίκα τὸ ὀστέον πρὸς τὴν μήνιγγα, θαμινὰ δὲ ἐξαιρεῦντα τὸν πρίονα σκοπεῖσθαι καὶ ἄλλως καὶ τῇ μήλῃ πέριξ κατὰ τὴν ὁδὸν τοῦ πρίονος· καὶ γὰρ πολὺ θᾶσσον διαπρίεται τὸ ὀστέον, ἢν ὑπόπυόν τε ἐὸν ἤδη καὶ διάπυον πρίῃς. καὶ πολof bone and exposed to morbid influences for a long time, since it may eventually become fungoid. There is also another danger if you remove (sc. the bone) at once by sawing right to the membrane, namely that in the process of trephining you may accidently injure the membrane. Rather, when the bone is sawn almost through and can be moved, you must leave off trephining and let the bone separate by itself, since a bone that is sawn almost through but left still attached will cause no trouble, because the remaining piece is already thin. Otherwise therapy should be managed as seems best for the wound.

During the procedure of trephining, you must remove the saw frequently because it heats the bone, and immerse it in cold water; for as the trephine is heated by turning, it burns the bone by heating and drying it, leading to a loss of more bone around the trephination than would otherwise occur. Also, if you prefer to drill through to the membrane at once, in order to remove the bone, still remove the trephine frequently and immerse it in cold water.

If you have not been managing a patient's case from the beginning, but took it over later from someone else, so that you are behind schedule in the treatment, you must immediately trephine the bone with a sharp-toothed trephine right to the membrane, frequently removing the saw to examine it, and especially inspecting all around its path with a probe: for the bone will be sawn through much more quickly, if while you are trephining it is already suppurating, and pus is present all through it. Often the bone λάκις τυγχάνει ἐπιπόλαιον ἐὸν τὸ ὀστέον, ἄλλως τε καὶ ἢν ταύτῃ τῆς κεφαλῆς ᾗ τὸ τρῶμα ᾗ τυγχάνει λεπτότερον ἐὸν τὸ ὀστέον ἢ παχύτερον. ἀλλὰ φυλάσσεσθαι χρὴ ὡς μὴ λάθῃς προσβαλὼν τὸν πρίονα, ἀλλ' ὅπῃ δοκέει πάχιστον εἶναι τὸ ὀστέον, ἐς τοῦτο αἰεὶ ἐνστηρίζειν τὸν πρίονα, θαμινὰ σκοπούμενος, καὶ πειρᾶσθαι ἀνακινέων τὸ ὀστέον | ἀναβάλλειν, ἀφελὼν δὲ τὰ λοιπὰ ἰητρεύειν ὡς ἂν δοκέῃ ξυμφέρειν τῷ ἕλκει,[42] πρὸς τὸ γινόμενον ὁρῶν.

Καὶ ἤν, ἐξ ἀρχῆς λαβὼν τὸ ἴημα, αὐτίκα βούλῃ ἐκπρίσας τὸ ὀστέον ἀφελεῖν ἀπὸ τῆς μήνιγγος, ὡσαύτως χρὴ πυκινά τε σκοπεῖσθαι τῇ μήλῃ τὴν περίοδον τοῦ πρίονος, καὶ ἐς τὸ παχύτατον αἰεὶ τοῦ ὀστέου τὸν πρίονα ἐνστηρίζειν, καὶ ἀνακινέων βούλεσθαι ἀφελέειν τὸ ὀστέον. ἢν δὲ τρυπάνῳ χρῇ, πρός γε[43] τὴν μήνιγγα μὴ ἀφικνέεσθαι, ἢν ἐξ ἀρχῆς λαμβάνων τὸ ἴημα τρυπᾷς, ἀλλ' ἐπιλιπεῖν τοῦ ὀστέου λεπτόν, ὥσπερ καὶ ἐν τῇ πρίσει γέγραπται [ἢν ἐξ ἀρχῆς λαμβάνων τὸ ἴημα πρίεις].[44]

42 Two passages from *Airs Waters Places* are inserted into the text here due to the inversion of a bifolium in one of V's ancestors: see Loeb *Hippocrates* vol. 1 (LCL 147), pp. 67f.

43 γε Ermerins: δὲ V

44 ἢ ἐξ ἀρχῆς λαμβάνων τὸ ἰήμα πρίες V: del. Ho^2

may chance to be shallow, especially if the wound is in an area of the skull where the bone happens (sc. naturally) to be thinner rather than thicker. You should also be careful not to advance the trephine without realizing it, and always fix the trephine at the place in the bone that appears to be thickest; examine the saw regularly, and try, by moving the bone back and forth, to lift it up. After you remove the bone, manage the subsequent therapy as seems best for the wound, in view of the circumstances.

If you are trephining in a case you have managed since its onset, you may wish immediately to saw the bone out away from the membrane and remove it: in this case, too, you should frequently remove the trephine from the membrane and examine its circular path with a probe; always fix the trephine in the bone where it is thickest, and by moving the bone back and forth remove it. If you use a perforator in a case you have managed from the beginning, do not advance all the way (sc. through the bone) as far as the membrane with the perforator, but leave a thin piece of bone next (sc. to the membrane) intact, the way I described under trephination above.

IN THE SURGERY

INTRODUCTION

This loose collection of notes on how to arrange a practitioner's office and perform various surgical practices, in particular forms of bandaging, is organized by subject in twenty-five chapters:

1	Sources of clinical knowledge
2	The components of surgical practice
3–4	The operator and his hands
5	The instruments
6	The assistants
7–10	Bandages and their application
11–12/14	Compresses and splints
13	Preparation of water
15/19	Positions of the limbs in bandaging
16	Traction
17	Friction
18	Complex bandages
20–21	Two surgical aphorisms
22	The protection of damaged tissues
23	The adjustment of dislocations
24	The restoration of atrophied parts
25	The management of specific conditions

The text of *In the Surgery* was widely known among ancient medical writers and lexicographers, including Bacchius of Tanagra in the third century BC, and is so titled by Erotian, Galen, and Oribasius.[1] Erotian lists the treatise among the "therapeutic works devoted to surgery" in his census of the Hippocratic Collection, and also includes in his *Collection of Hippocratic Words* about twenty terms drawn certainly or possibly from its text.[2] Galen glosses words from the text in his *Examination of Difficult Words in Hippocrates*,[3] devotes a *Commentary* in three books to the treatise,[4] and quotes from it in several of his other writings.[5]

In the ninth and tenth centuries, *In the Surgery* along with Galen's commentary received attention among various Arabic medical writers and translators, such as Ḥunain ibn Isḥāq and Rhazes.[6] Parts (K. 18B 738.6–67.9; 863.11–88.17; 894.11–925.6) of the commentary, including the Hippocratic lemmata, are preserved in Arabic translation in the manuscript Scorial. Arab. 845 dated to 1190, fols. 128ʳ–55ᵛ; in addition, an almost complete version of the Hippocratic text (without Galen's commentary) based on the same Arabic translation is extant in the manuscript

1 Cf. *Testimonien*, vol. I, pp. 379–91.

2 See Erotian, p. 9; Nachmanson, pp. 342–46.

3 See Perilli, p. 416 s.v. Off. Med.

4 Cf. Boudon, p. 160; Manetti/Ros., pp. 1546f. The Greek text of Galen's *Commentary* is edited from the sole independent witness, Parisinus Graecus 1849 (XIV c.) fols. 175ᵛ–206ᵛ, by C. H. Th. Reinhold in Ἱπποκράτης (Athens, 1866), vol. 3, pp. 47–233. Important for the study of the Galenic text is S. Vogt, *De Galeni in libellum κατ' ἰητρεῖον commentariis* (diss., Marburg, 1910).

5 Cf. *Testimonien*, vol. II,1, pp. 375–91; vol. II,2, pp. 290–99.

6 Cf. Ullmann, pp. 30f. and 50f.; Sezgin, pp. 36 and 123; *Testimonien*, vol. III, pp. 325–27.

Aya Sofya 3632 (XIII c.), fols. 110v–17v. Both these Arabic translations have been edited and translated into English:

Lyons, M. C. *Galeni in Hippocratis de officina medici commentariorum versionem Arabicam, quoad exstat, ex codice Scorialensi . . . edidit et in linguam anglicam vertit.* CMG Suppl. Orientale I. Berlin, 1963.

———. *Kitāb Buqrāṭ al-maʿrūf bi-Qāṭīṭrīyūn ay ḥānūt al-ṭabīb* (*Hippocrates: In the Surgery*). Cambridge, 1968. (= Ar.)

The only trace the treatise has left in medieval Latin literature is as titles (*chatha iatrion/catdiatrion*) in two slightly divergent lists of Hippocratic works accompanying the Hippocratic biography *Yppocratis genus, vita, dogma* preserved in the twelfth-century manuscript (Brussels, 1342–1350).[7] With the Renaissance, *In the Surgery* became a standard part of the Hippocratic Corpus, appearing in all the collected editions and translations, including Adams, Kühlewein, and Withington, as well as in the two important surgical collections by Vidius and Petrequin.

In preparing the present edition, I have read the three independent Greek manuscripts B, M, and V from microfilms and consulted Lyons' English translation of the Arabic version. The recently published work, J. Jouanna, *Hippocrate. L'Officine du médecin*, Budé VII (3) (Paris, 2024), provides a thorough discussion of the text, its complex transmission, and many important points of interpretation.

[7] Pinault, p. 132, line 62.

ΚΑΤ' ἸΗΤΡΕΙΟΝ

III 272 Littré 1. Ἢ ὅμοια ἢ ἀνόμοια, ἐξ ἀρχῆς, ἀπὸ τῶν μεγίστων, ἀπὸ τῶν ῥηΐστων, ἀπὸ τῶν πάντῃ πάντως γινωσκομένων, ἃ καὶ ἰδεῖν καὶ θιγεῖν καὶ ἀκοῦσαι ἔστιν· ἃ καὶ τῇ ὄψει καὶ τῇ ἁφῇ καὶ τῇ ἀκοῇ καὶ τῇ ῥινὶ καὶ τῇ γλώττῃ καὶ τῇ γνώμῃ ἔστιν αἰσθέσθαι· ἅ, οἷς γινώσκομεν, ἅπασιν ἔστι γνῶναι. |

274 2. Τὰ δὲ ἐς χειρουργίην κατ' ἰητρεῖον· ὁ ἀσθενέων,
276 ὁ δρῶν, οἱ | ὑπηρέται, τὰ ὄργανα, τὸ φῶς—ὅκου, ὅκως, ὅσα, οἷς, ὡς, ὁκότε· τὸ σῶμα, τὰ ἄρμενα· ὁ χρόνος, ὁ τρόπος, ὁ τόπος. |

278 3. Ὁ δρῶν, ἢ καθήμενος ἢ ἑστεώς, συμμέτρως πρὸς ἑωυτόν, πρὸς τὸ χειριζόμενον, πρὸς τὴν αὐγήν. αὐγῆς μὲν οὖν δύο εἴδεα, τὸ μὲν κοινόν, τὸ δὲ τεχνητόν· τὸ μὲν οὖν κοινὸν οὐκ ἐφ' ἡμῖν, τὸ δὲ τεχνητὸν καὶ ἐφ' ἡμῖν. ὧν ἑκατέρου δύο αἱ χρήσιες, ἢ πρὸς αὐγὴν ἢ ὑπ'[1] αὐγήν. ὑπ' αὐγὴν μὲν οὖν ὀλίγη τε ἡ χρῆσις καταφανής τε ἡ μετριότης· τὰ δὲ πρὸς αὐγήν, ἐκ τῶν

[1] πρὸς αὐγὴν ἢ ὑπ' M Ar.: ὑπ' αὐγὴν ἢ πρὸς BV

IN THE SURGERY

1. Either alike or not alike; from the beginning; from the greatest, from the easiest, from those known altogether and in every respect, which are to see, to touch, and to hear. And those that are to be perceived by sight, touch, hearing, the nose, the tongue, and contemplation; that can be known by all the means by which we know.[1]

2. The makeup of surgical practice in the physician's office: the patient, the operator, the assistants, the instruments, the light—where, how, how many, by which means, as, when.[2] The body, the surgical apparatus, the time, the manner, the place.

3. The operator, either seated or standing, must be positioned suitably for himself, in relation to the part being operated on, and to the light. Of light there are two kinds, the natural and the artificial: the natural is not within our power, the artificial is within our power. Each of them has two kinds of application, as direct light and as oblique light: oblique light is little used, and the suitable amount is obvious. With direct light coming from sources

[1] The beginning of the text is apparently mutilated, leaving its interpretation at first very hypothetical. [2] Various guesses about what this series of questions is referring to are recorded as far back as Galen's *Commentary*, but none inspires much credit.

παρεουσέων. ἐκ τῶν ξυμφερουσέων αὐγέων πρὸς τὴν λαμπροτάτην τρέπειν τὸ χειριζόμενον, πλὴν ὁκόσα λαθεῖν δεῖ ἢ ὁρῆν αἰσχρόν, οὕτω δὲ τὸ μὲν χειριζόμενον ἐναντίον τῇ αὐγῇ, τὸν δὲ χειρίζοντα ἐναντίον τῷ
280 χειριζομένῳ, πλὴν ὥστε μὴ ἐπισκοτάζειν· οὕτως | γὰρ ἂν ὁ μὲν δρῶν ὁρῴη, τὸ δὲ χειριζόμενον οὐχ ὁρῷτο.

Πρὸς ἑωυτὸν δέ, καθημένῳ μὲν πόδες ἐς τὴν ἄνω ἴξιν κατ' ἰθὺ γούνασι· διάστασιν δὲ ὀλίγον συμβεβῶτες. γούνατα δὲ ἀνωτέρω βουβώνων σμικρόν, διάστασιν δέ, ἀγκῶσιν θέσει, καὶ παραθέσει· ἱμάτιον εὐσταλέως, εὐκρινέως, ἴσως, ὁμοίως ἀγκῶσιν, ὤμοισιν.

Πρὸς δὲ τὸ χειριζόμενον, τοῦ μὲν πρόσω καὶ ἐγγὺς καὶ τοῦ ἄνω καὶ τοῦ κάτω, καὶ ἔνθα ἢ ἔνθα ἢ μέσον. τοῦ μὲν πρόσω καὶ ἐγγὺς ὅριον, ἀγκῶνας ἐς μὲν τὸ πρόσθεν γούνατα μὴ ἀμείβειν, ἐς δὲ τὸ ὄπισθεν πλευ-
282 ράς· τοῦ δὲ ἄνω μὴ ἀνωτέρω μαζῶν ἄκρας | χεῖρας ἔχειν· τοῦ δὲ κάτω, μὴ κατωτέρω ἢ ὥστε στῆθος ἐπὶ γούνασιν ἔχοντα, ἔχειν χεῖρας ἄκρας ἐγγωνίους πρὸς βραχίονας. τὰ μὲν κατὰ μέσον οὕτως· τὰ δὲ ἔνθα ἢ ἔνθα, μὴ ἔξω τῆς ἕδρης, κατὰ λόγον δὲ τῆς ἐπιστροφῆς προσβαλλόμενον τὸ σῶμα, καὶ τοῦ σώματος τὸ ἐργαζόμενον.

Ἑστεῶτα δέ, ἰδεῖν μὲν καὶ ἐπ' ἀμφοτέρων βεβῶτα ἐξ ἴσου τῶν ποδῶν ἅλις, δρῆν δὲ τῷ ἑτέρῳ ἐπιβεβαῶτα, μὴ τῷ κατὰ τὴν δρῶσαν χεῖρα· ὕψος γούνατα[2]

[2] γούνατα BM: γουνάτων V

that are available and useful, the operated part must be turned toward what is brightest, except for any part that should remain unexposed or is shameful to see: the part operated on faces the light, the operator faces the part, but in such a way as not to cast any shadow on it. In this way the operator will have a good view, but the part operated on will not be exposed to sight.

As far as the operator himself is concerned, when he is seated his feet should be directly below his knees, together but leaving a small space between them. His knees should be a little higher than the groin, slightly separated, and provide support for the elbows to rest on them. His cloak should be orderly, without creases, even, and well fitting on his elbows and shoulders.

Concerning the part operated on: its farness and nearness, its highness and lowness, and its being on the one side or the other, or in the middle. The limits of far and near are defined by where the (sc. operator's) elbows pass either in front of his knees or behind his ribs; the limits of high and low are reached when either his forearms are above his breasts or—with his chest lowered on to his knees—his forearms go beyond forming a right angle with his upper arms. This applies when the operation is in the middle, but when it is on one side or the other, the surgeon, while remaining seated where he is, moves according to the inclination needed with his whole body or the limb carrying out the procedure.

In the standing position, the surgeon should conduct his examination while treading evenly on both his feet, but he should operate with his weight on one foot, namely on the foot opposite to the arm he is operating with: in height his knees should be in the same relationship to his groins

πρὸς βουβῶνας ὡς ἐν ἕδρῃ· καὶ τὰ ἄλλα ὅρια τὰ αὐτά. ὁ δὲ χειριζόμενος τῷ χειρίζοντι τῷ ἄλλῳ μέρει τοῦ σώματος ὑπηρετείτω, ἢ ἑστεὼς ἢ καθήμενος ἢ κείμενος, ὡς ἂν ῥήϊστα ὃ δεῖ σχῆμα ἔχων διατελῇ,
284 φυλάσσων ὑπόρρυσιν, | ὑπόστασιν, ἔκστρεψιν, καταντίαν, ὡς ὃ δεῖ σώζηται καὶ σχῆμα καὶ εἶδος τοῦ χειριζομένου ἐν παρέξει, ἐν χειρισμῷ, ἐν τῇ ἔπειτα ἕξει.

4. Ὄνυχας μήτε ὑπερέχειν μήτε ἐλλείπειν δακτύλων
286 κορυφάς· | ἐς χρῆσιν ἀσκέειν δακτύλοισι μὲν ἄκροις, τὰ πλεῖστα λιχανῷ πρὸς μέγαν· ὅλῃ δὲ καταπρηνεῖ, ἀμφοτέρῃσι δὲ ἐναντίῃσι. δακτύλων εὐφυΐα· μέγα τὸ ἐν μέσῳ τῶν δακτύλων, καὶ ἀπεναντίον τὸν μέγαν τῷ λιχανῷ. νοῦσος δέ, δι᾽ ἣν καὶ βλάπτονται, τοῖσι δ᾽ ἐκ γενεῆς ἢ ἐν τροφῇ εἴθισται ὁ μέγας ὑπὸ τῶν ἄλλων |
288 δακτύλων κατέχεσθαι δῆλον. τὰ ἔργα πάντα ἀσκέειν ἑκατέρῃ δρῶντα, καὶ ἀμφοτέρῃσιν ἅμα—ὅμοιαι γάρ εἰσιν ἀμφότεραι—στοχαζόμενον ἀγαθῶς, καλῶς, ταχέως, ἀπόνως, εὐπόρως, εὐρύθμως.

5. Ὄργανα μὲν καὶ ὅτε, καὶ οἵως, εἰρήσεται. ὅκου[3] δεῖ μὴ ἐμποδὼν τῷ ἔργῳ μηδὲ ἐκποδὼν τῇ ἀναιρέσει, παρὰ τὸ ἐργαζόμενον δὲ τοῦ σώματος· ἄλλος δὲ ἣν διδῷ, ἕτοιμος ὀλίγῳ πρότερον ἔστω, ποιείτω δέ, ὅταν κελεύῃς.

[3] ὅκου M "where" Ar.: οὐ BV

as when he is sitting in a chair. The other limits are similar. The patient should help the operator with the rest of his body, by standing, sitting or lying in such a way as easily to maintain the posture he should have, being careful not to slip, drop down, twist, or hang down, so that the position and alignment of the part being operated on will be preserved as it should be in the stage of presentation, during the operation, and afterward.

4. The operator's nails should neither grow beyond the ends of his fingers, nor fall short of them. He should practice exercising the tips of the fingers, usually in opposing the forefinger to the thumb, sometimes holding the whole hand with the palm downward, and at other times with the two hands held against one another. It is a favorable dispensation of the fingers that there are wide spaces between them, and that it is possible to oppose the thumb to the forefinger: but it is obviously an impairment for those who, either congenitally or through their nurture, are in the habit of holding their thumb down below their fingers. You should practice all the movements I have described with each hand separately and with both hands at the same time—for the hands are both alike—aiming to do them well, gracefully, quickly, effortlessly, readily, and rhythmically.

5. The instruments, both when they are to be used and how, will be discussed. Their proper place is neither to be so close as to get in the way of the procedure being performed, nor to be so far away as to be out of reach, but rather they should be close to what is being done with the body. If someone is to hand them over, he should be ready with them a little beforehand, and obey when they are requested.

6. Οἱ δὲ περὶ τὸν ἀσθενέοντα, τὸ μὲν χειριζόμενον παρεχόντων, ὡς ἂν δοκῇ·[4] τὸ δὲ ἄλλο σῶμα κατεχόντων, ὡς ἀτρεμῇ, σιγῶντες, ἀκούοντες τοῦ ἐφεστεῶτος. |
290 7. Ἐπιδέσιος δύο εἴδεα, εἰργασμένον καὶ ἐργαζόμενον. ἐργαζόμενον μὲν ταχέως, ἀπόνως, εὐπόρως, εὐρύθμως. ταχέως μὲν ἀνύειν τὰ ἔργα· ἀπόνως δὲ ῥηϊδίως δρῆν· εὐπορίη δὲ ἐς πᾶν ἑτοίμως· εὐρυθμίη δὲ ὁρῆσθαι ἡδέως· ἀφ' ὧν δὲ ταῦτα ἀσκημάτων εἴρηται. εἰργασμένον δὲ ἀγαθῶς, καλῶς· καλῶς μὲν ἁπλῶς εὐκρινέως· ἢ ὅμοια καὶ ἴσα, ἴσως καὶ ὁμοίως· ἢ ἄνισα καὶ ἀνόμοια ἀνίσως καὶ ἀνομοίως. τὰ δὲ εἴδεα ἁπλοῦν
292 | εὔκυκλον, σκέπαρνον, σιμόν, ὀφθαλμὸς καὶ ῥόμβος καὶ ἡμίτομον· ἁρμόζον τὸ εἶδος τῷ εἴδει καὶ τῷ πάθει τοῦ ἐπιδεομένου.

8. Ἀγαθῶς δὲ δύο εἴδεα τοῦ ἐπιδεομένου· ἰσχύος
294 μὲν | ἢ πιέξει, ἢ πλήθει ὀθονίων. τὸ μὲν οὖν αὐτὴ ἡ ἐπίδεσις ἰῆται, τὸ δὲ τοῖς ἰωμένοισιν ὑπηρετέει. ἐς μὲν οὖν ταῦτα νόμος· ἐν δὲ τούτοισι μέγιστα ἐπιδέσιος· πίεξις μὲν ὥστε τὰ ἐπικείμενα μὴ ἀφεστάναι, μηδὲ ἐρηρεῖσθαι, ἀλλ' ἡρμόσθαι μέν, προσηναγκάσθαι δὲ μή, ἧσσον τὰ μέσα, ἥκιστα δὲ τὰ ἔσχατα.[5] ἅμμα καὶ

[4] δοκῇ V: δοθῇ BM: "as the latter (the physician) thinks best" Ar.

[5] μέσα, ἥκιστα δὲ τὰ ἔσχατα Jouanna: ἧσσον MV: εισσον B . . . μέσα BMV

6. Of the assistants around the patient, some should present the part being operated on as required, and others should hold the rest of the body securely so that it is still, themselves remaining silent and obeying the chief.

7. Of bandaging, there art two aspects: the completed bandage, and the application of the bandage: the process of application should be quick, effortless, ready, and rhythmical: quick (means) to bring the work rapidly to its conclusion; effortless to do this with ease; ready to fulfill every requirement; and rhythmical in such a way that it is a pleasure to watch. The appropriate procedures have been described above. The completed bandage should be well and correctly produced: correct in that it is simple and has no creases. If the parts are symmetrical and equal, the bandage should be equally and symmetrically executed; if unequal and unsymmetrical, then unequally and unsymmetrically. As to their forms, there are the simple circular, the oblique (= adze-like), the very oblique, the "eye," the rhomb, and the hemi-rhomb. The form of the bandage should fit the form of the damage in the part being bandaged.

8. There are two ways in which things are bandaged "well," either in their tension resulting from strength, or by a quantity of dressings. Now sometimes it is the bandage itself that provides the treatment, and at other times the bandage is only assisting other healing measures. The following is the rule, and these are the most important factors in bandaging: pressure should be sufficient to prevent what is being attached from falling off, but not so great as to cause excessive tightness: rather the bandage should fit properly without doing any violence, and exert less pressure in the center, and least of all at the ends. The

296 ῥάμμα νεμόμενον | μὴ κάτω, ἀλλ᾽ ἄνω, ἐν παρέξει καὶ
σχέσει καὶ ἐπιδέσει καὶ πιέξει. ἀρχὰς βάλλεσθαι μὴ
ἐπὶ τὸ ἕλκος, ἀλλ᾽ ἔνθα τὸ ἅμμα. τὸ δὲ ἅμμα μήτε ἐν
298 τρίβῳ μήτε ἐν ἔργῳ, | μήτε ἐκεῖσε ὅκου ἐνεόν, ὡς μὴ
ἐς τὸ ἐνεὸν κείσεται. τὸ δὲ ῥάμμα μαλθακόν, μὴ μέγα.

9. Εὖ γε μήν γνόντα ὅτι ἐς τὰ κατάντη καὶ ἀπόξη
300 φεύγει | πᾶς ἐπίδεσμος, οἷον κεφαλῆς μὲν τὸ ἄνω,
κνήμης δὲ τὸ κάτω, ἐπιδεῖν δεξιὰ ἐπ᾽ ἀριστερὰ ἢ ἀριστερὰ ἐπὶ δεξιά, πλὴν ἐπὶ κεφαλῆς, ταύτην δὲ κατ᾽ ἴξιν. τὰ δὲ ἐναντία ἀπὸ δύο ἀρχέων· ἢν δὲ ἀπὸ μιῆς, ἐφ᾽ ἑκάτερα ὅπερ[6] ὅμοιον ἐς τὸ μόνιμον, οἷον τὸ μέσον τῆς κεφαλῆς, ἢ ὅ τι ἄλλο τοιοῦτο. τὰ δὲ κινεύμενα, οἷον ἄρθρα, ὅπῃ μὲν συγκάμπτεται, ὡς ἥκιστα καὶ εὐσταλέστατα περιβάλλειν, οἷον ἰγνύη· ὅπῃ δὲ περιτείνεται, ἁπλᾶ τε καὶ πλατέα, οἷον μύλη· προσπεριβάλλειν δὲ καταλήψιος μὲν τῶν περὶ ταῦτα εἵνεκα, ἀναλήψιος δὲ τοῦ ξύμπαντος ἐπιδέσμου, ἐν τοῖσιν
302 ἀτρεμέουσι | καὶ λαπαρωτέροισι τοῦ σώματος, οἷον τὸ ἄνω καὶ τὸ κάτω τοῦ γούνατος· ὁμολογέει δέ, ὤμου μὲν ἡ περὶ τὴν ἑτέρην μασχάλην περιβολή, βουβῶνος δὲ ἡ περὶ τὸν ἕτερον κενεῶνα, καὶ κνήμης ἡ ὑπὲρ γαστροκνήμης. ὁπόσοισι μὲν ἄνω ἡ φυγή, κάτωθεν ἡ ἀντίληψις, οἷσι δὲ κάτω, τοὐναντίον· οἷσι δὲ μὴ ἔστιν, οἷον κεφαλή, τούτων ἐν τῷ ὁμαλωτάτῳ τὰς καταλή-

[6] ἑκάτερα ὅπερ B: ὅπερ V: om. M

knot and stitching should not be drawn downward, but upward, taking into account the presentation, the positioning, the bandaging, and the compression. The ends of the bandage are not placed over the lesion, but on one side or the other where the knot is to be, and the knot should not be where there is friction or movement, or where it would be useless, so that the knot will not be useless. The knot and stitches should be soft, and not bulky.

9. Being well aware that every bandage tends to slip toward parts that hang down and along those that have a tapering form, such as upward on the head, and downward on the calf, bind right parts (sc. of the body) toward the left, and left parts toward the right, except on the head, where the direction should be vertical. Parts which have two opposite sides (sc. should be bandaged) from two starting points, but if from only one, extend the bandage both ways so that it will have an identical relationship to the fixed point, as for example to the center of the head or anything else like this. Moving parts, like joints: where there is flexion, the turns should be made as few and as compact as possible, as for example the back of the knee; where a part spreads out, the bandage should be simple and broad, as for example the kneecap. Make additional turns to provide a hold for what is around the bandage, and to secure the whole bandage to immobile and concave parts of the body, such as above and below the knee. Suitable for the shoulder is a loop around the opposite axilla, for the groin around the opposite flank, and for the lower leg around the top of the calf. In cases where there is slippage upward, a downward counterforce must be provided, where the slippage is downward, the reverse. In cases where this is not feasible, as in the head, make holdfasts

ψιας ποιέεσθαι,[7] καὶ ἥκιστα λοξῷ τῷ ἐπιδέσμῳ χρέεσθαι,[8] ὡς τὸ μονιμώτατον ὕστερον περιβληθὲν τὰ πλανωδέστατα κατέχῃ. ὁπόσοισι δὲ τοῖσιν ὀθονίοισι μὴ εὐκαταλήπτως, μηδὲ εὐαναλήπτως ἔχει, ῥάμμασι τὰς ἀναλήψιας[9] ποιέεσθαι ἐκ καταβολῆς ἢ ξυρραφῆς.

10. Ἐπιδέσματα καθαρά, κοῦφα, μαλθακά, λεπτά. ἑλίσσειν ἀμφοτέρῃσιν ἅμα, καὶ ἑκατέρῃ χωρὶς ἀσκέειν. τῇ πρεπούσῃ δὲ ἐς τὰ πάχη καὶ τὰ πλάτη τῶν
304 μορίων[10] τεκμαιρόμενον | χρῆσθαι. ἑλίξιος κεφαλαὶ σκληραί, ὁμαλαί, εὐκρινέες. τὰ δὲ δὴ μέλλοντα ἀποπίπτειν κακίω[11] ταχέως ἀποπεσόντων· τὰ δὲ ὡς μὴ πιέζειν μήτε ἀποπίπτειν τὰ εἰρημένα. |

306 11. Ὧν δὲ ἔχεται ἢ ἐπίδεσις ἢ ὑπόδεσις ἢ ἀμφότερα· ὑπόδεσις μὲν αἰτίη ὥστε ἢ ἀφεστεῶτα προ<σ>στεῖλαι,[12] ἢ ἐκπεπταμένα συστεῖλαι, ἢ τὰ ξυνεσταλμένα διαστεῖλαι, ἢ διεστραμμένα διορθῶσαι, ἢ τἀναντία. παρασκευὴ δέ· ὀθόνια μαλθακά, λεπτά, καθαρά, κοῦφα, πλατέα, μὴ ἔχοντα ξυρραφάς, μηδ' ἐξάστιας, καὶ ὑγιᾶ ὥστε τάνυσιν φέρειν καὶ ὀλίγῳ κρέσσω, μὴ ξηρά, ἀλλ' ἔγχυμα χυμῷ ᾧ ἕκαστα σύντροφα. ἀφεστεῶτα μὲν προ<σ>στεῖλαι[13] ὥστε τὰ μετέωρα τῆς ἕδρης ψαύειν μέν, πιέζειν δὲ μή· ἦρχθαι
308 δὲ ἐκ τοῦ ὑγιέος, τελευτᾶν | πρὸς τὸ ἕλκος, ὡς τὸ μὲν ὑπεὸν[14] ἐξαθέλγηται, ἕτερον δὲ μὴ ἐπισυλλέγηται·

[7] ποιέεσθαι BV: om. M
[8] χρέεσθαι V: ποιέεσθαι BM
[9] ἀναλήψιας BV: καταλήψιας M

on the evenest parts, and employ as little oblique traction as possible, so that the most stable and outermost turn will secure the less stable ones. For dressings that cannot be properly secured and supported, make supports with threaded sutures in loops, or a continuous suture.

10. Bandages—clean, light, soft, thin: practice rolling them with both hands at the same time, and with each hand separately. Use one that you have determined to be suitable to the thickness and breadth of the parts; the margins of the roll (sc. should be) firm, even, and free of creases. Bandages that might fall off are worse than those that fall off quickly. Bandages that do not cause compression or fall off have been described above.

11. The use of bandages above or below compresses, or both: an underbandage is used a) to bringing together what is separated, b) to reduce what is everted, c) to separate what has adhered, and d) to straighten what is distorted—or the reverse. Preparation: dressings that are soft, thin, clean, light, broad, free of joins or selvages, sound enough to bear tension and a little more, and not dry, but damp with their appropriate moisture. a) Close an open (sc. lesion) so that the high parts touch the base, but do not press on it: begin from the sound tissue and end at the open lesion, so that the contents will be pressed out, but no new humor will collect. Bind vertical

---

10 μορίων Littré Petrequin (from Galen *Comm.*, p. 772) Ar.: ὀθονίων codd.

11 κακίω BV: καλῶς M

12 προ‹σ›στεῖλαι Jouanna: προστεῖλαι BMV

13 προ‹σ›στεῖλαι Jouanna: προστιλαι B: om. MV

14 ὑπεὸν MV: "suppuration" Ar.: ὑγιέον B

ἐπιδεῖν τὰ μὲν ὀρθὰ ἐς ὀρθόν, τὰ δὲ λοξὰ λοξῶς, ἐν σχήματι ἀπόνῳ, ἐν ᾧ μήτε ἀπόσφιγξις μήτε ἀπόστασις ἔσται, ἐξ οὗ ὅταν μεταλλάσσῃ, ἢ ἐς ἀνάληψιν ἢ ἐς θέσιν, μὴ μεταλλάξουσιν, ἀλλ᾽ ὅμοια ταῦθ᾽ ἕξουσι μύας, φλέβας, νεῦρα, ὀστέα. ἀναλελάμφθαι δὲ ἢ κέ-
310 εσθαι ἐν σχήματι ἀπόνῳ κατὰ φύσιν· ὧν δὲ | ἂν μὴ ἀποστῇ,[15] τἀναντία· ὧν δὲ ἐκπεπταμένα συστεῖλαι, τὰ μὲν ἄλλα ταὐτά, ἐκ πολλοῦ δέ τινος δεῖ τὴν συναγωγήν, καὶ ἐκ προσαγωγῆς τὴν πίεξιν, τὸ πρῶτον ἥκιστα, ἔπειτα ἐπὶ μᾶλλον, ὅριον τοῦ μάλιστα τὸ συμψαύειν. ὧν δὲ συνεσταλμένα διαστεῖλαι, ξὺν μὲν φλεγμονῇ, τἀναντία. ἄνευ δὲ ταύτης, παρασκευῇ μὲν τῇ αὐτῇ, ἐπιδέσει δὲ ἐναντίῃ. διεστραμμένα δὲ διορθῶσαι, τὰ μὲν ἄλλα κατὰ ταὐτά· δεῖ δὲ τὰ μὲν ἀπεληλυθότα ἐπάγειν, τὰ δὲ ἐπεληλυθότα ἀπαγαγεῖν, ἐπιδέσει, παρακολλήσει, ἀναλήψει· τὰ δὲ ἐναντία, ἐναντίως.

12. Κατήγμασι δὲ σπληνῶν μήκεα, πλάτεα, πάχεα,
312 πλήθεα. | μῆκος ὅση ἡ ἐπίδεσις· πλάτος, τριῶν ἢ τεσσάρων δακτύλων· πάχος, τρίπτυχα ἢ τετράπτυχα· πλῆθος, κυκλεῦντας μὴ ὑπερβάλλειν, μηδὲ ἐλλείπειν· οἷσι δὲ ἐς διόρθωσιν, μῆκος κυκλεῦντας· πάχος καὶ πλάτος τῇ ἐνδείῃ τεκμαίρεσθαι, μὴ ἀθρόα πληρεῦν-

[15] μὴ ἀποστῇ BV: "have been joined" Ar.: ἀποστῇ M

---

[3] Galen cannot understand this statement, offering four different possibilities from various commentators.

wounds vertically, and oblique ones obliquely, in a position causing no pain, avoiding both tight compression and laxness, so that when you change to a sling or to fixation, patients do not suffer any disturbance, but maintain their parts as they are, namely the muscles, vessels, cords, and bones; the part should be lifted up and supported comfortably in its natural position. Where the lesion is not open, do the opposite.[3] b) When a wound that you are bringing together gapes open, follow the same practices, but you must begin the approximation at a good distance (sc. from the wound), and then gradually increase the pressure, beginning with only a little, but then adding more and more, the limit being reached when the parts touch. c) In cases where you are separating parts that have adhered, if there is inflammation, the opposite policy is best; if, however, there is no inflammation, use the same apparatus (sc. as for drawing parts together), but pull in the opposite direction. d) If you are straightening what is distorted, follow the same principles: what is turned out must be brought in, and what has turned in must be brought out, by bandaging, agglutination, and suspension—the reverse, reversely.

12. Compresses used on fractures: their length, breadth, thickness, and number: length will be the same as that of the bandage, the breadth three or four fingers, the thickness of three or four folds, the number sufficient to go around the limb without hanging over or falling short. If they are being used to level out a wound, they must be long enough to go around, and their thickness and width should be judged by the gap being filled, without

τας. τῶν δὲ ὀθονίων ὑποδεσμίδες εἰσὶ δύο· τῇ πρώτῃ ἐκ τοῦ σίνεος ἐς τὸ ἄνω τελευτώσῃ· τῇ δὲ δευτέρῃ ἐκ τοῦ σίνεος ἐς τὸ κάτω, ἐκ τοῦ κάτω ἐς τὸ ἄνω τελευτώσῃ· κατὰ τὸ σίνος πιέζειν μάλιστα, ἥκιστα τὰ ἄκρα, τὰ δὲ ἄλλα κατὰ λόγον.

Ἡ δὲ ἐπίδεσις πολὺ τοῦ ὑγιέος προσλαμβανέτω. ἐπιδέσμων δὲ πλῆθος, μῆκος, πλάτος· πλῆθος μὲν μὴ ἡσσᾶσθαι τοῦ σίνεος, μηδὲ νάρθηξιν ἐνέρεισιν εἶναι, μηδὲ ἄχθος, μηδὲ περίρρεψιν, μηδὲ ἐκθήλυνσιν· μῆ-
314 *κος δὲ καὶ πλάτος, τριῶν ἢ τεσσάρων* | *ἢ πέντε ἢ ἓξ πήχεων μὲν μῆκος, δακτύλων δὲ πλάτος. καὶ παραιρέματος περιβολαὶ τοσαῦται ὥστε μὴ πιέζειν· μαλθακὰ δέ, μὴ παχέα· ταῦτα πάντα ὡς ἐπὶ μήκει καὶ πλάτει καὶ πάχει τοῦ παθόντος.*

Νάρθηκες δὲ λεῖοι, ὁμαλοί, σιμοὶ κατ᾽ ἄκρα, σμικρῷ μείους ἔνθεν καὶ ἔνθεν τῆς ἐπιδέσιος, παχύτατοι δὲ ᾗ ἐξήριπε τὸ κάτηγμα. ὁπόσα δὲ κυρτὰ καὶ ἄσαρκα φύσει, φυλασσόμενον τῶν ὑπερεχόντων, οἷον τὸ κατὰ δακτύλους ἢ σφυρά, ἢ τῇ θέσει ἢ τῇ βραχύτητι.
316 *παραιρέμασι δὲ ἑρμάζειν,*[16] | *μὴ πιέζειν· τὸ πρῶτον κηρωτῇ μαλθακῇ*[17] *λείῃ καὶ καθαρῇ ἑλισσέτω.*

13. *Ὕδατος θερμότης, πλῆθος· θερμότης μὲν κατὰ τῆς ἑωυτοῦ χειρὸς καταχεῖν, πλῆθος δὲ χαλάσαι μὲν*
318 *καὶ ἰσχνῆναι τὸ* | *πλεῖστον ἄριστον, σαρκῶσαι δὲ καὶ ἁπαλῦναι τὸ μέτριον· μέτρον δὲ τῆς καταχύσιος, ἔτι*

[16] *ἑρμάζειν* BV: "use as a prop" Ar.: *ἁρμόζειν* M
[17] Add. *ὁμαλῇ* M

their bunching together in a mass. There are two underbandages of the dressings; the first goes upward from the wound and ends at the top, while the second begins from the wound in a downward direction, and then turning back upward ends at the top. Apply the maximum pressure over the lesion, and the least at its edges, with the rest in proportion.

The overbandage should extend far over the healthy tissue. The number, length, and breadth of bandages over dressings: the number must be adequate for the lesion, but should not press on the splints, be a burden, slip askew, or cause atrophy. Length and width: three, four, five or six cubits for the length, and the same number of fingers (i.e., inches) for the breadth. Let there be as many coils of the supporting bands as not to exert compression; they are to be soft and not very thick. All these details should be adapted to the length, breadth and thickness of the wounded part.

Splints (sc. should be) smooth, even, tapered toward the ends, a little shorter than the bandage in each direction, and thickest where the broken part of the bone projects. In places that are convex in structure and naturally bare of flesh, care must be taken of the prominences, such as on the digits and the ankles, by their position or by shortening them: use the supporting bandages as a prop, but do not exert pressure. Prepare the dressing first by rolling the bandages in soft, smooth, clean cerate.

13. Of water (sc. consider) the temperature and amount: to determine the temperature pour it over your own hand; the best quantity for relaxing and reducing a swelling is a very generous one; for promoting the growth of flesh and softening, it is a moderate amount. The

μετεωριζομένου δεῖ, πρὶν συμπίπτειν, παύεσθαι· τὸ μὲν γὰρ πρῶτον αἴρεται, ἔπειτα δὲ ἰσχναίνεται.

14. Θέσις δὲ μαλθακή, ὁμαλή, ἀνάρροπος τοῖσιν ἐξέχουσι τοῦ σώματος, οἷον πτέρνῃ καὶ ἰσχίῳ, ὡς μήτε ἀνακλᾶται μήτε ἀποκλᾶται μήτε ἐκτρίβηται,[18] σωλῆνα παντὶ τῷ σκέλει ἢ ἡμίσει· ἐς τὸ πάθος δὲ βλέπειν καὶ τὰ ἄλλα ὁκόσα βλάπτει δῆλα.

15. Πάρεξις δέ, καὶ διάτασις, καὶ ἀνάπλασις[19] καὶ τἆλλα κατὰ φύσιν. φύσις δὲ ἐν μὲν ἔργοις, τοῦ ἔργου τῇ πρήξει, ὃ βούλεται τεκμαρτέον· ἐς δὲ ταῦτα, ἐκ τοῦ
320 ἐλινύοντος, ἐκ τοῦ κοινοῦ, | ἐκ τοῦ ἔθεος· ἐκ μὲν τοῦ ἐλινύοντος καὶ ἀφειμένου τὰς ἰθυωρίας σκέπτεσθαι, οἷον τὸ τῆς χειρός· ἐκ δὲ τοῦ κοινοῦ, ἔκτασιν, σύγκαμψιν, οἷον τὸ ἐγγὺς τοῦ ἐγγωνίου πήχεος πρὸς βραχίονα· ἐκ δὲ τοῦ ἔθεος, ὅτι οὐκ ἄλλα σχήματα φέρειν δυνατώτερα, οἷον σκέλεα ἔκτασιν. ἀπὸ τούτων γὰρ ῥήϊστα πλεῖστον χρόνον ἔχοι ἂν μὴ μεταλλάσσοντα. ἐν δὲ τῇ μεταλλαγῇ ἐκ διατάσιος ὅμοια ταῦθ' ἔχουσιν ἐς ἕξιν ἢ θέσιν μύες, φλέβες, νεῦρα, ὀστέα, ᾗ μάλιστα εὔθετα καὶ εὔσχετα.

16. Διάτασις, μάλιστα τὰ μέγιστα καὶ πάχιστα, καὶ ὅκου ἀμφότερα· δεύτερα, ὧν τὸ ὑποτεταγμένον,

[18] ἐκτρίβηται M: ἐκτρέπειται B: ἐκτρέπηται V

[19] ἀνάπλασις Froben from Galen *Comm. lemma* (p. 851): κατάτασις codd.: "stretching downward" Ar.

amount of the affusion being supplied while a part is still swollen must cease before it collapses, for first it swells, and then it recedes.

14. The support (on which a lower limb is set should be) soft, even and sloping upward, in particular for the projecting parts of the body such as with the heel or the hip, in order to avoid their being bent up, turned down, or bruised. Apply a hollow splint to the whole leg rather than to half; observe any possible damage or other obvious disadvantages (sc. of this splint).

15. The presentation, degree of extension, reduction, and everything else (sc. regarding a wounded limb) should be done according to nature. What nature is must be deduced in these procedures from the execution itself of the procedure that is desired—in such cases, from the limb's position at rest, its normal position, and its habitual position. From its position in rest and relaxation, you must determine a limb's natural direction, as for example for the arm; from its normal position, judge its natural extension and flexion, as for example the almost right-angle relationship between the forearm and the upper arm; from the habitual position notice that certain limbs are not more able to tolerate other positions, as for example the thighs with extension. For from this knowledge, a patient will be able to keep his position most easily and for the longest time without change; and in the change that follows surgical extension, the muscles, vessels, cords and bones will have the most likeness in their direction and placement, so that they will be very easily secured and maintained.

16. Extension (sc. will be applied) strongest when the largest and thickest bones (sc. of the arm) are both broken: secondly when it is the one positioned below (ulna); least

322 ἥκιστα ὧν τὸ ἄνω· | μᾶλλον δὲ τοῦ μετρίου βλάβη,
πλὴν παιδίων· ἔχειν ἀνάντη σμικρόν· διορθώσιος
παράδειγμα, τὸ ὁμώνυμον, τὸ ὁμόζυγον, τὸ ὅμοιον, τὸ
ὑγιές.

17. Ἀνάτριψις δύναται λῦσαι, δῆσαι, σαρκῶσαι, μινυθῆσαι· ἡ σκληρὴ δῆσαι· ἡ μαλακὴ λῦσαι· ἡ πολλὴ μινυθῆσαι· ἡ μετρίη παχῦναι.

18. Ἐπιδεῖν δὲ τὸ πρῶτον· ὁ μὲν ἐπιδεδεμένος μάλι-
στα φάτω πεπιέχθαι κατὰ τὸ σίνος· ἥκιστα τὰ ἄκρα·
ἡρμάσθαι δέ, μὴ πεπιέχθαι, πλήθει, μὴ ἰσχύι· τὴν δὲ
ἡμέρην ταύτην καὶ νύκτα, ὀλίγῳ μᾶλλον, τὴν δὲ
ὑστέρην, ἧσσον· τρίτῃ, χαλαρά. εὑρεθήτω δὲ τῇ μὲν
ὑστεραίῃ ἐν ἄκροις οἴδημα μαλθακόν. τῇ τρίτῃ δὲ τὸ
324 ἐπιδεθὲν λυθέν, ἰσχνότερον, παρὰ πάσας τὰς | ἐπιδέ-
σιας τοῦτο. τῇ δὲ ὑστεραίῃ ἐπιδέσει, ἢν διηνεκέως
ἐπιδεδεμένον φανῇ, μαθεῖν δεῖ· ἐντεῦθεν δὲ μᾶλλον
καὶ ἐπὶ πλέοσι πιεχθήτω· τῇ δὲ τρίτῃ ἐπὶ μᾶλλον καὶ
ἐπὶ πλέοσι. τῇ δὲ ἑβδόμῃ ἀπὸ τῆς πρώτης ἐπιδέσιος
λυθέντα εὑρεθήτω ἰσχνά, χαλαρὰ τὰ ὀστέα. ἐς δὲ
νάρθηκας δεθέντα, ἢν ἰσχνὰ καὶ ἄκνησμα καὶ ἀνέλ-
κεα ᾖ, ἐᾶν μέχρις εἴκοσιν ἡμερέων ἀπὸ τοῦ σίνεος· ἢν
δέ τι ὑποπτεύηται, λῦσαι ἐν μέσῳ· νάρθηκας διὰ
τρίτης ἐρείδειν.

19. Ἡ ἀνάληψις, ἡ θέσις,[20] ἡ ἐπίδεσις, ὡς ἐν τῷ

[20] θέσις BV: αἴσθησις M

when it is the upper one (radius). More than moderate stretching causes harm, except in children. Hold the limb slightly raised. The model for making your adjustment should be the homonymous, corresponding, similar, sound limb.

17. Friction can relax, contract, promote the growth of flesh, or thin: if firmly applied it contracts, if gently, it relaxes, if repeatedly, it thins, and if moderately, it builds flesh.

18. The first bandaging: the person being bandaged should report that there is pressure mainly at the lesion and least at the ends; also, that the bandage provides support with its bulk or tension, but is without compression. On the first day and night the pressure reported should increase a little, on the second day be less, and on the third day the bandage should be loose. On the second day, a mild edema should be found at the extremities; on the third day when the bandage is removed, this edema should have gone down. The same occurs with all succeeding bandages. For the second bandaging, you must determine whether the wound appears to have been held firm for the whole time, and then employ more dressings to create more pressure. At the third bandaging use even more dressings to produce yet more pressure. On the seventh day after the first bandaging, with the removal of the bandage no swelling should be found, and the bones should be mobile. When the limb is put up in splints, if it is free of swelling, itching, or any lesion, leave it in peace until the twentieth day from the injury, but if you suspect any trouble, remove them in the middle of that period. Tighten the splints every other day.

19. In suspension, putting up, and bandaging, be care-

αὐτῷ σχήματι ᾖ διαφυλάσσειν. κεφάλαια σχημάτων, ἔθη, φύσιες ἑκάστου τῶν μελέων· τὰ δὲ εἴδεα,[21] ἐκ τοῦ τρέχειν, ὁδοιπορέειν, ἑστάναι, κατακεῖσθαι, ἐκ τοῦ ἔργου, ἐκ τοῦ ἀφεῖσθαι.

20. Το δέ,[22] ὅτι χρῆσις κρατύνει, ἀργείη τήκει. |

326 21. Ἡ πίεξις πλήθει ἢ ἰσχύι.[23]

22. Οκόσα δὲ ἐκχυμώματα, ἢ φλάσματα, ἢ σπάσματα, ἢ οἰδήματα ἀφλέγμαντα, ἐξαρείαται[24] αἷμα ἐκ τοῦ τρώματος, ἐς μὲν τὸ ἄνω τοῦ σώματος τὸ πλεῖστον, βραχὺ δέ τι καὶ ἐς τὸ κάτω· μὴ κατάντεα τὴν
328 χεῖρα ἔχοντα ἢ τὸ σκέλος· τιθέμενον τὴν ἀρχὴν | κατὰ τὸ τρῶμα καὶ μάλιστα ἐρείδοντα, ἥκιστα τὰ ἄκρα, μέσως τὰ διὰ μέσου· τὸ ἔσχατον πρὸς τὸ ἄνω τοῦ σώματος νεμόμενον· ἐπιδέσει, πιέξει, ἄταρ καὶ ταῦτα πλήθει μᾶλλον ἢ ἰσχύι. μάλιστα δὲ τούτοισιν ὀθόνια, λεπτά, κοῦφα, μαλθακά, καθαρά, πλατέα, ὑγιᾶ, ὡς ἂν ἄνευ ναρθήκων· καὶ καταχύσει χρῆσθαι πλέονι.

23. Τὰ δὲ ἐκπτώματα, ἢ στρέμματα, ἢ διαστήματα, ἢ ἀποσπάσματα, ἢ ἀποκλάσματα, ἢ διαστρέμματα, οἷα τὰ κυλλά,[25] τὰ ἑτερόρροπα, ἔνθεν μὲν ἐξέστη, συνδιδόντα, ὅπῃ δέ, συντείνοντα, ὡς ἐς τἀναντία ῥέπῃ, ἐπιδεθέντα ἢ πρὶν ἐπιδεθῆναι, σμικρῷ μᾶλλον ἢ ὥστε ἐξ ἴσου εἶναι· καὶ τοῖσιν ἐπιδέσμοισι, καὶ τοῖσι

21 Add. τῶν μελέων V Ar.
22 τὸ δὲ B: om. MV
23 ἢ ἰσχύι V: ἰσχύι BM: om. Ar.
24 ἐξαρείαται BM: ἐξάρη ἅτε V
25 κυλλά M: κοιλα B: κυλὰ V

ful that each of the limbs maintains the same position, namely its natural, habitual position. The habitual position depends upon whether the person is running, walking, standing, lying down, working, or relaxing.

20. Next, usage strengthens, idleness debilitates.

21. Pressure: number of dressings or amount of tension.

22. In all cases of bruisings, crushings, sprains or swellings without inflammation, blood is driven away from the wound (sc. by bandaging) mainly toward the upper parts of the body, but also a small part downward. Prevent the arm or the leg from hanging down; begin by applying the bandage at the site of the wound, and exert the most pressure there, least at the extremities, and a moderate amount in between; bring the final turns toward the upper part of the body. Employ bandaging and compression, but here too pressure is to be generated by the number of dressings rather than by tension. In these cases especially, the dressings should be thin, light, soft, clean, wide, and sound, the same way as when no splints are used. Also employ copious irrigation.

23. Dislocations, sprains, separations, avulsions, splinterings, or distortions such as deformities to one side: the bandage should yield on the side from which the parts deviate and exert force on the side to which they deviate, so that when they are put up, or before they are put up, that are not straight, but incline slightly to the opposite

σπλήνεσι, καὶ τοῖσιν ἀναλήμμασι, καὶ τοῖσι σχήμασι, κατατάσει, ἀνατρίψει, διορθώσει,[26] καταχύσει[27] πλέονι. |

330 24. Τὰ δὲ μινυθήματα, πολὺ προσλαμβάνοντα τοῦ ὑγιέος, ἐπιδεῖν ὡς ἂν ἐξ ἐπιδρομῆς τὰ συντακέντα
332 πλέον ἢ αὐτόματα ἐμινύθει, | ἀλλοίῃ τῇ ἐπιδέσει παραλλάξαντα, ἐκκλίνῃ ἐς τὴν αὔξησιν καὶ τὴν ἀνάπλασιν τῶν σαρκῶν ποιήσηται. βέλτιον δὲ καὶ τὰ ἄνωθεν, οἷον κνήμης καὶ τὸν μηρόν, καὶ τὸ ἕτερον σκέλος τὸ ὑγιὲς συνεπιδεῖν, ὡς ὁμοιότερον ᾖ καὶ ὁμοίως ἐλινύῃ, καὶ ὁμοίως τῆς τροφῆς ἀποκλείηται καὶ δέχηται. ὀθονίων πλήθει, μὴ πιέξει· ἀνιέντα πρῶτον τὸ μάλιστα δεόμενον, καὶ ἀνατρίψει χρώμενον σαρκούσῃ καὶ καταχύσει· ἄνευ ναρθήκων. |

334 25. Τὰ δὲ ἑρμάσματα καὶ ἀποστηρίγματα, οἷον στήθει, πλευρῇσι, κεφαλῇ, καὶ τοῖσιν ἄλλοισιν, ὅσα τοιαῦτα· τὰ μὲν σφυγμῶν εἵνεκεν, ὡς μὴ ἐνσείηται· τὰ δὲ καὶ τῶν διαστασίων τῶν κατὰ τὰς ἁρμονίας ἐν τοῖσι κατὰ τὴν κεφαλὴν ὀστέοις ἐρεισμάτων χάριν· ἐπί τε βηχέων ἢ πταρμῶν, ἢ ἄλλης κινήσιος, οἷον τὰ κατὰ θώρηκα καὶ κεφαλὴν ἀποστηρίγματα γίνεται. τούτων ἁπάντων αἱ αὐταὶ συμμετρίαι τῆς ἐπιδέσιος· ᾗ γὰρ τὰ σίνη μάλιστα πεπιέχθαι· ὑποτιθέναι οὖν εἴριον μαλθακὸν ἅρμοζον τῷ πάθει· ἐπιδεῖν δὲ μὴ μᾶλλον πιεζεῦντα <ἢ>[28] ὥστε τοὺς σφυγμοὺς μὴ ἐν-

[26] Add. σπλήνεσι, ἐπιδέσμοισι ἀνάληψι θέσι καὶ B

[27] καταχύσει B: ταῦτα καὶ ταχύσει M: κατὰ ταῦτα καὶ

side: employ bandages, compresses, slings, postures, extension, friction, straightening, and copious irrigation.

24. Atrophied parts: in bandaging include a good amount of healthy limb so that the wasted tissues will acquire more through afflux than they lose by themselves: by the alteration of a bandage to another kind, a part may be compelled to incline more toward growth and the addition of flesh. It is also advantageous to extend the bandage over the upper parts, such as the top of the leg and the thigh, and to connect the other (i.e., atrophied part of the) leg to the sound part, so that it will come to become like it and to share its rest, and its deprivation or reception of nutriment. Use many dressings, and not tension; first release the part that especially requires it, and then employ friction to create flesh, and also irrigation. Stay away from splints.

25. Supports and props for the chest, ribs, head, and other such parts are used as follows: sometimes against throbbing to prevent the part from shaking; at other times as supports when there is a separation of the commissures in the bones of the head. Also, when there is coughing, sneezing, or any other movement, they act as supports for the chest and head. The appropriate modes of bandaging apply equally in all these cases, since the chief pressure should always be applied to the lesion. Underneath place a piece of soft wool fitted to the injured part. Do not make the bandage tighter than suffices to prevent the throbbing

---

καταχύσει ταῦτα V: "we also have the same aims when we pour on a considerable quantity (of water)" Ar.

[28] ἢ add. recc.

σείειν, μηδὲ μᾶλλον ἢ ὥστε τῶν διεστηκότων τὰ ἔσχατα τῶν ἁρμονιῶν συμψαύειν ἀλλήλων, μηδὲ τὰς
336 | *βῆχας καὶ τοὺς πταρμοὺς ὥστε κωλύειν,*[29] *ἀλλ' ὥστε ἀποστήριγμα εἶναι ὡς μήτε διαναγκάζηται, μήτε ἐνσείηται.*[30]

[29] *ἐνσείεω* (line xx) . . . *κωλύειν* BV: *ἐνσείεσθαι, ἀλλὰ τὰς διαστάσιας τῶν ἁρμονίων ψαύειν τὰ ἔσχατα ἀλλήλων· μήτε ἀμφὶ τοῖσι πταρμοῖσι καὶ βηξίν* M

[30] Add. BV *Ὀστέων φύσις· δακτύλων μὲν ἁπλὰ καὶ ὀστέα καὶ ἄρθρα* (= the first line of *Instruments of Reduction*).

from shaking the part, nor than is necessary to bring the edges of the separated commissures into contact with one another; nor so tight that it would inhibit coughs and sneezes, but rather just such as to be a support against violent separation or internal shaking.

FRACTURES AND JOINTS

INTRODUCTION

Fractures and *Joints* together provide an extensive collection of surgical knowledge about the injuries of the bones of the body, including their etiology, diagnosis, and methods of treatment. The material in the two treatises is arranged somewhat unsystematically in chapters as follows:

Fractures

1–3	Polemic regarding the positioning of the arm for bandaging in fractures and dislocations.
4–7	Fractures of the forearm.
8	Fractures of the upper arm.
9–14	Dislocations in the foot and ankle.
15–18	Fractures of the bones of the lower leg.
19–23	Fractures of the thigh.
24–30	Compound fractures in general with polemical remarks.
31–32	Use of the lever in the reduction of fractures.
33–34	Management of incurable fractures; resection of bone.
35–36	Compound fractures of the thigh and upper arm.
37	Dislocation of the knee compared to that of the elbow.

Fractures

38–44	Dislocations of the elbow.
45–46	Fracture of the olecranon and the trochlea of the humerus.
47–48	Treatments of injuries to the elbow; bandaging.[1]

Joints

1–12	Dislocations of the shoulder joint.	
13	Evulsion of the collarbone from the acromion.	
14–16	Fractures of the collarbone.	
17–25	Dislocations of the elbow.	[interpolated
26–28	Dislocations at the wrist.	from *Instruments*
29	Dislocation of the finger joints.	*of Reduction* 7–19]
30–32	Dislocations of the lower jaw.	
33–34	Fractures of the lower jaw.	
35–39	Fractures and contusions of the nose.	
40	Bruising of the ear.	
41–48	Curvation of the spine.	
49–50	Fractures and contusions of the ribs.	
51–61	Dislocations of the hip joint.	
62–63	Congenital dislocations in the ankle (clubfoot).	
64–68	Compound dislocations of the wrist, knee, elbow, and fingers.	
69	Gangrene supervening in cases of fracture.	
70–78	Reductions of a dislocated hip joint.	
79	General observations on joints and their dislocations.	

[1] See Jouanna *Frac.*, pp. xii–xxxiii.

Joints

80	Dislocations of the finger joints.	
81–87	Injuries of the knee, ankle, and foot.	[interp. from *Inst. Red.* 26–31]

Both treatises have numerous references backward and forward to subjects handled in their texts, and references in chapters 67 and 72 of *Joints* refer to statements made in chapters 31 and 13, respectively, of *Fractures*, raising the possibility that the two works were once joined.[2] Furthermore, *Fractures* seems to have provided the text from which most of *Instruments of Reduction* was excerpted.[3]

An important witness for both the Hippocratic text and the Hellenistic reception of *Joints* is Apollonius of Citium's (1 BC) monograph *On the Subject of Joints*, which consists of a systematic account of dislocations and their treatment built around the text of the Hippocratic work: *Joints* is frequently quoted verbatim, so that altogether approximately one quarter of its text is included. Apollonius, however, does not always take over the Hippocratic text as he finds it, but freely shortens passages he judges to be too diffuse, lengthens those he finds too jejune, and adds material from other writers on subjects "Hippocrates" has omitted. A series of thirty ancient drawings preserved in the manuscript B,[4] to which Apollonius specifically refers,

[2] Cf. Schleiermach.

[3] See Withington, pp. 85–89, for a discussion of the possible relationships among the treatises.

[4] See Schöne and Apoll. Cit.: both these editions contain black-and-white reproductions of the colored illustrations in B.

illustrate various injuries, procedures, and instruments mentioned in the text.

Fractures and *Joints* are listed among the "therapeutic works pertaining to surgery" in the census in Erotian's *Collection of Hippocratic Words*,[5] and together they contain almost one hundred terms certainly or possibly glossed by the lexicographer.[6] Galen wrote commentaries on them, both of which have survived, and he refers to the texts many times in his other writings.[7] Two papyri fragments provide evidence that the text of *Fractures* was in circulation in Egypt during Galen's lifetime,[8] and both treatises are known by many other ancient authors, in particular lexicographers and medical writers.[9] In the medieval period, however, besides being quoted a few times by Rhazes, they have left little trace.[10]

The Greek texts of *Fractures* and *Joints* are transmitted in the three independent manuscripts B, M, and V, which form the basis of volume 2 of H. Kühlewein's *Hippocratis*

[5] See Erotian, p. 9.

[6] Cf. Nachmanson, pp. 367–93; Jouanna *Frac.*, pp. xcv–cxiv.

[7] Galen: *Fractures Commentary*, vol. 18B, pp. 318–628; see Jouanna *Frac.*, pp. cxv–cxxviii. Galen: *Joints Commentary*, vol. 18A, pp. 300–767; see Kudlien on the transmission of Galen's *Joints* commentary. Cf. Manetti/Ros., pp. 1531–35; *Testimonien*, vol. II,1, pp. 144–66 and 288–300; vol. II,2, pp. 106–16 and 230–35.

[8] PBrookl 94 (MP^3 538.02) and PAberd 124 (MP^3 538.03); see Andorlini I, pp. 3–8; Jouanna *Frac.*, pp. lxxix–lxxxi.

[9] See *Testimonien*, vol. I, pp. 110–84 and 258–73; vol. III, pp. 162–66 and 239–46.

[10] Cf. Ullmann, p. 31; Kibre, p. 93.

Opera (Leipzig, 1902), a volume not directly available to Withington when he was preparing the third Loeb *Hippocrates* volume (1928). Of particular importance in the preparation of the present volume has been

Jouanna, J. *Hippocrate. Sur les fractures.* Budé VII (1). Paris, 2022. (= Jouanna *Frac.*)

The present edition of *Joints* is based on the work of earlier scholars and a collation of the manuscripts B, M, and V from microfilms. Adams' and Withington's English translations have been invaluable for the interpretation of the text.

ΠΕΡΙ ΑΓΜΩΝ

III 412 Littré 1. Ἐχρῆν τὸν ἰητρὸν τῶν ἐκπτωσίων τε καὶ κατηγμάτων ὡς ἰθύτατα τὰς κατατάσιας[1] ποιέεσθαι· αὕτη γὰρ ἡ δικαιοτάτη φύσις. ἢν δέ τι ἐγκλίνῃ[2] ἢ τῇ ἢ τῇ, ἐπὶ τὸ πρηνὲς ῥέπειν· ἐλάσσων γὰρ ἡ ἁμαρτὰς ἢ ἐπὶ τὸ ὕπτιον. οἱ μὲν οὖν μηδὲν προβουλεύοντες[3] οὐδὲν ἐξαμαρτάνουσιν ὡς ἐπὶ τὸ πολύ· αὐτὸς γὰρ ὁ ἐπιδεό-
414 μενος τὴν χεῖρα ἀπορέγει οὕτως | ὑπὸ τῆς δικαίης φύσιος ἀναγκαζόμενος· οἱ δὲ ἰητροὶ σοφιζόμενοι δῆθεν ἔστιν οἳ ἁμαρτάνουσι.

Σπουδὴ μὲν οὖν οὐ[4] πολλὴ χεῖρα κατηγυῖαν χειρίσαι, καὶ παντὸς δὲ ἰητροῦ, ὡς ἔπος εἰπεῖν· ἀναγκάζομαι δὲ πλείω γράφειν περὶ αὐτοῦ ὅτι οἶδα ἰητροὺς σοφοὺς δόξαντας εἶναι ἀπὸ σχημάτων χειρὸς ἐν ἐπιδέσει, ἀφ᾽ ὧν ἀμαθέας αὐτοὺς ἐχρῆν δοκέειν εἶναι. ἄλλα γὰρ πολλὰ οὕτω ταύτης τῆς τέχνης κρίνεται· τὸ γὰρ ξενοπρεπὲς οὔπω ξυνιέντες, εἰ χρηστόν, καὶ[5] μᾶλλον ἐπαινέουσιν ἢ τὸ σύνηθες, ὃ ἤδη οἴδασιν ὅτι χρηστόν, καὶ τὸ ἀλλόκοτον μᾶλλον ἢ τὸ εὔδηλον. ῥητέον οὖν ὁκόσας ἐθέλω τῶν ἁμαρτάδων τῶν ἰητρῶν † τὰς

[1] κατατάσιας BM: καταστά. V
[2] ἐγκ. BM: ἐκκ.V

FRACTURES

1. In dislocations and fractures, the surgeon should make extensions in as straight a line as possible, for this is most conformable with nature; but if there has to be an inclination one way or the other, it should be toward pronation (palm down) rather than supination (palm up), since this will do less harm. As a rule, surgeons who have no preconceived idea make no mistake, for a patient himself holds out the arm for bandaging in the natural position, while theorizing practitioners are just the ones who go wrong.

Actually, the treatment of a fractured arm is not difficult, and to tell the truth is within almost any surgeon's capability; still I am compelled to write a good deal about it because I know practitioners who have gained credit for competence by bandaging arms in positions which should rather have given them a name for ignorance. Many other parts of this art are also misjudged in the same way, for people prefer what is novel before they know whether it will work, over what is customary and whose value is already established: the exotic over the obvious. To address, then, are the errors of practitioners that I recommend

---

3 -οντες B: -ονται MV
4 οὐ BM: om. V
5 καὶ B: om. MV

μὲν ἀποδιδάξαι, τὰς δὲ διδάξαι †περὶ τῆς φύσιος τῆς
χειρός· καὶ γὰρ ἄλλων ὀστέων τῶν κατὰ τὸ σῶμα
δίδαγμα ὅδε ὁ λόγος ἐστίν. |

416 2. Τὴν μὲν οὖν χεῖρα, περὶ οὗ ὁ λόγος, ἔδωκέ τις
418 ἐπιδῆσαι[6] | καταπρηνέα ποιήσας· ὁ δὲ ἠνάγκαζεν
οὕτως ἔχειν ὥσπερ οἱ τοξεύοντες, ἐπὴν τὸν ὦμον ἐμ-
βάλλωσι, καὶ οὕτως ἔχουσαν ἐπέδει, νομίζων ἑωυτῷ
εἶναι τοῦτο αὐτῇ τὸ κατὰ φύσιν· καὶ μαρτύριον ἐπή-
γετο τά τε ὀστέα ἅπαντα τὰ ἐν τῷ πήχει, ὅτι ἰθυωρίην
κατ᾽ ἄλληλα εἶχε, τήν τε ὁμοχροίην, ὅτι αὐτὴ καθ᾽
ἑωυτὴν τὴν ἰθυωρίην ἔχει οὕτω καὶ ἐκ τοῦ ἔξωθεν
μέρεος καὶ ἐκ τοῦ ἔσωθεν· οὕτω δὲ ἔφη καὶ τὰς σάρ-
κας καὶ τὰ νεῦρα πεφυκέναι, καὶ τὴν τοξικὴν ἐπήγετο
μαρτύριον. ταῦτα δὲ λέγων καὶ ταῦτα ποιέων σοφὸς
ἐδόκεεν εἶναι· τῶν δὲ ἄλλων τεχνέων ἐπελελήθει καὶ
ὁπόσα ἰσχύι ἐργάζονται καὶ ὅσα τεχνήμασιν, οὐκ εἰ-
δὼς ὅτι ἄλλο ἐν ἄλλῳ τὸ κατὰ φύσιν σχῆμά ἐστιν,
καὶ ἐν τῷ αὐτῷ ἔργῳ ἕτερα τῆς δεξιῆς χειρὸς σχήματα
κατὰ φύσιν ἐστί, καὶ ἕτερα τῆς ἀριστερῆς, ἢν οὕτω
420 τύχῃ. ἄλλο μὲν γὰρ σχῆμα ἐν | ἀκοντισμῷ κατὰ φύ-
σιν, ἄλλο δὲ ἐν σφενδονήσει, ἄλλο δὲ ἐν λιθοβολή-
σει,[7] ἄλλο ἐν πυγμῇ, ἄλλο ἐν τῷ ἐλινύειν. ὁκόσας δ᾽
ἄν τις τέχνας ἐρευνήσῃ,[8] οὐ τὸ αὐτὸ σχῆμα τῶν χει-
ρέων κατὰ φύσιν ἐστίν καὶ ἐν ἑκάστῃ τῶν τέχνων,
ἀλλὰ πρὸς τὸ ἄρμενον ὃ ἂν ἔχῃ ἕκαστος, καὶ πρὸς τὸ
ἔργον ὃ ἂν ἐπιτελέσασθαι θέλῃ, σχηματίζονται αἱ
χεῖρες· τοξικὴν δὲ ἀσκέοντι εἰκὸς τοῦτο τὸ σχῆμα
κράτιστον εἶναι τῆς ἑτέρης χειρός· τοῦ γὰρ βραχίο-

should be unlearned . . . learned about the structure of the arm: this discourse also teaches about other bones of the body.

2. To come to our subject, a patient once presented his arm to be dressed in the position of pronation, but the surgeon made him hold it the way archers do when they bring forward their (sc. left) shoulder, and he bandaged it in this posture, persuading himself that this is its natural position. He adduced as evidence that all the bones of the forearm are parallel to one another, and that their surfaces too lie in a direct line away from both their outer and inner parts; he declared this to be the natural disposition of the muscles and cords, and then brought in the art of the archer as evidence. This gave an appearance of learning to his discourse and practice, but he had forgotten the other arts and all the things that the arm executes by strength or artifice, not understanding that the natural position varies among them, and that in doing the same work it might even happen that the right arm has one natural position and the left a different one. For there is one natural position when throwing the javelin, another in using the sling, another in hurling a stone, another in boxing, and another in repose: how many arts could be found in which the natural position of the arms is not the same, but in which they assume positions in accordance with the apparatus being employed and the work being performed! As to the archer, clearly the above posture is the most effective one

---

6 ἐπιδῇ. B: καταδῇ. MV

7 -βολήσει B: -βόλησι M: -βόλησιν V

8 ἐρευνήσῃ Kühlewein: εὗροι ἐν ᾗσιν codd.

νος τὸ γιγγλυμοειδές, ἐν τῇ τοῦ πήχεος βαθμίδι ἐν
τούτῳ τῷ σχήματι ἐρεῖδον ἰθυωρίην ποιέει τοῖσιν
ὀστέοισι τοῦ πήχεος καὶ τοῦ βραχίονος, ὡς εἰ ἓν εἴη
τὸ πᾶν· καὶ ἡ ἀνάκλασις τοῦ ἄρθρου κέκλασται[9] ἐν
τούτῳ τῷ σχήματι. εἰκὸς οὖν οὕτως ἀκαμπτότατόν τε
καὶ τετανώτατον εἶναι τὸ χωρίον, καὶ μὴ ἡσσᾶσθαι,
μηδὲ συνδιδόναι, ἑλκομένης τῆς νευρῆς ὑπὸ τῆς δε-
422 ξιῆς χειρός, | καὶ οὕτως ἐπὶ πλεῖστον μὲν τὴν νευρὴν
ἑλκύσει, ἀφήσει δὲ ἀπὸ στερεωτάτου καὶ ἀθροωτάτου·
ἀπὸ τῶν τοιούτων γὰρ ἀφεσίων τῶν τοξευμάτων, τα-
χεῖαι καὶ αἱ ἰσχύες καὶ τὰ μήκεα γίνεται. ἐπιδέσει δὲ
καὶ τοξικῇ οὐδὲν κοινόν. τοῦτο μὲν γάρ, εἰ ἐπιδήσας
ἔχειν τὴν χεῖρα οὕτως ἔμελλε, πόνους ἂν ἄλλους πολ-
λοὺς προσετίθει μείζονας τοῦ τρώματος· τοῦτο δ᾽ εἰ
συγκάμψαι ἐκέλευεν, οὔτε τὰ ὀστέα οὔτε τὰ νεῦρα
οὔτε αἱ σάρκες ἔτι ἐν τῷ αὐτῷ ἐγίνοντο, ἀλλὰ ἄλλῃ
μετεκοσμεῖτο κρατέοντα τὴν ἐπίδεσι. καὶ τί ὄφελός
ἐστι τοξικοῦ σχήματος; καὶ ταῦτα ἴσως οὐκ ἂν ἐξ-
ημάρτανε σοφιζόμενος, εἰ εἴα τὸν τετρωμένον αὐτὸν
τὴν χεῖρα παρασχέσθαι.

3. Ἄλλος δ᾽ αὖ τις τῶν ἰητρῶν ὑπτίην τὴν χεῖρα
δούς, οὕτω κατατείνειν ἐκέλευε, καὶ οὕτως ἔχουσαν
ἐπέδει, τοῦτο νομίζων τὸ κατὰ φύσιν εἶναι, τῷ τε χροῒ
σημαινόμενος καὶ τὰ ὀστέα νομίζων κατὰ φύσιν εἶναι
οὕτως, ὅτι φαίνεται τὸ ἐξέχον ὀστέον τὸ παρὰ τὸν
καρπὸν ᾗ ὁ σμικρὸς δάκτυλος, κατ᾽ ἰθυωρίην εἶναι
424 τοῦ ὀστέου, ἀφ᾽ | ὁκοίου τὸν πῆχυν οἱ ἄνθρωποι με-

for the left arm; for the hinge-like end of the arm in this position, being supported on the base provided by the forearm, makes a straight line of the bones of the upper arm and forearm, as if the whole were one, and any flexure of the joint disappears in this attitude. Clearly, then, the parts in question are most inflexible and extended in this position, so as neither to be overcome or give way when the bow is drawn by the right arm. In this way, the archer will be able to draw the bowstring the farthest, and shoot with the greatest force and rapidity, for shafts launched in this way fly strongly, swiftly and far. But in reality, there is no reason to connect reducing fractures and archery. For, first, if the operator, after reducing an arm, fixed it in this position, he would inflict many additional harms, greater than those of the injury itself; and in addition, if he ordered the patient to bend the elbow, none of the bones, cords, or muscles would remain in their bandaged position, but would rearrange themselves in spite of the dressings. What, then, is the advantage of the archer's position? And perhaps our theorizer would not have committed these errors if he had let the injured person himself present the arm.

3. Again, another practitioner, placed an arm in the supine position, instructed an assistant to extend it, and then bandaged it in this position, supposing this to be the natural one, on the basis of the surface indications, and also believing that the bones were in their natural position because the prominent bone at the wrist on the little finger side appeared to be in line with the bone from which the cubit is measured (medial epicondyle of the humerus). He

---

9 κέκλασται MV: τετασθαι B

τρέουσι. ταῦτα μαρτύρια ἐπήγετο ὅτι κατὰ φύσιν
οὕτως ἔχει, καὶ ἐδόκεε εὖ λέγειν. ἀλλὰ τοῦτο μέν, εἰ
ὑπτίη ἡ χεὶρ κατατείνοιτο, ἰσχυρῶς πονοίη ἄν· γνοίη
δ' ἄν τις τὴν ἑωυτοῦ χεῖρα κατατείνας ὡς ἐπώδυνον
τὸ σχῆμα. ἐπεὶ καὶ ἀνὴρ ἥσσων κρέσσονα διαλαβὼν
οὕτως ἐν τῇσιν ἑωυτοῦ χερσίν, ὡς κλᾶται ὁ ἀγκὼν
ὕπτιος, ἄγοι ἂν ὅπῃ ἐθέλοι· οὔτε γὰρ εἰ ξίφος ἐν
ταύτῃ τῇ χειρὶ ἔχοι, ἔχοι ἂν ὅ τι χρήσαιτο τῷ ξίφει·
οὕτω βίαιον τοῦτο τὸ σχῆμά ἐστιν. τοῦτο δέ, εἰ ἐπι-
δήσας τις ἐν τούτῳ τῷ σχήματι ἐῴη, μέζων μὲν πό-
νος, εἰ περιΐοι, μέγας δὲ καὶ εἰ κατακέοιτο. τοῦτο δέ,
εἰ συγκάμψει τὴν χεῖρα, ἀνάγκη τούς τε μύας καὶ τὰ
ὀστέα ἄλλο σχῆμα ἔχειν. ἠγνόει δὲ καὶ τάδε τὰ ἐν τῷ
σχήματι χωρὶς τῆς ἄλλης λύμης· τὸ γὰρ ὀστέον τὸ
παρὰ τὸν καρπὸν ἐξέχον, τὸ κατὰ τὸν σμικρὸν δάκτυ-
λον, τοῦτο μὲν τοῦ πήχεός ἐστι· τὸ δὲ ἐν τῇ συγκάμ-
ψει ἐὸν ἀπ' ὅτευ τὸν πῆχυν οἱ ἄνθρωποι μετρέουσι,
426 τοῦτο δὲ | τοῦ βραχίονος ἡ κεφαλή ἐστιν. ὁ δὲ ᾤετο
τωὐτὸ ὀστέον εἶναι τοῦτό τε κἀκεῖνο, πολλοὶ δὲ καὶ
ἄλλοι· ἔστι δὲ ἐκείνῳ τῷ ὀστέῳ τωὐτὸ ὁ ἀγκὼν κα-
λεόμενος, ὃν ποτὶ στηριζόμεθα. οὕτως οὖν ὑπτίην
ἔχοντι τὴν χεῖρα, τοῦτο μὲν τὸ ὀστέον διεστραμμένον
φαίνεται, τοῦτο δὲ τὰ νεῦρα τὰ ἀπὸ τοῦ καρποῦ τεί-
νοντα ἐκ τοῦ εἴσω μέρεος καὶ ἀπὸ τῶν δακτύλων,
ταῦτα ὑπτίην ἔχοντι τὴν χεῖρα διεστραμμένα γίνεται·
τείνει γὰρ ταῦτα τὰ νεῦρα πρὸς τὸ τοῦ βραχίονος
ὀστέον, ὅθεν ὁ πῆχυς μετρεῖται. αὗται τοσαῦται καὶ
τοιαῦται αἱ ἁμαρτάδες καὶ ἀγνοίαι τῆς φύσιος τῆς

adduced this as evidence for the naturalness of the position, and seemed to be speaking the truth. But, to begin with, if the arm is extended in the supine position, it will be very painful; in fact, anyone who holds his arm extended in this position will discover how painful it is. Indeed, a weaker man grasping a stronger one with his hands, with the latter's elbow extended in the supine position, would be able to lead him wherever he chose; for not even if the stronger man had a sword in this hand would he be able to use it: this is how constraining the position is. Furthermore, if someone bandaged a patient's arm in this position and left him that way, his pain would be very great when he walked about, and great enough even when he was recumbent. Again, if he flexed the arm, it would be absolutely necessary for both the muscles and the bones to change their positions. Besides, the practitioner was also ignorant of the following facts about this position: the projecting bone at the wrist on the side of the little finger (ulna) belongs to the forearm, whereas the one at the bend of the elbow (from which men measure the cubit) is the head of the arm (medial epicondyle of the humerus): he however thought they both belong to the same bone, as many other people do too. The former (ulna) is in fact the same bone which is called the elbow (olecranon), on which we lean. In a patient with the forearm supinated like this, first, the bone appears to be distorted; and secondly, the cords stretching from the wrist on its inner side and from the fingers also become distorted when the arm is in this supine position, for these cords extend to the bone of the upper arm from which the cubit is measured. This is the number and character of these errors and misunderstandχειρός. εἰ δ᾽ ὡς[10] ἐγὼ κελεύω, χεῖρα κατηγυῖαν κατατείνοι τις, ἐπιστρέψαι μὲν τὸ ὀστέον ἐς ἰθύ, τὸ κατὰ τὸν σμικρὸν δάκτυλον, τὸ ἐς τὸν ἀγκῶνα τεῖνον· ἰθυωρίην δὲ ἕξει τὰ νεῦρα τὰ ἀπὸ τοῦ καρποῦ πρὸς τοῦ βραχίονος τὰ ἄκρα τείνοντα. ἀναλαμβανομένη δὲ ἡ χεὶρ ἐν παραπλησίῳ σχήματι ἔσται, ἐν ᾧπερ καὶ ἐπιδεομένη, ἄπονος μὲν ὁδοιπορέοντι, ἄπονος δὲ κατακειμένῳ καὶ ἀκάματος. καθίννυσθαι δὲ χρὴ τὸν ἄνθρωπον οὕτως, ὅκως ᾖ τὸ ἐξέχον τοῦ ὀστέου πρὸς τὴν λαμπροτάτην τῶν παρεουσῶν αὐγέων, ὡς μὴ λάθῃ τὸν χειρίζοντα ἐν τῇ κατατάσει, εἰ[11] ἱκανῶς ἐξίθυνται. τοῦ γε μὴν ἐμπείρου οὐδ᾽ ἂν τὴν χεῖρα λάθοι ἐπαγομένην τὸ ἐξέχον· ἀτὰρ καὶ ἀλγέει μάλιστα κατὰ τὸ ἐξέχον[12] ψαυόμενον. |

428 4. Τῶν δὲ ὀστέων τοῦ πήχεος, ὧν μὴ ἀμφότερα κατέηγεν, ῥᾴων ἡ ἴησις, ὧν[13] τὸ ἄνω ὀστέον τέτρωται, καίπερ παχύτερον ἐόν, ἅμα μὲν ὅτι τὸ ὑγιὲς ὑποτεταγμένον[14] γίνεται ἀντὶ θεμελίου, ἅμα δὲ ὅτι εὐκρυπτότερον γίνεται, πλὴν ᾖ[15] τὸ ἐγγὺς τοῦ καρποῦ· παχείη γὰρ ἡ τῆς σαρκὸς ἐπίφυσις ἡ ἐπὶ τὸ ἄνω. τὸ δὲ κάτω ὀστέον ἄσαρκον καὶ οὐκ εὐξύγκρυπτον, καὶ κατατάσιος ἰσχυροτέρης δεῖται. ἢν δὲ μὴ τοῦτο συντριβῇ, ἀλλὰ τὸ ἕτερον, ἐλαφροτέρη ἡ κατάτασις ἀρκέει, ἢν δὲ ἀμφότερα κατεηγῇ, ἰσχυροτάτης κατατάσιος δεῖται. παιδίου μὲν γὰρ ἤδη εἶδον καταταθέντα

[10] εἰ δ᾽ ὡς Aldina: εἰδὼς BMV
[11] εἰ B: om. MV
[12] ἀτὰρ . . . ἐξέχον B: om. MV
[13] ὧν B: ἢν B^2MV

ings concerning the structure of the arm. But if anyone recommends extension of a fractured arm, as I do, he will turn the bones stretching from the extremity of the little finger to the elbow into a straight line, and the cords stretching from the wrist to the (sc. lower) end of the upper arm will be in a direct line. Further, the arm when put in a sling will be in the same position as when it is bandaged, and will cause the patient no pain when he walks, no pain when he lies down, and call forth no sense of weariness. You should seat the patient in such a way that the projecting part of the bone is turned toward the brightest light available, in order that the operator will not fail to see whether the extent of the extension is producing an adequate straightening. Of course, the hand of an experienced practitioner will not fail to recognize the prominence (sc. at the fracture) by touch; also, there is a special tenderness at the prominence when palpated.

4. When one but not both of the bones of the forearm is fractured, treatment is easier if the upper bone (radius) is injured, even though it is the thicker of the two, both because the sound bone lying beneath it acts as a support, and because the injury is easier to hide, except in the part near the wrist, since the musculature covering the upper bone is thick. But the lower bone (ulna) lacks a muscular covering; it also requires stronger extension. If it is not this bone but the other one (ulna) that is broken, a lighter extension will suffice: if both are broken, very strong extension is required. In the case of a child, I have seen the

---

14 -ταγμένον B: -ταμένον MV
15 ᾗ Kühlewein: η B: εἰ MV

μᾶλλον ἢ ὡς ἔδει,[16] οἱ δὲ πλεῖστοι ἧσσον τείνονται ἢ
ὡς δεῖ. χρὴ δ᾽ ἐπὴν τείνωσι, τὰ θέναρα προσβάλ-
430 λοντα | διορθοῦν· ἔπειτα χρίσαντα κηρωτῇ (μὴ πάνυ
πολλῇ, ὡς μὴ περιπλέῃ τὰ ἐπιδέσματα) οὕτως ἐπιδεῖν
ὅπως μὴ κατωτέρω ἄκρην τὴν χεῖρα ἕξει τοῦ ἀγκῶ-
νος, ἀλλὰ σμικρῷ τινι[17] ἀνωτέρω, ὡς μὴ τὸ αἷμα ἐς
ἄκρον ἐπιρρέῃ, ἀλλὰ ἀπολαμβάνηται. ἔπειτα ἐπιδεῖν
τῷ ὀθονίῳ, τὴν ἀρχὴν βαλλόμενος κατὰ τὸ κάτηγμα·
ἐρείδων μὲν οὖν, μὴ πιέζων δὲ κάρτα. ἐπὴν δὲ περι-
βάλλῃ κατὰ τωὐτὸ δὶς ἢ τρίς, ἐπὶ τὸ ἄνω νεμέσθω
ἐπιδέων, ἵνα αἱ ἐπιρροαὶ τοῦ αἵματος ἀπολαμβάνων-
ται, καὶ τελευτησάτω κεῖθι. χρὴ δὲ μὴ μακρὰ εἶναι τὰ
πρῶτα ὀθόνια· τῶν δὲ δευτέρων ὀθονίων τὴν μὲν ἀρ-
χὴν βάλλεσθαι ἐπὶ τὸ κάτηγμα· περιβαλὼν δὲ ἅπαξ
ἐς τωὐτό, ἔπειτα νεμέσθω ἐς τὸ κάτω καὶ ἐπὶ ἧσσον
πιέζων, καὶ ἐπὶ μέζον διαβιβάσκων, ὡς ἂν αὐτῷ
ἱκανὸν γένηται τὸ ὀθόνιον ἀνάπαλιν δρομῆσαι κεῖθι
ἵνα περ τὸ ἕτερον ἐτελεύτησεν. ἐνταῦθα μὲν οὖν τὰ
ὀθόνια ἐπ᾽ ἀριστερὰ ἢ ἐπὶ δεξιὰ ἐπιδεδέσθω, ἢ ἐπὶ
ὁκότερα ἂν ξυμφέρῃ πρὸς τὸ σχῆμα τοῦ κατάγματος
καὶ ἐφ᾽ ὁκότερα ἂν περιρρέπειν ξυμφέρῃ. μετὰ δὲ
ταῦτα, σπλῆνας κατατείνειν χρὴ κεχρισμένους κη-
ρωτῇ ὀλίγῃ· καὶ γὰρ προσηνέστερον καὶ εὐθετώτερον.
432 ἔπειτα οὕτως ἐπιδεῖν τοῖσιν ὀθονίοισιν[18] | ἐναλλάξ, ὁτὲ
μὲν ἐπὶ δεξιά, ὁτὲ δὲ ἐπ᾽ ἀριστερά· καὶ τὰ μὲν πλείω
κάτωθεν ἀρχόμενος ἐς τὸ ἄνω ἄγειν, ἔστι δ᾽ ὅτε καὶ
ἄνωθεν ἐς τὸ κάτω. τὰ δὲ ὑπόξηρα ἀκέεσθαι τοῖσι
σπλήνεσι κυκλεῦντα· τῷ δὲ πλήθει τῶν περιβολέων

bones extended more than was necessary, but most patients receive less extension than necessary. During extension you must use the palms of your hands to press the parts into position; then after anointing with cerate (in no great amount in order to keep the dressings from slipping), proceed to bandage it in such a position that the patient's hand will not be lower than his elbow, but a little higher, in order that blood will not flow to the extremity, but be kept back. Then apply a linen bandage, putting its beginning at the fracture: provide support, but without pressing too hard. After two or three turns are made at the same place, the bandage should be carried upward, so that the afflux of blood will be kept back, and it should terminate there. The first bandages should not be long. The start of the second bandage should be at the fracture, making one turn there, and then be carried downward, with decreasing pressure and at wider intervals, so that enough of the bandage will be left for it to run back again to the place where the other one ended. The bandages in this part of the dressing should be applied either to the left or to the right, whichever suits the form of the fracture and the direction toward which the limb ought to incline. After this, compresses anointed with a little cerate should be laid along the arm, which will make the bandages more accommodating and easier to place. Then position the bandages crosswise, alternately to the right and the left, beginning in most cases from below and moving upward, but on other occasions from above downward. Treat lean parts by surrounding them with compresses; bring them

16 ἔ̓δει MV: ειδέι B   17 *τινι* B[3]: *καὶ* BMV
18 ὡ̔ς add. MV

μὴ πᾶν ἀθροὸν ξυνδιορθοῦντα, ἀλλὰ κατὰ μέρος. προσπεριβάλλειν δὲ χρὴ χαλαρὰ καὶ περὶ τὸν καρπὸν τῆς χειρὸς ἄλλοτε καὶ ἄλλοτε. πλῆθος δὲ τῶν ὀθονίων ἱκανὸν τὸ πρῶτον αἱ δύο μοῖραι.

5. Σημεῖα δὲ τοῦ καλῶς ἰητρευομένου ταῦτα, καὶ ὀρθῶς[19] ἐπιδεομένου· εἰ ἐρωτῴης αὐτὸν εἰ πεπίεκται, καὶ εἰ φαίη μὲν πεπιέχθαι, ἡσύχως δέ, καὶ μάλιστα εἰ κατὰ τὸ κάτηγμα φαίη· τοιαῦτα μὲν οὖν[20] φάναι χρὴ πεπρηγμένα διὰ τέλεος τὸν ὀρθῶς ἐπιδεόμενον. σημεῖα δὲ τάδε τῆς μετριότητος· τὴν μὲν ἡμέρην, ἣν ἂν ἐπιδεθῇ, καὶ τὴν νύκτα δοκείτω αὐτὸς ἑωυτῷ μὴ ἐπὶ ἧσσον πεπιέχθαι, ἀλλ᾽ ἐπὶ μᾶλλον· τῇ δὲ ὑστεραίῃ οἰδημάτιον ἐλθεῖν ἐς χεῖρα ἄκρην μαλθακόν· μετριότητος γὰρ σημεῖον τῆς πιέξιός.[21] τελευτώσης δὲ τῆς ἡμέρης, ἐπὶ ἧσσον δοκείτω πεπίεχθαι· τῇ δὲ τρίτῃ χαλαρά οἱ[22] δοκείτω εἶναι τὰ ἐπιδέσματα. καὶ ἢν μέν τι τούτων τῶν εἰρημένων ἐλλείπῃ, γινώσκειν χρὴ ὅτι χαλαρωτέρη ἡ ἐπίδεσις τοῦ μετρίου· ἢν δέ τι
434 | τῶν εἰρημένων πλεονάζῃ, γινώσκειν χρὴ ὅτι μᾶλλον ἐπιέχθη τοῦ μετρίου· καὶ τούτοισι σημαινόμενος τὸ ὕστερον ἐπιδέων ἢ χαλᾶν μᾶλλον ἢ πιέζειν. ἀπολύσαντα δὲ χρὴ τριταῖον ἐόντα κατατεινάμενον καὶ διορθωσάμενον <αὖθις ἐπιδῆσαι>·[23] κἢν μετρίως τὸ πρῶτον τετυχήκῃς ἐπιδήσας, ταύτην τὴν ἐπίδεσιν χρὴ ὀλίγῳ μᾶλλον ἢ ἐκείνην πιέσαι. βάλλεσθαι δὲ χρὴ τὰς ἀρχὰς κατὰ τὸ κάτηγμα, ὥσπερ καὶ τὸ

[19] ὀρθῶς B^2: ὅρος BMV   [20] μὲν οὖν B: τοίνυν MV

to a level not all at once, but gradually, by the number of circumvolutions. You should put additional loose turns now and then at the wrist. Two sets of the bandages will be sufficient for the first dressing.

5. Here are the signs of good treatment and correct bandaging: if you ask the patient whether the part is compressed, he will say that it is, but moderately, and that this is mainly at the fracture; a properly bandaged patient must give a report like this throughout the course of the treatment. The following are indications of a correct moderation (sc. in tightness): during the day of the dressing and the following night, the pressure should seem to the patient not to diminish, but rather to increase; on the next day, a slight and soft swelling will appear in the hand: this is a sign of moderate pressure. At the end of that day, the pressure should seem less, and on the third day you should find the bandages loose. If, then, any of the conditions mentioned are absent, you can conclude that the bandaging was slacker than moderate, but if any of them is excessive, you must conclude that the pressure was more than moderate; take this evidence as your guide to make the next dressing looser or tighter. You should remove the dressing on the third day after the extension and adjustment, <and bind it again:> if your previous bandaging was satisfactory, make the next one a little tighter. The beginning of the bandages should be placed over the fracture,

---

21 *σου* add. MV 22 *χαλαρά οἱ* B[3]: *χαλαρὰς οἱ* M: *χαλαρασοι* B: *χαλαράσοι* V

23 *αὖθις ἐπιδῆσαι* add Littré: see Jouanna's discussion, pp. 109–11

πρότερον· ἢν μὲν γὰρ τοῦτο πρότερον ἐπιδέῃς, †ἐξαρ(ε)ιαται†[24] ἐκ τούτου οἱ ἰχῶρες ἐς τὰς ἐσχατιὰς ἔνθα καὶ ἔνθα· ἢν δέ τι ἄλλο πρότερον πιέξῃς, ἐς τοῦτο †ἐξαρ(ε)ιαται† ἐκ τοῦ πιεχθέντος, ἐς πολλὰ δὲ εὔχρηστον τοῦτο ξυνιέναι. οὕτως οὖν ἄρχεσθαι μὲν ἀεὶ χρὴ τὴν ἐπίδεσιν καὶ τὴν πίεξιν ἐκ τούτου τοῦ χωρίου, τὰ δὲ ἄλλα κατὰ λόγον, ὡς ἂν προσωτέρω ἀπὸ τοῦ κατήγματος ἀγάγῃς, ἐπὶ ἧσσον τὴν πίεξιν ποιέεσθαι. χαλαρὰ δὲ παντάπασι μηδέποτε περιβάλ-
436 λειν, ἀλλὰ προσπεπτωκότα. | ἔπειτα δὲ πλείοσιν ὀθονίοισι χρὴ ἐπιδεῖν ἑκάστην τῶν ἐπιδεσίων· ἐρωτώμενος δὲ φάτω ὀλίγῳ μᾶλλόν οἱ πεπίεχθαι ἢ τὸ πρότερον, καὶ μάλιστα φάτω κατὰ τὸ κάτηγμα καὶ τὰ ἄλλα κατὰ λόγον· καὶ ἀμφὶ τῷ οἰδήματι, καὶ ἀμφὶ τῷ πονέειν, καὶ ἀμφὶ τῷ ῥηΐζειν, κατὰ λόγον τῆς προτέρης ἐπιδέσιος γενέσθω. ἐπὴν δὲ τριταῖος ᾖ, χαλαρώτερά οἱ δοκείτω εἶναι τὰ ἐπιδέσματα. ἔπειτα ἀπολύσαντα χρὴ αὖθις ἐπιδῆσαι, ὀλίγῳ μᾶλλον πιέζοντα, καὶ ἐν πᾶσι τοῖσιν ὀθονίοισιν οἷσί περ ἤμελλεν ἐπιδεῖσθαι· καὶ ἔπειτα πάντα αὐτὸν ταὐτὰ[25] καταλαβέτω, ἅπερ καὶ ἐν τῇσι πρώτῃσι περιόδοισι τῶν ἐπιδεσίων.

6. Ἐπὴν δὲ τριταῖος γένηται, ἑβδομαῖος δὲ ἀπὸ τῆς πρώτης ἐπιδέσιος, ἢν ὀρθῶς ἐπιδέηται, τὸ μὲν οἴδημα ἐν ἄκρῃ τῇ χειρὶ ἔσται, οὐδὲ τοῦτο λίην μέγα· τὸ δ' ἐπιδεόμενον χωρίον ἐν πάσῃσι τῇσιν ἐπιδέσεσιν ἐπὶ τὸ[26] λεπτότερον καὶ ἰσχνότερον εὑρεθήσεται, ἐν δὲ τῇ ἑβδόμῃ καὶ πάνυ λεπτὸν καὶ τὰ ὀστέα τὰ κατεηγότα

as before, since if you did this before, the serous effusions will have been driven away from there into the outer parts on both sides; but if you formerly exerted pressure anywhere else, they will have been driven toward this place (the fracture) away from the part compressed. It is useful for many reasons to understand this. First, it demonstrates the necessity always to begin the bandaging and compression at this place, and to reduce the pressure proportionally elsewhere as you get further from the point of the fracture. Never make the turns altogether slack, but rather closely adherent. Further, you should use more bandages at each dressing; and when the patient is asked, he should say that the pressure is a little more than it was before, especially at the point of fracture, and proportionately in the other parts. And as regards the swelling, and the feeling of pain and relief, things should develop the same way as with the previous dressing. When the third day comes, the patient should find the dressings quite loose. Then, after undoing them, the dresser should bandage again with a little more pressure, and with all the bandages that he is intending to use. Afterward the patient should experience all the same symptoms he had in the first cycles of bandaging.

6. When the third day is reached (the seventh from the first dressing), if the bandaging has been properly done, there will be swelling in the hand, but it will not be very great. As to the part bandaged, it will be found to be thinner and more shrunken at each dressing, and on the seventh day it will be quite thin, while the fractured bones

[24] Cf. Jouanna, pp. 111–13. [25] ταὐτὰ Jouanna: ταῦτα BM: om. V [26] τὸ MV: om. B

ἐπὶ μᾶλλον κινεύμενα καὶ εὐπαράγωγα ἐς κατόρθωσιν. καὶ ἢν ᾖ ταῦτα τοιαῦτα, κατορθωσάμενον χρὴ ἐπιδῆσαι ὡς ἐς νάρθηκας, ὀλίγῳ μᾶλλον πιέσαντα[27] ἢ τὸ πρότερον, ἢν μὴ πόνος τις πλείων ᾖ ἀπὸ τοῦ οἰδήματος τοῦ ἐν ἄκρῃ τῇ χειρί. ἐπὴν δ᾽ ἐπιδήσῃς τοῖσιν ὀθονίοισι, τοὺς νάρθηκας περιθεῖναι χρὴ καὶ περιλαβεῖν ἐν τοῖσι δεσμοῖσιν ὡς χαλαρωτάτοισιν, ὁκόσον ἠρεμέειν, ὥστε μηδὲν συμβάλλεσθαι ἐς τὴν πίεξιν τῆς χειρὸς τὴν τῶν ναρθήκων πρόσθεσιν. |
438 μετὰ δὲ ταῦτα, ὅ τε πόνος αἵ τε ῥᾳστῶναι αἱ αὐταὶ γινέσθωσαν αἵπερ καὶ ἐν τῇσι προτέρῃσι περιόδοισι τῶν ἐπιδεσίων. ἐπὴν δὲ τριταῖος ἐὼν φῇ χαλαρὸν εἶναι, τότ᾽ ἔπειτα χρὴ τοὺς νάρθηκας ἐρείσασθαι, μάλιστα μὲν κατὰ τὸ κάτηγμα, ἀτὰρ καὶ τἆλλα κατὰ λόγον, ᾗπερ καὶ ἡ ἐπίδεσις ἐχάλα μᾶλλον ἢ ἐπίεζεν. παχύτατον δὲ χρὴ εἶναι τὸν νάρθηκα ᾗ ἐξέστη τὸ κάτηγμα, μὴ μὴν πολλῷ. ἐπιτηδεύειν δὲ χρὴ μάλιστα μὲν κατ᾽ ἰθυωρίην τοῦ μεγάλου δακτύλου, ὡς μὴ κείσεται ὁ νάρθηξ, ἀλλὰ τῇ ἢ τῇ, μηδὲ κατὰ τὴν τοῦ σμικροῦ δακτύλου ἰθυωρίην, ᾗ τὸ ὀστέον ὑπερέχει ἐν τῷ καρπῷ, ἀλλὰ τῇ ἢ τῇ. ἢν δὲ ἄρα πρὸς τὸ κάτηγμα συμφέρῃ κεῖσθαι κατὰ ταῦτά τινας τῶν ναρθήκων, βραχυτέρους αὐτοὺς χρὴ τῶν ἄλλων ποιέειν, ὡς μὴ ἐξικνέωνται πρὸς τὰ ὀστέα τὰ ὑπερέχοντα παρὰ τὸν καρπόν· κίνδυνος γὰρ ἑλκώσιος καὶ νεύρων ψιλώσιος. χρὴ δὲ διὰ τρίτης ἐρείδειν τοῖσι νάρθηξι πάνυ ἡσυχῇ, οὕτω τῇ γνώμῃ ἔχοντα, ὡς οἱ νάρθηκες φυλακῆς εἵ-

will be more mobile and readier for adjustment. If this is so, after making the adjustment, you should bandage as for splints, applying a little more pressure than before, unless there is an increase of pain from the swelling in the hand. When you have applied the bandages, attach the splints around the limb and secure them in ligatures as loose as possible while still providing adequate firmness, so that the addition of splints will contribute nothing to the compression of the arm. After this, the pain and the relief following it should be the same as in the previous courses of bandaging. When, on the third day, the patient says his bandage is loose, you should tighten the splints, especially at the site of the fracture, but also in the other places proportionately, wherever the dressing has become too slack. The splint should be thicker where the fracture projects, but not too much so: you should also see that the splint is not lying in the line of the thumb, but on one side or the other, nor in the line of the little finger where the bone projects at the wrist, but on one side or the other. If in spite of this, it is still of benefit for the fracture that some of the splints be placed here, you should make them shorter than the other ones, so that they do not reach to the bones that project at the wrist, which would bring a risk of ulceration and denudation of the cords. You should tighten the splints every third day, very slightly, bearing in

---

27 *πιέσαντα* Froben: -*οντας* BMV

*νεκα τῆς ἐπιδέσιος προσκέονται, ἀλλʼ οὐ τῆς πιέξιος
εἵνεκεν ἐπιδέδενται.*[28]

7. *Ἦν μὲν οὖν εὖ εἰδῇς ὅτι ἱκανῶς τὰ ὀστέα ἀπί-
θυνται ἐν τῇσι προτέρῃσιν ἐπιδέσεσι, καὶ μήτε κνη-
σμοὶ λυπέωσι, μήτε τις ἕλκωσις μηδεμία ὑποπτεύηται
440 εἶναι, ἐὰν χρὴ | ἐπιδεδέσθαι ἐν τοῖσι νάρθηξιν, ἔστʼ
ἂν ὑπὲρ εἴκοσιν ἡμέρας γένηται. ἐν γὰρ τριήκοντα
μάλιστα τῇσι συμπάσῃσι κρατύνεται ὀστέα τὰ ἐν
πήχει τὸ ἐπίπαν· ἀτρεκὲς δὲ οὐδέν· μάλα γάρ τε καὶ
φύσις φύσιος καὶ ἡλικίη ἡλικίης διαφέρει. ἐπὴν δὲ
λύσῃς, ὕδωρ θερμὸν καταχέαι χρὴ καὶ μετεπιδῆσαι,
ἧσσον μὲν ὀλίγῳ πιέσαντα ἢ τὸ πρόσθεν, ἐλάσσοσι
δὲ τοῖσιν ὀθονίοισιν ἢ τὸ πρότερον· καὶ ἔπειτα διὰ
τρίτης ἡμέρης λύσαντα ἐπιδεῖν, ἐπὶ μὲν ἧσσον πιέ-
ζοντα, ἐπὶ δὲ ἐλάσσοσι τοῖσιν ὀθονίοισιν. ἐπὴν δέ,
ὅταν τοῖσι νάρθηξι δεθῇ, ὑποπτεύῃς τὰ ὀστέα μὴ
ὀρθῶς κεῖσθαι, ἢ ἄλλο τι ὀχλέῃ τὸν τετρωμένον, λύ-
σαντα ἐν τῷ μεσηγὺ*[29] *τοῦ χρόνου ἢ ὀλίγῳ πρόσθεν
μετεπιδῆσαι. δίαιτα δὲ τούτοισιν οἷσιν ἂν μὴ ἕλκεα
ἐξ ἀρχῆς γένηται ἢ ὀστέα ἔξω ἐξίσχῃ, ἀρκέει ὑπο-
φαύλη. ἐνδεέστερον δὲ χρὴ διαιτᾶν ἄχρις ἡμερέων
δέκα, ἅτε δὴ καὶ ἐλινύοντας, καὶ ὄψοισιν ἁπαλοῖσι
χρῆσθαι ὁκόσα τῇ διεξόδῳ μετριότητα παρασχήσει·
οἴνου δὲ καὶ κρεηφαγίης ἀπέχεσθαι· ἔπειτα μέντοι ἐκ
442 προσαγωγῆς | ἀνακομίζεσθαι. οὗτος ὁ λόγος ὥσπερ
νόμος κεῖται δίκαιος περὶ κατηγμάτων ἰήσιος, ὥς τε
χειρίζειν χρή, ὥς τε ἀποβαίνειν ἀπὸ τῆς δικαίης χει-
ρίξιος. ὅ τι δʼ ἂν μὴ οὕτως ἀποβαίνῃ, εἰδέναι χρὴ ὅτι*

mind that they are put there to protect the dressing, but not attached for the sake of pressure.

7. If you see then, that the bones are sufficiently realigned by the previous dressings, and there is neither any painful irritation nor any suspicion of a sore, you must leave the part bandaged with splints until beyond the twentieth day, since generally it takes thirty days in total for the bones of the forearm to consolidate; but there is nothing exact about this, since both constitutions and ages differ greatly in their effects on recovery. When you remove the dressing, make an affusion of warm water and then apply a new bandage, using a little less pressure and fewer bandages than before. After this, remove and reapply every other day with less pressure and fewer bandages. If, in cases where splints are used, you suspect that the bones are not properly adjusted, or that something else is troubling the patient, remove the dressing and replace it in the middle of the interval or a little sooner. A normal diet will suffice in cases where no lesion arises at the onset, or any protrusion of the bone; this diet should be slightly restricted for the first ten days, seeing that the patients are resting; and soft dishes should be favored that promote a moderate amount of evacuation: avoid wine and meat, but then gradually build the patient up. This account gives a sort of normal rule for the treatment of fractures, both how they should be handled surgically, and the results of correct handling. If any of the results are not as has been described, you may be sure that there has been

---

28 ἐπιδέδενται M: -δεδεανται B: ἐπιδέονται V
29 μεσηγὺ B: ἡμίσει MVB²

ἐν τῇ χειρίξει τι ἐνδεὲς πεποίηται ἢ πεπλεόνασται. ἔτι δὲ τάδε χρὴ προσξυνιέναι ἐν τούτῳ τῷ ἁπλῷ τρόπῳ, ἃ οὐ κάρτα ἐπιμελέονται οἱ ἰητροί, καίτοι πᾶσαν μελέτην καὶ πᾶσαν ἐπίδεσιν οἷά τε διαφθείρειν[30] ἐστί, μὴ ὀρθῶς ποιεύμενα. ἢν γὰρ τὰ μὲν ὀστέα ἄμφω κατηγῇ, ἢ τὸ κάτω μοῦνον, ὁ δὲ ἐπιδεδεμένος ἐν ταινίῃ τινὶ τὴν χεῖρα ἔχῃ ἀναλελαμμένος, τυγχάνῃ δὲ ἡ ταινίη κατὰ τὸ κάτηγμα πλείστη ἐοῦσα, ἔνθεν δὲ καὶ ἔνθεν ἡ χεὶρ ἀπαιωρῆται, τοῦτον ἀνάγκη τὸ ὀστέον εὑρεθῆναι διεστραμμένον ἔχοντα πρὸς τὸ ἄνω μέρος· ἢν δέ, κατεηγότων τῶν ὀστέων οὕτως, ἄκρην τε τὴν χεῖρα ἐν τῇ ταινίῃ ἔχῃ καὶ παρὰ τὸν ἀγκῶνα, ὁ δὲ ἄλλος πῆχυς μετέωρος ᾖ, οὕτως εὑρεθήσεται τὸ ὀστέον ἐς τὸ κάτω μέρος διεστραμμένον ἔχων. χρὴ οὖν ἐν ταινίῃ πλάτος ἐχούσῃ μαλθακῇ, τὸ πλεῖστον τοῦ πήχεος καὶ τὸν καρπὸν τῆς χειρὸς ὁμαλῶς αἰωρέεσθαι. |

444 8. Ἢν δὲ ὁ βραχίων καταγῇ, ἢν μέν τις ἀποτανύσας τὴν χεῖρα ἐν τούτῳ τῷ σχήματι διατείνῃ, ὁ μῦς τοῦ βραχίονος κατατεταμένος ἐπιδεθήσεται· ἐπὴν δ' ἐπιδεθεὶς συγκάμψῃ τὸν ἀγκῶνα, ὁ μῦς τοῦ βραχίονος ἄλλο σχῆμα σχήσει. δικαιοτάτη οὖν βραχίονος κατάτασις ἥδε· ξύλον πηχυαῖον ἢ ὀλίγῳ βραχύτερον, ὁκοῖοι οἱ στειλεοί εἰσι τῶν σκαφίων, κρεμάσαι χρὴ ἔνθεν καὶ ἔνθεν σειρῇ δήσαντα. καθίσαντα δὲ χρὴ[31] τὸν ἄνθρωπον ἐπὶ ὑψηλοῦ τινος, τὴν χεῖρα ὑπερκεῖσθαι, ὡς ὑπὸ τῇ μασχάλῃ γένηται ὁ στειλεὸς ἔχων ξυμμέτρως, ὥστε μόλις δύνασθαι καθίννυσθαι τὸν

some defect or excess in the application. Furthermore, you should acquaint yourself with the following points in this basic method, points to which some practitioners do not pay much attention, but which, if not properly appreciated, can spoil the results of all your careful bandaging. If both bones are broken, or only the lower one (ulna), and the patient, after bandaging, has his arm held in some sort of sling which happens to be chiefly at the point of fracture, while the (sc. part of the) arm on either side is unsupported, the bone will necessarily be distorted toward the upper side; while if, when the bones are broken in the same way, his hand and the part near the elbow are supported in the sling, but the rest of the forearm is unsupported, this bone will be found distorted toward the lower side. It follows that as much as possible of the forearm and wrist should be supported evenly in a soft wide sling.

8. If the upper arm is fractured, if you extend the whole arm and keep it in this position, the muscle of the upper arm (biceps) will be bandaged in a state of extension, but when the bandaged patient flexes his elbow, the muscle will assume a different position. The most correct mode of extension of the arm is as follows: take a piece of wood shaped like a spade handle, a cubit in length or a little shorter, and suspend it by a cord tied at each end. Seat the patient on some high object, and pass his arm over the piece of wood so that it comes evenly under his axilla, such

---

30 διαφθείρειν Aldina: διαφθειριν B: δεῖ φθείρειν MV

31 χρὴ B: om. MV

ἄνθρωπον, σμικροῦ δέοντα μετέωρον εἶναι. ἔπειτα θέντα τι ἄλλο ἔφεδρον, καὶ ὑποθέντα σκύτινον ὑποκεφάλαιον, ἢ ἓν ἢ πλείω, ὅπως ξυμμέτρως σχήσει ὕψεος τοῦ πήχεος πλαγίου πρὸς ὀρθὴν γωνίην· ἄριστον μὲν σκύτος πλατὺ καὶ μαλθακὸν ἢ ταινίην πλατέην ἀμφιβάλλοντα, τῶν μεγάλων τι σταθμίων ἐξαρτῆσαι, ὅ τι μετρίως ἕξει κατατείνειν. εἰ δὲ μή, τῶν ἄνδρων ὅστις ἐρρωμένος, ἐν τούτῳ τῷ σχήματι τοῦ
446 πήχεος παρὰ τὸν ἀγκῶνα καταναγκαζέτω | ἐς τὸ κάτω. ὁ δὲ ἰητρὸς ὀρθὸς μὲν ἐὼν χειριζέτω, τὸν ἕτερον πόδα ἐπὶ ὑψηλοτέρου τινὸς ἔχων, κατορθώσας δὲ τοῖσι θέναρσι τὸ ὀστέον· ῥηϊδίως δὲ κατορθώσεται· ἀγαθὴ γὰρ ἡ κατάτασις, ἤν τις καλῶς παρασκευάσηται. ἔπειτα ἐπιδείτω, τάς τε ἀρχὰς βαλλόμενος ἐπὶ τὸ κάτηγμα, καὶ τἆλλα πάντα ὥσπερ πρότερον παρῃνέθη, χειριζέτω. καὶ ἐρωτήματα ταὐτὰ[32] ἐρωτάτω καὶ σημείοισι χρήσθω τοῖσιν αὐτοῖσι, εἰ μετρίως ἔχει, ἢ οὔ. καὶ διὰ τρίτης ἐπιδείτω, καὶ ἐπὶ μᾶλλον πιεζείτω, καὶ ἑβδομαῖον ἢ ἐναταῖον ἐν νάρθηξι δησάτω· καὶ ἢν ὑποπτεύῃ μὴ καλῶς κεῖσθαι τὸ ὀστέον, μεσηγὺ τούτου τοῦ χρόνου, λυσάτω, καὶ εὐθετισάμενος μετεπιδησάτω.

Κρατύνεται δὲ μάλιστα βραχίονος ὀστέον ἐν τεσσαράκοντα ἡμέρῃσιν. ἐπὴν δὲ ταύτας ὑπερβάλλῃ, λύειν χρή, καὶ ἐπὶ ἧσσον πιέζειν τοῖσιν ὀθονίοισι καὶ ἐπὶ ἐλάσσοσιν ἐπιδεῖν. δίαιταν δὲ ἀκριβεστέρην τινὰ ἢ τὸ πρότερον διαιτᾶν, καὶ πλείω χρόνον· τεκμαίρεσθαι δὲ πρὸς τοῦ οἰδήματος τοῦ ἐν ἄκρῃ τῇ χειρί, |

that he can hardly sit down and is almost suspended. Then, set out another stool with one or more leather cushions on it, so that it will be level with the patient's forearm flexed at a right angle. It is best to pass some broad piece of soft leather or a wide piece of cloth around the arm, and suspend large weights from it that will produce a moderate extension. Alternately, with the arm in this position, have a strong man grasp the elbow and force it downward. As to the surgeon, he should operate standing with one foot on some quite raised support, and adjust the bone with the palms of his hands. The adjustment will be easy, as a result of the good extension, if it is properly managed. Then have him do the bandaging, starting the bandages at the fracture, and performing all the rest of the operation as previously directed; he should ask the same questions and use the same indications to judge whether the bandage is correct or not. He should bandage again every other day, with increasing pressure, and on the seventh or ninth day bandage the arm in splints. If he suspects the bone is not well aligned, he should loosen the dressings toward the middle of this period, and after setting it right, reapply them.

The bone of the upper arm (humerus) generally consolidates in forty days; when these are exceeded you should undo the dressings and diminish the pressure by rebandaging with fewer bandages. Employ a somewhat stricter diet than before, over a longer period: judge this from the swelling in the hand, keeping an eye on its sever-

---

[32] ταὐτὰ Littré: ταῦτα codd.

448 τὴν ῥώμην ὁρέων. προσξυνίεναι δὲ χρὴ καὶ τάδε, ὅτι βραχίων κυρτὸς πέφυκεν ἐς τὸ ἔξω μέρος· ἐς τοῦτο τοίνυν τὸ μέρος καὶ διαστρέφεσθαι θέλει,[33] ἐπὴν μὴ καλῶς ἰητρεύηται· ἀτὰρ καὶ τἆλλα πάντα ὀστέα, ἐς[34] ὅπερ πέφυκε διεστραμμένα, ἐς τοῦτο καὶ ἰητρευόμενα φιλέει διαστρέφεσθαι, ἐπὴν κατεαγῇ. χρὴ τοίνυν, ἐπὴν τοιοῦτόν τι ὑποπτεύηται, ταινίῃ πλατείῃ προσεπιλαμβάνειν τὸν βραχίονα κύκλῳ περί τε τὸ στῆθος περιδέοντα· καὶ ἐπὴν ἀναπαύεσθαι μέλλῃ, μεσηγὺ τοῦ ἀγκῶνος καὶ τῶν πλευρέων σπλῆνά τινα πολύπτυχον πτύξαντα ὑποτιθέναι, ἢ ἄλλο τι ὃ τούτῳ ἔοικεν· οὕτω γὰρ ἂν ἀλορδότατον[35] τὸ κύρτωμα τοῦ ὀστέου γένοιτο· φυλάσσεσθαι δὲ ἀεὶ[36] χρή, ὅπως μὴ ἄγαν διαστρέψῃ[37] ἐς τὸ ἔσω μέρος.

9. Ποὺς δὲ ἀνθρώπου ἐκ πολλῶν καὶ μικρῶν ὀστέων συγκεῖται, ὥσπερ καὶ χεὶρ ἄκρη· κατάγνυται μὲν οὐ πάνυ τι ταῦτα τὰ ὀστέα, ἢν μὴ σὺν τῷ χρωτὶ[38] τιτρωσκομένῳ ὑπὸ ὀξέος τινὸς ἢ βαρέος· τὰ μὲν οὖν τιτρωσκόμενα, ἐν ἑλκωσίων μέρει εἰρήσεται, ὡς χρὴ ἰητρεύειν. ἢν δέ τι κινηθῇ ἐκ τῆς χώρης, ἢ τῶν δακτύλων ἄρθρον ἢ ἄλλο τι τῶν ὀστέων τοῦ ταρσοῦ καλεομένου, ἀναγκάζειν μὲν χρὴ ἐς τὴν ἑωυτοῦ χώρην ἕκα-
450 στον, ὥσπερ | καὶ τὰ ἐν τῇ χειρὶ εἴρηται· ἰητρεύειν δὲ καὶ κηρωτῇ καὶ σπλήνεσι καὶ ὀθονίοισιν, ὥσπερ καὶ τὰ κατήγματα, πλὴν ναρθήκων, τὸν μὲν αὐτὸν τρόπον πιεζεῦντα, διὰ τρίτης δὲ ἐπιδέοντα· ὑποκρινέσθω δὲ ὁ ἐπιδεόμενος παραπλήσια, οἷά περ καὶ ἐν τοῖσι κατήγμασι, καὶ περὶ τοῦ πεπιέχθαι καὶ περὶ τοῦ χαλαρὸν

ity. It is also necessary to bear in mind that the upper arm inclines naturally outward, and therefore tends to get distorted in this direction when it is improperly managed. In fact, all bones when fractured tend to become distorted during treatment toward the side to which they naturally incline. So, whenever you suspect anything of this kind, you should pass an additional broad band around the arm, and continue it around the chest; when the patient goes to bed, put a compress with many folds or something of the kind between the elbow and the ribs; in this way the curvature of the bone will be rectified. You must always take care, however, that it is not bent too far inward.

9. The human foot, like the hand, is composed of many small bones. These bones are not often broken, unless the skin is also wounded by something sharp or heavy. The proper treatment of the wounded parts will be discussed in the section on lesions of soft parts. But if any of the bones is displaced, whether at a joint of the toes or some bone of what is called the tarsus, you should force each of them back into its proper place as has been described for the bones of the hand. Treat as in cases of fracture with cerate, compresses, and bandages, but without splints, using pressure in the same way and changing the dressings every other day; a patient's answers as to pressure and relaxation should be like those in all cases of fracture. All

---

33 θέλει B: φιλέει MV

34 ἐς B: om. MV

35 αλ. B: ἰθὺ MV

36 ἀεὶ B: μέντοι MV

37 διαστρ. B: om. MV

38 χρωτὶ B: χρώματι MV

εἶναι.[39] ὑγιέα δὲ γίνεται ἐν εἴκοσιν ἡμέρῃσι τελέως ἅπαντα, πλὴν ὁκόσα κοινωνέει τοῖσι τῆς κνήμης ὀστέοισι κατ᾽ αὐτὴν τὴν ἴξιν.[40] συμφέρει δὲ κατακεῖσθαι τοῦτον τὸν χρόνον. ἀλλὰ γὰρ οὐ τολμέουσιν ὑπερορῶντες τὰ νοσήματα, ἀλλὰ περιέρχονται πρὶν ὑγιέες γενέσθαι. διὰ τοῦτο καὶ οἱ πλεῖστοι οὐκ ἐξυγιαίνουσι τελέως, ἀλλὰ πολλάκις αὐτοὺς ὁ πόνος ὑπομιμνήσκει εἰκότως· ὅλον γὰρ τὸ ἄχθος οἱ πόδες τοῦ σώματος ὀχέουσιν. ὁκόταν οὖν μήπω ὑγιέες ἐόντες ὁδοιπορέωσι, φλαύρως ξυναλθέεται τὰ ἄρθρα τὰ κινηθέντα· διὰ τοῦτο ἄλλοτε καὶ ἄλλοτε ὀδυνῶνται τὰ πρὸς τῇ κνήμῃ.

10. Τὰ δὲ κοινωνέοντα τοῖσι τῆς κνήμης ὀστέοισι μέζω τε τῶν ἑτέρων ἐστί, καὶ κινηθέντων τούτων πολυχρονιωτέρη ἡ ἄλθεξις. ἴησις μὲν οὖν ἡ αὐτή·
452 ὀθονίοισι δὲ πλείοσι | χρῆσθαι καὶ σπλήνεσι, καὶ ἐπὶ πᾶν ἔνθεν καὶ ἔνθεν ἐπιδέειν· πιέζειν δὲ ὥσπερ καὶ τὰ ἄλλα πάντα, ταύτῃ μάλιστα ᾗ ἐκινήθη, καὶ τὰς πρώτας περιβολὰς τῶν ὀθονίων κατὰ ταῦτα ποιέεσθαι, ἐν δὲ ἑκάστῃ τῶν ἀπολυσίων[41] ὕδατι πολλῷ θερμῷ χρέεσθαι· ἐν πᾶσι δὲ πολλὸν ὕδωρ θερμὸν καταχέειν τοῖσι κατ᾽ ἄρθρα σίνεσιν. αἱ δὲ πιέξιες καὶ αἱ χαλάσιες ἐν τοῖσιν αὐτοῖσι χρόνοισι τὰ αὐτὰ σημεῖα δεικνυόντων ἅπερ ἐπὶ τοῖσι πρόσθεν· καὶ τὰς μετεπιδέσιας ὡσαύτως χρὴ ποιέεσθαι. ὑγιέες δὲ τελέως οὗτοι γίνονται ἐν τεσσεράκοντα ἡμέρῃσι μάλιστα, ἢν τολμέωσι κατακεῖσθαι· ἢν δὲ μή, πάσχουσι τὰ αὐτὰ ἃ καὶ πρότερον καὶ ἐπὶ μᾶλλον.

these bones are completely healed in twenty days, except those which are attached to the bones of the leg in a vertical line. It is good to stay in bed during this period, but some patients, denying the injury, cannot bring themselves to do this, and get up before they are well. This is why most of them do not make a complete recovery; but often the pain reminds them of the real situation, since the feet carry the whole weight of the body. Thus, if they walk about when they are not yet well, the unstable joints do not heal very well; on this account there are occasional pains in the area of the lower leg.

10. The (sc. tarsal) bones connected with those of the lower leg are larger than the others (sc. in the wrist), and when they are displaced healing takes much longer. Treatment, in fact, is the same, but more bandages and compresses should be used, and the dressings should be extended completely in both directions; use pressure as in all the other cases, especially at the point of the displacement, and make the first turns of the bandage there. At each change of dressing use plenty of warm water; in truth, much warm water should be poured over all injuries of the joints. The same signs indicating pressure and slackness reveal themselves in the same periods as in the former cases, and the change of dressings should be made in the same way. These patients generally recover completely in forty days, if they can bring themselves to stay in bed; failing this, they suffer the same as the former cases, but to a greater degree.

---

39 *χαλαρὸν εἶναι* Petrequin: *χαλᾶν* B: *χαλαρὰν εἶναι* MV

40 *κατ' αὐτὴν τὴν ἴξιν* Ermerins: *καὶ αυτη τη ειξη* B: *καὶ αὐτῇ τῇ ἴξει* MV

41 *ἀπολυσ.* MV: *ἐπιλυς.* B

11. Ὅσοι δὲ πηδήσαντες ἀπὸ ὑψηλοῦ τινὸς ἐστηρίξαντο τῇ πτέρνῃ ἰσχυρῶς, τούτοισι διίσταται μὲν τὰ ὀστέα, φλέβια δὲ ἐκχυμοῦνται ἀμφιφλασθείσης τῆς σαρκὸς ἀμφὶ τὸ ὀστέον· οἴδημα δὲ ἐπιγίνεται καὶ πόνος πολύς· τὸ γὰρ ὀστέον τοῦτο οὐ σμικρόν ἐστι, καὶ ὑπερέχει μὲν ὑπὸ τὴν ἰθυωρίην τῆς κνήμης, κοινωνέει δὲ φλεψὶν καὶ νεύροις ἐπικαίροις· ὁ τένων δὲ ὁ ὀπίσθιος τούτῳ προσήρτηται τῷ ὀστέῳ. τούτους χρὴ ἰητρεύειν μὲν κηρωτῇ καὶ σπλήνεσι καὶ ὀθονίοισιν· ὕδατι δὲ θερμῷ πλείστῳ ἐπὶ τούτοισι χρῆσθαι καὶ ὀθονίων πλειόνων ἐπὶ τούτοισι δεῖ καὶ ἄλλως ὡς βελ-
454 τίστων καὶ προσηνεστάτων. καὶ ἢν μὲν τύχῃ | ἁπαλὸν τὸ δέρμα φύσει ἔχων τὸ ἀμφὶ τὴν πτέρνην, ἐᾶν οὕτως· ἢν δὲ παχὺ καὶ σκληρόν, οἷα μετεξέτεροι ἴσχουσι, κατατάμνειν χρὴ ὁμαλῶς καὶ διαλεπτύνειν, μή[42] διατιτρώσκοντα. ἐπιδεῖν δὲ ἀγαθῶς οὐ πάντος ἀνδρός ἐστι τὰ τοιαῦτα· ἢν γάρ τις ἐπιδέῃ, ὥσπερ καὶ τὰ ἄλλα τὰ κατὰ τὰ σφυρὰ ἐπιδεῖται, ὁτὲ μὲν περὶ τὸν πόδα περιβαλλόμενος, ὁτὲ δὲ περὶ τὸν τένοντα, αἱ ἀποσφίγξιες αὗται χωρίζουσι τὴν πτέρνην ᾗ τὸ φλάσμα ἐγένετο. καὶ οὕτω κίνδυνος σφακελίσαι τὸ ὀστέον τὸ τῆς πτέρνης· καίτοι ἢν σφακελίσῃ, τὸν αἰῶνα πάντα ἱκανὸν ἀντίσχειν τὸ νόσημα. καὶ γὰρ τἆλλα ὅσα μὴ ἐκ τοιούτου τρόπου σφακελίζει, ἀλλ᾽ ἐν κατακλίσει μελανθείσης τῆς πτέρνης ὑπὸ ἀμελείης τοῦ σχήματος, ἢ ἐν κνήμῃ τρώματος γενομένου ἐπικαίρου καὶ χρονίου καὶ κοινοῦ τῇ πτέρνῃ, ἢ ἐν μηρῷ ἢ ἐπὶ ἄλλῳ νοσήματι ὑπτιασμοῦ χρονίου γενομένου. ὁμῶς

11. In those who, in leaping from a height, come down violently on their heel, the bones are separated, and blood is extravasated from the small vessels due to contusion of the flesh around the bone; swelling supervenes and severe pain. The bone involved (calcaneum) is not small, it extends beyond the line of the leg, and it is connected with critical vessels and cords; the posterior tendon of the leg (Achilles tendon) is inserted into it. You should treat these patients with cerate, compresses, and bandages, using an abundance of hot water; and they require plenty of bandages, in particular the best and softest ones you can find. If the skin about the heel is naturally smooth, leave it alone, but if it is thick and hard as in some persons, you should pare it evenly down, and thin it without, however, going through to the flesh. It is not within every man's capacity to bandage such cases properly, for if someone applies the bandage as is done in other cases at the ankle, sometimes taking one turn around the foot and another time around the tendon, the bandage compresses the part and excludes the heel where the contusion is, so that there is a risk of necrosis of the heel bone; and if there is necrosis, the malady may last the patient's whole life. In fact, other kinds of necrosis not arising from this cause, as for example when the heel blackens in a bedridden patient due to the carelessness of his position, or when there is a serious and chronic wound in the lower leg communicating with the heel, or in the thigh, or during another malady involving prolonged rest on his back; all these cases of

---

42 μὴ MV: om. B

καὶ τοῖσι τοιούτοισι χρόνια καὶ ὀχλώδεα καὶ πολ-
λάκις ἀναρρηγνύμενα, ἢν μὴ χρηστῇ μὲν μελέτῃ θε-
456 ραπευθῇ, πολλῇ δὲ ἡσυχίῃ, ὡς τά | γ᾿ ἐκ τοῦ τοιούτου
τρόπου σφακελίζοντα, καὶ κινδύνους μεγάλους τῷ
σώματι παρέχει πρὸς τῇ ἄλλῃ λύμῃ. καὶ γὰρ πυρετοὶ
ὑπεροξέες, συνεχέες, τρομώδεες, λυγγώδεες, γνώμης
ἁπτόμενοι, καὶ ὀλιγήμεροι κτείνοντές. γένοιντο δ᾿ ἂν
καὶ φλεβῶν αἱμορρόων πελιώσιες, ναρκώσιες καὶ
γαγγραινώσιες ὑπὸ τῆς πιέξιος· γένοιτο δ᾿ ἂν ταῦτα
ἔξω τοῦ ἄλλου σφακελισμοῦ. ταῦτα μὲν οὖν εἴρηται
οἷα τὰ ἰσχυρότατα φλάσματα γίνεται· τὰ μέντοι πλεῖ-
στα ἡσυχαίως ἀμφιφλᾶται καὶ οὐδεμία πολλὴ
σπουδὴ τῆς μελέτης, ἀλλ᾿ ὅμως ὀρθῶς γε χρὴ χειρί-
ζειν. ἐπὴν μέντοι ἰσχυρὸν δόξῃ εἶναι τὸ ἔρεισμα, τά
τε εἰρημένα ποιέειν χρή καὶ τὴν ἐπίδεσιν τὴν πλεί-
στην ποιέεσθαι ἀμφὶ τὴν πτέρνην, περιβάλλοντα ἄλ-
λοτε πρὸς τὰ ἄκρα τοῦ ποδὸς ἀντιπεριβάλλοντα, ἄλ-
λοτε πρὸς τὰ μέσα, ἄλλοτε πρὸς τὰ περὶ τὴν κνήμην.
προσεπιδεῖν δὲ καὶ τὰ πλησίον πάντα καὶ ἔνθεν καὶ
ἔνθεν, ὥσπερ καὶ πρόσθεν εἴρηται. καὶ ἰσχυρὴν μὲν
μὴ ποιέεσθαι τὴν πίεξιν, πολλοῖσι δὲ τοῖς ὀθονίοισιν.
458 ἄμεινον δὲ καὶ ἐλλέβορον| πῖσαι ἢ[43] αὐθημερὸν ἢ τῇ
ὑστεραίῃ· ἀπολῦσαι δὲ τριταῖον καὶ αὖτις μετεπιδῆ-
σαι. σημεῖα δὲ τάδε, εἰ παλιγκοταίνει ἢ οὔ· ἐπὴν μὲν
τὰ ἐκχυμώματα τῶν φλεβῶν καὶ τὰ μελάσματα καὶ
τὰ ἐγγὺς ἐκείνων ὑπέρυθρα γίνηται καὶ ὑπόσκληρα,
κίνδυνος παλιγκοτῆσαι· ἀλλ᾿ ἢν μὲν ἀπύρετος ᾖ,
φαρμακεύειν ἄνω χρὴ ὥσπερ εἴρηται, καὶ ὅσα ἂν μὴ

necrosis are chronic and troublesome, and often relapse if not treated with the most skillful attention and long rest. Necroses of this sort, indeed, besides other harm, also bring great dangers to the body: there may be very acute fevers, continuous and attended by tremblings, hiccups, and affections of the mind, and death in a few days. There may also be lividity of the large blood vessels, loss of sensation, and gangrene due to compression; however, these signs may also occur without any necrosis. The above remarks apply to very severe contusions, but often the parts are only moderately contused, and require no very great care, although they do require competent treatment. But when the crushing seems to be severe, the above directions must be followed, the greater part of the bandaging being about the heel, taking turns sometimes around the front end of the foot, sometimes about the middle part, and sometimes carrying on around the leg. All the neighboring parts in both directions should be included in the bandage, as explained above; and do not exert strong pressure, but use many bandages; it is also good to give a dose of hellebore on the first or second day. Remove the bandage and reapply it on the third day. The following are signs whether or not there will be complications: when there are extravasations from the blood vessels, and dark spots, and the neighboring parts become reddish and rather hard, there is a danger of aggravation. If there is no fever, you should give an emetic as was directed; also, in

---

43 ἢ B: om. MV

ξυνεχεῖ πυρεταίνῃ· ἢν δὲ ξυνεχεῖ πυρεταίνῃ, μὴ φαρμακεύειν· ἀπέχειν δὲ σιτίων καὶ ῥυφημάτων, ποτῷ δὲ χρῆσθαι ὕδατι καὶ μὴ οἴνῳ, ἀλλὰ ὀξυγλυκεῖ. ἢν δὲ μὴ μέλλῃ παλιγκοταίνειν τὰ ἐκχυμώματα καὶ τὰ μελάσματα καὶ τὰ περιέχοντα ὑπόχλωρα γίνεται καὶ οὐ σκληρά· ἀγαθὸν τοῦτο τὸ μαρτύριον ἐν πᾶσι τοῖσιν ἐκχυμώμασι, τοῖσι μὴ μέλλουσι παλιγκοταίνειν· ὅσα δὲ σὺν σκληρύσμασι πελιοῦται, κίνδυνος μὲν μελανθῆναι. τὸν δὲ πόδα ἐπιτηδεύειν χρὴ ὅπως ἀνωτέρω τοῦ ἄλλου σώματος ἔσται τὰ πλεῖστα ὀλίγον. ὑγιὴς δ᾽ ἂν γένοιτο ἐν ἑξήκοντα ἡμέρῃσιν, εἰ ἀτρεμέοι. |

460 12. Ἡ δὲ κνήμη δύο ὀστέα ἔχει, τῇ μὲν συχνῷ λεπτότερον τὸ ἕτερον τοῦ ἑτέρου, τῇ δὲ οὐ πολλῷ λεπτότερον· ξυνέχεται δὲ ἀλλήλοισι τὰ πρὸς τοῦ ποδός, καὶ ἐπίφυσιν κοινὴν ἔχει· ἐν ἰθυωρίῃ δὲ τῆς κνήμης οὐ ξυνέχεται· τὰ δὲ πρὸς τοῦ μηροῦ ξυνέχεται, καὶ ἐπίφυσιν ἔχει καὶ ἡ ἐπίφυσις διάφυσιν. μακρότερον δὲ τὸ ἕτερον ὀστέον σμικρῷ τὸ κατὰ τὸν σμικρὸν δάκτυλον· ἡ μὲν φύσις τοιαύτη τῶν ὀστέων τῶν ἐν τῇ κνήμῃ.

13. Ὀλισθάνει δὲ ἔστιν ὅτε τὰ πρὸς τοῦ ποδός, ὅτε μὲν σὺν τῇ ἐπιφύσει ἀμφότερα τὰ ὀστέα, ὁτὲ δὲ ἡ ἐπίφυσις ἐκινήθη, ὁτὲ δὲ τὸ ἕτερον ὀστέον. ταῦτα δὲ
462 ὀχλώδεα μὲν ἧσσον ἢ τὰ ἐν τῷ | καρπῷ τῶν χειρέων, εἰ τολμῷεν ἀτρεμέειν οἱ ἄνθρωποι. ἴησις δὲ παραπλησίη, οἵη περ ἐκείνων· τήν τε γὰρ ἐμβολὴν χρὴ ποιέεσθαι ἐκ κατατάσιος ὥσπερ ἐκείνων, ἰσχυροτέρης δὲ

cases where the fever is not continuous. If, however, there is continuous fever, do not give an evacuant, but avoid foods and soups, and as drink use water and not wine; but hydromel may be taken. If there are not going to be complications, the extravasations and dark spots, and the parts around them, become yellowish and are not hard: this is good evidence in all extravasations that they are not going to develop complications, but in those which turn livid and hard there is a danger of gangrene. Care must be taken that the foot is, as a rule, kept a little higher than the rest of the body. The patient will recover in sixty days if he remains at rest.

12. The lower leg has two bones, one much slenderer (fibula) than the other (tibia) at one end, but not so much slenderer at the other end. The parts near the foot are joined together and have a common epiphysis, but along the length of the leg they are not united; the parts next the thigh are united and have an epiphysis, and this epiphysis has a diaphysis. The bone on the side of the little toe is slightly longer. This is the disposition of the leg bones.

13. Such bones are occasionally dislocated at the foot end, sometimes both bones together with the epiphysis, sometimes (sc. just) the epiphysis, and sometimes only one of the bones. These dislocations cause less trouble than dislocations of the wrist, if the patients can bring themselves to stay in bed. The treatment is similar to that of the wrist, with reduction being made by extension just

δεῖται τῆς κατατάσιος, ὅσῳ καὶ ἰσχυρότερον τὸ σῶμα ταύτῃ. ἐς τὰ πλεῖστα μὲν γὰρ ἀρκέουσιν ἄνδρες δύο, ὁ μὲν ἔνθεν, ὁ δὲ ἔνθεν τείνοντες. ἢν δὲ μὴ ἰσχύωσιν, ἰσχυροτέρην ῥηΐδιόν ἐστι ποιέειν τὴν κατάτασιν· ἢ γὰρ πλήμνην κατορύξαντα χρή, ἢ ἄλλο τι ὅ τι τούτῳ ἔοικεν, μαλθακόν τι περὶ τὸν πόδα περιβάλλειν· ἔπειτα πλατέσι βοείοισιν ἱμᾶσιν περιδήσαντα τὸν πόδα τὰς ἀρχὰς τῶν ἱμάντων ἢ πρὸς ὕπερον ἢ πρὸς ἕτερον ξύλον προσδήσαντα, τὸ ξύλον πρὸς τὴν πλήμνην ἄκρον ἐνθέντα ἀνακλᾶν, τοὺς δὲ ἀντιτείνειν ἄνωθεν, τῶν τε ὤμων ἐχομένους καὶ τῆς ἰγνύης. ἔστι δὲ καὶ τὸ ἄνω τοῦ σώματος ἀνάγκῃ[44] προσλαβεῖν. τοῦτο
464 μὲν ἢν βούλῃ, | ξύλον στρογγύλον λεῖον κατορύξας βαθέως, μέρος τι αὐτοῦ ὑπερέχον τοῦ ξύλου μεσηγὺ τῶν σκελέων ποιήσασθαι παρὰ τὸν περίνεον, ὡς κωλύῃ ἀκολουθέειν τὸ σῶμα τοῖσι πρὸς ποδῶν τείνουσιν· ἔπειτα πρὸς τὸ τεινόμενον σκέλος μὴ ῥέπειν· τὸν δέ τινα πλάγιον παρακαθήμενον ἀπωθέειν τὸν γλουτόν, ὡς μὴ περιέλκηται τὸ σῶμα. τοῦτο δὲ καὶ ἢν παρὰ[45] τὰς μασχάλας ἔνθεν καὶ ἔνθεν τὰ ξύλα παραπεπήγῃ, αἱ δὲ χεῖρες παρατεταμέναι φυλάσσωνται· προσεπιλαμβάνηται[46] δέ τις κατὰ τὸ γόνυ, καὶ οὕτως ‹ἂν›[47] ἀντιτείνοιτο. τοῦτο δ' ἢν παρὰ τὸ γόνυ βούλῃ,[48] ἄλλους ἱμάντας περιδήσας ἢ περὶ τὸν μηρόν, πλήμνην ἄλλην ὑπὲρ κεφαλῆς κατορύξας, ἐξαρτήσας τοὺς ἱμάντας ἔκ τινος ξύλου, τὸ ξύλον στηρίζων ἐς

as there, but stronger extension is required since the body is stronger in this part. As a rule, two men will suffice, one pulling one way and one the other, but if they are not strong enough, it is easy to make the extension more powerful. Fix either a wheel nave or something like that in the ground, put a soft wrapping around the patient's foot, and then binding broad straps of oxhide to it, attach the ends of the straps to a pestle or some other piece of wood. Put the end of the wood into the wheel nave and pull it back, while assistants provide counterextension from above by holding the patient by the shoulders and hollow of the knee. It is also possible to exert force on the upper part of the body by some apparatus. For example, A. You may fix a smooth, rounded piece of wood deeply in the ground and place its upper projecting part between the legs at the perineum, so as to prevent the body from following when the assistants make the extension by the feet; and also, in order that the leg which is being extended does not incline downward, an assistant seated at the side should press back on the hip so that the body is not drawn sideways. B. If you like, pegs may be fixed under both axillae, and the arms kept extended along the sides, while someone else takes hold at the knee to make counterextension. C. If you prefer, fasten straps in the same way around the knee or the thigh; and fixing another wheel nave in the ground above the patient's head, attach the straps to a staff of wood and by levering the wood in the wheel nave, exert

---

44 ἀνάγκῃ Littré: ἀνάγκην BM: ἀνάγκη V

45 παρὰ B: περὶ MV

46 -βάνηται B: -βανέτω MVB[2]

47 ἂν add. Jouanna

48 βούλῃ B: βούληται MVB[2]

τὴν πλήμνην τἀναντία τῶν πρὸς ποδῶν ἕλκειν. τοῦτο δ᾽ ἢν βούλῃ, ἀντὶ τῶν πλημνέων δοκίδα ὑποτείνας ὑπὸ τὴν κλίνην μετρίην, ἔπειτα πρὸς τῆς δοκίδος ἔν-
466 θεν καὶ ἔνθεν τὴν κεφαλὴν στηρίζων,[49] | ἀνακλῶν τὰ ξύλα κατατείνειν τοὺς ἱμάντας. ἢν δὲ θέλῃς, ὀνίσκους καταστήσας ἔνθεν καὶ ἔνθεν, ἐπ᾽ ἐκείνων τὴν κατάτασιν ποιέεσθαι. πολλοὶ δὲ καὶ ἄλλοι τρόποι κατατασίων· ἄριστον δέ, ὅστις ἐν πόλει μεγάλῃ ἰητρεύει, κεκτῆσθαι ἐσκευασμένον ξύλον, ἐν ᾧ πᾶσαι αἱ ἀνάγκαι ἔσονται πάντων μὲν κατηγμάτων, πάντων δὲ ἄρθρων ἐμβολῆς ἐκ κατατάσιος καὶ μοχλεύσιος. ἀρκέει δὲ τὸ ξύλον, ἢν ᾖ τοιοῦτον οἷον οἱ τετράγωνοι στῦλοι δρύϊνοι[50] γίνονται, μῆκος καὶ πλάτος καὶ πάχος.

14. Ἐπὴν δὲ ἱκανῶς κατατανύσῃς, ῥηΐδιον ἤδη τὸ ἄρθρον ἐμβαλεῖν· ὑπεραιωρέεται γὰρ ἐς ἰθυωρίην ὑπὲρ τῆς ἀρχαίης ἕδρης. κατορθοῦσθαι οὖν χρὴ τοῖσι θέναρσι τῶν χειρῶν, τοῖσι μὲν ἐς τὸ ἐξεστηκὸς ἐρείδοντα, τοῖσι δὲ ἐπὶ θάτερα κατώτερον τοῦ σφυροῦ ἀντερείδοντα. ἐπὴν δ᾽ ἐμβάλῃς, ἢν μὲν οἷόν τε ᾖ, κατατεταμένον ἐπιδεῖν χρή· ἢν δὲ κωλύηται ὑπὸ τῶν ἱμάντων, ἐκείνους λύσαντα ἀντικατατείνειν, ἔστ᾽ ἂν ἐπιδήσῃς. ἐπιδεῖν δὲ τὸν αὐτὸν τρόπον καὶ τὰς ἀρχὰς ὡσαύτως βαλλόμενον κατὰ τὸ ἐξεστηκός, καὶ τὰς περιβολὰς τὰς πρώτας πλείστας κατὰ τοῦτο ποιέεσθαι, καὶ τοὺς σπλῆνας πλείστους κατὰ τοῦτο, καὶ
468 τὴν πίεξιν μάλιστα κατὰ | τὠυτόν· προσεπιδεῖν δὲ καὶ ἔνθεν καὶ ἔνθεν ἐπὶ συχνόν. μᾶλλον δέ τι τοῦτο τὸ

counterextension to the extension at the feet. D. If you like, instead of the wheel naves, stretch a plank of appropriate length under the bed, and then levering the head of the plank at each end, draw back on the wood and exert extension with the straps. E. If you prefer, set up windlasses at either end of the plank and make the extension with these. There are also many other methods of extension. The best thing for anyone who practices in a large city is to acquire a wooden apparatus capable of carrying out all the mechanical methods useful for all kinds of fractures, and for reduction of all the joints by extension and leverage. This wooden apparatus will suffice if it is like the quadrangular planks regularly made of oak, in length, breadth and thickness.

14. When you have made sufficient extension, it will then be easier to reduce the joint, since the bone will already be elevated in a direct line above its old position. It must be adjusted with the palms of the hands, pressing upon the projecting part with one palm, and with the other making counterextension on the opposite side below the ankle. After reduction, you should, if possible, apply a bandage, with the limb kept extended. If this is prevented by the straps, remove them and keep up counterextension until the bandaging is completed. Bandage in the same way (as for fractures) starting the bandages on the projecting part, and making the first and many more turns there; most of the compresses should also be there, and the pressure should fall especially on this part. Also extend the dressing for a good distance on either side. This joint re-

---

49 καὶ add. MV 50 στῦλοι δρ. Jouanna: στυλεοὶ δρύϊνοι στύλοι B: στύλοι οἷοι δρύϊνοι M

ἄρθρον πεπίεχθαι χρὴ ἐν τῇ πρώτῃ ἐπιδέσει ἢ τὸ ἐν τῇ χειρί. ἐπὴν δὲ ἐπιδήσῃς, ἀνωτέρω μὲν τοῦ ἄλλου σώματος ἐχέτω τὸ ἐπιδεθέν, τὴν δὲ θέσιν δεῖ ποιέεσθαι οὕτως, ὅπως ἥκιστα ἀπαιωρηθήσεται ὁ πούς. τὸν δὲ ἰσχνασμὸν τοῦ σώματος οὕτως ποιέεσθαι, ὁποίην τινὰ δύναμιν ἔχει καὶ τὸ ὀλίσθημα· τὰ μὲν γὰρ σμικρόν, τὰ δὲ μέγα ὀλισθάνει. τὸ ἐπίπαν δὲ[51] ἰσχναίνειν μᾶλλον καὶ ἐπὶ πλείω χρόνον χρὴ ἐν τοῖσι κατὰ τὰ σκέλεα τρώμασι ἢ ἐν τοῖσι κατὰ χεῖρα· καὶ γὰρ μέζω καὶ παχύτερα ταῦτα ἐκείνων. καὶ δὴ καὶ ἀναγκαῖον ἐλινύειν τὸ σῶμα καὶ κατακεῖσθαι. μετεπιδῆσαι δὲ τὸ ἄρθρον οὔτε τι κωλύει τριταῖον οὔτε κατεπείγει· καὶ τἆλλα πάντα παραπλησίως χρὴ ἰητρεύειν, ὥσπερ καὶ τὰ παροιχόμενα. καὶ ἢν μὲν τολμᾷ ἀτρέμα κατακεῖσθαι, ἱκαναὶ τεσσαράκοντα ἡμέραι, ἢν μοῦνον ἐς τὴν ἑωυτῶν χώρην τὰ ὀστέα αὖθις καθίζηται. ἢν δὲ μὴ θέλῃ ἀτρεμέειν, χρῷτο μὲν ἂν οὖν[52] βραδέως τῷ σκέλει, ἐπιδεῖσθαι δὲ ἀναγκάζοιτ' ἂν πολὺν χρόνον. |
470 ὁπόσα μέντοι τῶν ὀστέων μὴ τελέως ἵζει ἐς τὴν ἑωυτῶν χώρην, ἀλλά τι ἐπιλείπει, τῷ χρόνῳ λεπτύνεται ἰσχίον καὶ μηρὸς καὶ κνήμη. καὶ ἢν μὲν εἴσω ὀλίσθῃ, τὸ ἔξω μέρος λεπτύνεται, ἢν δὲ ἔξω, τὸ εἴσω· τὰ πλεῖστα δὲ ἐς τὸ ἔσω ὀλισθάνει.

15. Ἐπὴν δὲ κνήμης ὀστέα ἀμφότερα καταγῇ ἄνευ ἑλκώσιος, κατατάσιος ἰσχυροτέρης δεῖται. κατατείνειν[53] δὲ τούτων τῶν τροπων ἐνίοισι τῶν προειρημένων τισί,[54] ἢν μεγάλαι αἱ παραλλάξιες ἔωσιν. ἱκα-

quires somewhat greater pressure at the first bandaging than the wrist does. After dressing, the bandaged part should be higher than the rest of the body, and positioned so that the foot is left unsupported as little as possible. The patient should undergo a reduction corresponding to his strength and to the displacement, for the displacement may be small or great. As a rule, the reducing treatment should be stricter and more prolonged in injuries in the leg than in those about the arm, since the former parts are larger and stouter than the latter. And it is especially needful for the person to be at rest and stay in bed. As to rebandaging the joint on the third day, there is neither any contraindication nor any urgency, and you should conduct all the other treatment as in the previous cases. If the patient manages to keep himself at rest and stay in bed, forty days will be sufficient, provided only that the bones have settled back again into their places. If he fails to keep still, he will recover use of the leg only slowly, and will have to be kept bandaged for a long time. Any of the bones that do not return completely to their original place, but have some inadequacy, in time cause atrophy of the hip, thigh, and leg. If the dislocation is inward, the outer part becomes atrophied; if outward, the inner part: and most dislocations are inward.

15. When both bones of the lower leg are fractured without any external wound, stronger extension is required. If there is much overlapping (sc. of the bones), carry out extension by some of the methods that have been

51 δὲ B: δεῖ MV 52 οὖν Jouanna: οὐ BMV
53 κατατείνειν B: τείνειν MV
54 τισί Aldina: τησι B: τινὶ MV

ναὶ δὲ καὶ αἱ ἀπὸ τῶν ἀνδρῶν κατατάσιες· ἐς[55] τὰ πλεῖστα γὰρ ἀρκέοιεν ἂν δύο ἄνδρες ἐρρωμένοι, ὁ μὲν ἔνθεν, ὁ δὲ ἔνθεν ἀντιτείνοντες. τείνειν δὲ ἐς τὸ ἰθὺ χρὴ κατὰ φύσιν καὶ κατὰ τὴν ἰθυωρίην τῆς κνήμης καὶ τοῦ μηροῦ, καὶ ἢν κνήμης ὀστέα κατηγυίης καταταννύῃς,[56] καὶ ἢν μηροῦ. καὶ ἐπιδεῖν οὕτως ἐκτεταμένων ἀμφοτέρων· ὁκότερον ἂν τούτων ἐπιδέῃς· οὐ γὰρ ταὐτὰ συμφέρει σκέλει τε καὶ χειρί· πήχεος μὲν γὰρ καὶ βραχίονος ἐπὴν ἐπιδεθῶσιν ὀστέα κατεηγότα, ἀναλαμβάνεται ἡ χείρ, καὶ ἢν ἐκτεταμένα ἐπιδέῃς, τὰ σχήματα τῶν σαρκῶν ἑτεροιοῦται ἐν τῇ συγκάμψει τοῦ ἀγκῶνος· ἀδύνατος γὰρ ὁ ἀγκὼν ἐκτετάσθαι πο-
472 λὺν χρόνον· οὐ γὰρ πολλάκις ἐν τοιούτῳ | εἴθισται ἐσχηματίσθαι, ἀλλ᾽ ἐν τῷ ξυγκεκάμφθαι· καὶ δὴ καὶ ἅτε δυνάμενοι οἱ ἄνθρωποι περιιέναι, ἐπὴν κατὰ χεῖρα τρωθῶσι, ξυγκεκάμφθαι κατὰ τὸν ἀγκῶνα δέονται. σκέλος δὲ ἔν τε τῇσιν ὁδοιπορίῃσι καὶ ἐν τῷ ἑστάναι εἴθισται ὁτὲ μὲν ἐκτετάσθαι, ὁτὲ δὲ σμικροῦ δεῖν ἐκτέτασθαι· καὶ εἴθισται καθεῖσθαι[57] ἐς τὸ κάτω κατὰ φύσιν καὶ δὴ καὶ πρὸς τὸ ὀχέειν τὸ ἄλλο σῶμα· διὰ τοῦτο εὔφορον αὐτῷ ἐστὶ τὸ ἐκτετάσθαι, ὅταν ἀνάγκην ἔχῃ, καὶ δὴ καὶ ἐν τῇσι κοιτῇσι πολλάκις ἐν τῷ σχήματι τούτῳ ἐστὶν ἐν τῷ ἐκτετάσθαι· ἐπὴν δὲ δὴ τρωθῇ, καὶ ἡ[58] ἀνάγκη καταδουλοῦται τὴν γνώμην, ὅτι ἀδύνατοι μετεωρίζεσθαι γίνονται, ὥστε οὐδὲ μέμνηνται περὶ τοῦ ξυγκαμφθῆναι καὶ ἀναστῆναι, ἀλλὰ τολμέουσιν ἐν τούτῳ τῷ σχήματι κείμενοι. διὰ οὖν

described. But extensions made by manpower are also sufficient, and in most cases two strong men pulling in opposite directions are sufficient. The traction must be directed in a straight line in accordance with the natural direction of the leg and thigh, both when it is being made for fractures of the leg bones, and of the thigh; apply the bandage while both are extended, to whichever of the two leg bones you are dressing. The same treatment does not suit both the leg and the arm: when fractures of the forearm and upper arm are bandaged, the arm is put in a sling, and if you bandage it while it is in the extended position, the relationships of the fleshy parts will be altered by bending the elbow. Further, the elbow cannot be kept extended for a long time, since it is not used to being held often in that posture, but to being flexed. And furthermore, since patients are able to walk about after injuries of the arm, they will need to flex it at the elbow. But the leg, both in walking and standing, is accustomed to be sometimes extended and sometimes nearly so, and naturally directed downward; and what is more, its function is to support the rest of the body. Extension therefore is easily borne when necessary, and indeed the leg frequently has this position even in bed. But if it is injured, necessity compels the mind to submit, because patients are incapable of getting up, so that they do not even think of bending their legs and standing up, but are able to lie at rest in this posture. For these reasons, then, the same position in either making extension or bandaging is not

55 ἐς B: om. MV 56 -τανύῃς B²: -νύῃς B: -τείνῃς MV
57 καὶ εἴθισται καθεῖσθαι B: om. MV
58 καὶ ἡ B: om. MV

ταύτας τὰς προφάσιας χειρὸς καὶ σκέλεος οὔτε ἡ κατάτασις οὔτε ἡ ἐπίδεσις τοῦ σχήματος ξυμφέρει ἡ αὐτή. ἢν μὲν οὖν ἱκανὴ ἡ κατάτασις ἡ ἀπὸ τῶν ἀνδρῶν ᾖ, οὐδὲν δεῖ μάτην ὀνεύεσθαι·[59] καὶ γὰρ σολοικότερον μηχανοποιέειν μηδὲν δέον· ἢν δὲ μὴ ἱκανὴ ἡ κατάτασις ἡ ἀπὸ τῶν ἀνδρῶν,[60] καὶ τῶν ἄλλων τινὰ
474 τῶν ἀναγκέων προσφέρειν, | ἣν τινά γε προσχωρέῃ. ὅταν δὲ δὴ ἱκανῶς καταταθῇ, ῥηΐδιον ἤδη κατορθώσασθαι τὰ ὀστέα καὶ ἐς τὴν φύσιν ἀγαγεῖν, τοῖσι θέναρσι τῶν χειρῶν ἀπιθύνοντα[61] καὶ ἐξευκρινέοντα.

16. Ἐπὴν δὲ κατορθώσῃς, ἐπιδεῖν τοῖσιν ὀθονίοισι κατατεταμένον, ἤν τ' ἐπὶ δεξιὰ ἤν τ' ἐπ' ἀριστερὰ περιφέρειν ξυμφέρῃ αὐτοῖσι τὰ πρῶτα ὀθόνια· βάλλεσθαι δὲ τὴν ἀρχὴν τοῦ ὀθονίου κατὰ τὸ κάτηγμα, καὶ περιβάλλεσθαι κατὰ τοῦτο τὰς πρώτας περιβολάς· κἄπειτα νέμεσθαι ἐπὶ τὴν ἄνω κνήμην ἐπιδέων, ὥσπερ καὶ ἐν τοῖσιν ἄλλοισι κατήγμασιν εἴρηται. τὰ δὲ ὀθόνια πλατύτερα χρὴ εἶναι καὶ μακρότερα καὶ πλείω πολὺ αὖ τὰ[62] κατὰ τὸ σκέλος τῶν ἐν τῇ χειρί. ἐπὴν δ' ἐπιδήσῃς, καταθεῖναι ἐφ' ὁμαλοῦ τινος καὶ μαλθακοῦ, ὥστε μὴ διεστράφθαι ἢ τῇ ἢ τῇ, μήτε λορδὸν μήτε κυφὸν εἶναι. μάλιστα δὲ ξυμφέρει προσκεφάλαιον ἢ λίνεον ἢ ἐρίνεον, μὴ σκληρόν, μέσον κατὰ μῆκος ποιήσαντα λαπαρὸν ὑποθεῖναι, ἢ ἄλλο τι ὃ τούτῳ ἔοικε.

Περὶ γὰρ τῶν σωλήνων τῶν ὑποτιθεμένων ὑπὸ τὰ σκέλεα τὰ κατεηγότα, ἀπορέω ὅ τι συμβουλεύσω, εἰ[63] ὑποτιθέναι χρὴ ἢ οὔ· ὠφελέουσι μὲν γάρ, οὐχ ὅσον applicable for both the arm and the leg. If, then, extension by manpower is enough, you should avoid useless trouble, for to have recourse to machines when it is not required is quite absurd. But if extension by manpower is not enough, bring in some of the mechanical aids, if any of them happens to seem useful. When once sufficient extension has been made, it becomes fairly easy to adjust the bones and bring them into their natural position, by straightening them and making coaptation with the palms of the hands.

16. After adjustment (sc. of the leg), apply the bandages while the limb is extended, making the turns with the first bandage either to right or left, as may be suitable. Start the bandage at the fracture and make the first turns there: then carry the bandaging up the leg as was directed for the other fractures. Bandages for the leg parts should be broader, longer, and much more numerous than those for the arm. On completing the dressing, place the limb on something even and soft, so that it does not get distorted to either side or become concave or convex. Most suitable is to place a pillow of linen or wool which is not very hard under the limb, making a median longitudinal depression in it, or something like this.

Concerning the hollow splints that are put under fractured legs, I feel at a loss whether or not to advise their use: for although they do some good, it is not so great as

---

59 *ὀνεύσθαι* B: *πονευεσθαι* M: om. V
60 *ᾗ, οὐ δεῖ μάτην . . . ἀνδρῶν* BM: om. V
61 *ἀπιθύ*. Jouanna: *ἀπειθύ* B: *ἄπευθύ* MV
62 *αὖ τὰ* Petrequin: *τὰ* B: *αὐτὰ* MV
63 *εἰ* Littré: *ἥ* B: *ἢ* MV

476 δὲ οἱ ὑποτιθέντες οἴονται· οὐ γὰρ | ἀναγκάζουσιν οἱ σωλῆνες ἀτρεμέειν, ὡς οἴονται· οὔτε γὰρ τῷ ἄλλῳ σώματι στρεφομένῳ ἢ ἔνθα ἢ ἔνθα ἐπαναγκάζει ὁ σωλὴν μὴ ἐπακολουθέειν τὸ σκέλος, ἢν μὴ ἐπιμελῆται αὐτὸς ὥνθρωπος· οὔτε αὖ τὸ σκέλος ἄνευ τοῦ σώματος κωλύει ὁ σωλὴν κινηθῆναι ἢ τῇ ἢ τῇ. ἀλλὰ μὴν ἀστεργέστερόν γε[64] ξύλον ὑποτετάσθαι, ἢν μὴ ὁμῶς αὖτις[65] μαλθακόν τι ἐς αὐτὸ ἐντεθῇ· εὐχρηστότατον δέ ἐστιν ἔν τε τῇσι μεθυποστρώσεσι καὶ ἐν τῇσιν ἐς ἄφοδον προσχωρήσεσιν.[66] ἔστιν οὖν σὺν σωλῆνι καὶ ἄνευ σωλῆνος, καὶ καλῶς καὶ αἰσχρῶς κατασκευάσασθαι. πιθανώτερον δὲ τοῖσι δημότῃσίν ἐστι καὶ τὸν ἰητρὸν ἀναμαρτητότερον εἶναι, ἢν σωλὴν ὑποκέηται· καίτοι ἀτεχνώτερόν γέ ἐστιν. δεῖ μὲν γὰρ ἐφ᾽ ὁμαλοῦ καὶ μαλθακοῦ κεῖσθαι πάντῃ πάντως ἐς ἰθύ· ἐπεί τοί γε ἀνάγκη κρατηθῆναι τὴν ἐπίδεσιν ὑπὸ τῆς διαστροφῆς τῆς ἐν τῇ θέσει,[67] ὅποι ἂν ῥέπῃ καὶ ὁκόσα ἂν ῥέπῃ. ὑποκρινέσθω δὲ ὁ ἐπιδεδεμένος ταὐτά, ἅπερ καὶ πρότερον εἴρηται· καὶ γὰρ τὴν ἐπίδεσιν χρὴ τοιαύτην εἶναι καὶ τὸ οἴδημα οὕτως ἐξαείρεσθαι ἐς τὰ ἄκρεα καὶ τὰς χαλάσιας οὕτω, καὶ τὰς μετεπιδέσιας διὰ
478 τρίτης· καὶ εὑρισκέσθω | ἐπὶ ἰσχνότερον τὸ ἐπιδεόμενον· καὶ τὰς ἐπιδέσιας ἐπὶ μᾶλλον ποιέεσθαι καὶ πλείοσι τοῖσιν ὀθονίοισι· περιλαμβάνειν τε καὶ τὸν πόδα χαλαρῶς, ἢν μὴ ἄγαν ἐγγὺς ᾖ τοῦ γούνατος τὸ τρῶμα. κατατείνειν δὲ μετρίως καὶ ἐπικατορθοῦν ἐφ᾽ ἑκάστῃ ἐπιδέσει χρὴ τὰ ὀστέα· ἢν γὰρ ὀρθῶς μὲν

those who use them suppose. The hollow splints do not secure immobility as their advocates imagine, since neither does the hollow splint forcibly prevent the limb from following the body when it is turned to one side or the other, unless the patient himself pays attention; nor does it hinder the leg itself apart from the body from moving this way or that. Besides, it is of course rather unpleasant to have wood under the limb, unless at the same time something soft is inserted in it. But it is very useful in changing the bedding, and in getting up to go to stool. It is thus possible either with or without the hollow splint to arrange the matter well or clumsily. Still, the laity has great faith in it, and the practitioner will be freer from blame if a hollow splint is applied; nonetheless, in my opinion, it remains quite a bad practice. In any case, the limb should lie on something even and soft, and be absolutely straight, since it necessarily follows that the bandaging is overcome by any deviation in posture, whatever the direction or extent of it may be. The patient should give the same replies as are recorded above, for the bandaging should be similar, and there should be the same swelling in the extremities, and likewise with the looseness and the changes of dressing every other day. So, too, the bandaged part should be found less swollen, and greater pressure should be used in the dressings, and more bandages. You should also make some slack turns around the foot if the injury is not too near the knee; make moderate extension and adjustment of the bones at each dressing; for if the treatment is correct

64 γε B: τε MV 65 αὖτις Kühlewein: ἄν τις BMV
66 προσχ. B: προχ. MV 67 θέσει B: διαθέσει MV

ἰητρεύηται, κατὰ λόγον δὲ τὸ οἴδημα χωρέῃ, ἐπὶ[68] μὲν λεπτότερον καὶ ἰσχνότερον τὸ ἐπιδεόμενον χωρίον ἔσται, ἐπὶ δ᾽ εὐπαραγωγότερα[69] τὰ ὀστέα, ἐνακούοντα δὲ τῆς κατατάσιος μᾶλλον. ἐπὴν δὲ ἑβδομαῖος ἢ ἐναταῖος ἢ ἑνδεκαταῖος γένηται, τοὺς νάρθηκας χρὴ[70] προστιθέναι, ὥσπερ καὶ ἐπὶ τοῖς ἄλλοισι κατήγμασιν εἴρηται. τῶν δὲ ναρθήκων τὰς ἐνέδρας χρὴ φυλάσσεσθαι κατά τε τῶν σφυρῶν τὴν ἴξιν καὶ κατὰ τὸν τένοντα τὸν ἐν τῇ κνήμῃ τοῦ ποδός. ὀστέα δὲ κνήμης κρατύνεται ἐν τεσσαράκοντα ἡμέρῃσιν, ἢν ὀρθῶς ἰητρεύηται. ἢν δὲ ὑποπτεύῃς τῶν ὀστέων τι δεῖσθαί τινος διορθώσιος ἤ τινα ἕλκωσιν ὀρρωδέῃς, ἐν τῷ μεσηγὺ χρόνῳ χρὴ λύσαντα καὶ εὐθετισάμενον μετεπιδῆσαι.

17. Ἢν δὲ τὸ ἕτερον ὀστέον κατεηγῇ ἐν κνήμῃ, κατατάσιος μὲν ἀσθενεστέρης δεῖται· οὐ μὴν ἐπιλείπειν χρὴ οὐδὲ βλακεύειν ἐν τῇ κατατάσει. μάλιστα μὲν ἐν τῇ πρώτῃ ἐπιδέσει κατατείνεσθαι ὅσον ἐφικνέεται ἀεί ποτε πάντα τὰ κατήγματα, εἰ δὲ μή, ὡς τάχιστα· ὅ τι γὰρ ἂν μὴ κατὰ τρόπον εὐθετισμένων τῶν ὀστέων ἐπιδέων τις πιέζῃ, ὀδυναίτερον τὸ χωρίον γίνεται. ἡ δὲ ἄλλη ἰητρείη ἡ αὐτή.

18. Τῶν δὲ ὀστέων, τὸ μὲν ἔσω τοῦ ἀντικνημίου
480 καλεομένου ὀχλωδέστερον | ἐν τῇ ἰητρείῃ ἐστί, καὶ κατατάσιος μᾶλλον δεόμενον· καὶ ἢν μὴ ὀρθῶς τὰ ὀστέα τεθῇ, ἀδύνατον κρύψαι· φανερὸν γὰρ καὶ ἄσαρκον πᾶν ἐστίν· καὶ ἐπιβαίνειν ἐπὶ τὸ σκέλος πολλῷ βραδύτερον δύναιντ᾽ ἄν, τούτου κατεηγότος. ἢν δὲ τὸ

and the edema subsides as it should, the bandaged part will be slenderer and more attenuated, while the bones on that side will be more mobile and lend themselves more readily to extension. On the seventh, ninth, or eleventh day splints should be applied as was directed in the case of other fractures, and special care must be taken as to the position of the splints, both in the line of the ankles, and about the back (Achilles) tendon running from the leg to the foot. The bones of the leg consolidate in forty days if properly treated. If you suspect that one of the bones requires some adjustment, or are afraid of ulceration, you should unbandage the part in the interval and reapply the bandage after setting it right.

17. If the other bone of the lower leg (fibula) is broken, the extension required is weaker: there should, however, be no omission or carelessness in the procedure. Especially at the first dressing sufficient extension should be made in all fractures to bring the bones together, or, failing this, as soon as possible; for when pressure is being applied in bandaging, if the bones have not been properly set, the part will be more painful. The rest of the treatment is the same.

18. The inner bone of what is called the shin (tibia) is the more troublesome to treat, requiring greater extension, and if the fragments are not correctly set, this cannot be hidden, for it is visible and entirely unhidden by flesh; when this bone is broken, the patient's ability to walk on it returns much later, while if the outer bone (fibula) is

---

68 ἐπὶ B: om. MV 69 δ' εὐπαραγω. B[2]: δὲ αὖ παραγω. MV 70 χρὴ B: om. MV

ἔξω ὀστέον κατεηγῇ, πολὺ μὲν εὐφορώτερον φέρουσι, πολὺ δὲ εὐκρυπτότερον, καὶ ἢν μὴ καλῶς συντεθῇ—ἐπίσαρκον γάρ ἐστιν—ἐπὶ πόδας τε ταχέως ἵστανται. τὸ πλεῖστον γὰρ τοῦ ἄχθεος ὀχέει τὸ ἔσωθεν τοῦ ἀντικνημίου ὀστέον. ἅμα μὲν γὰρ αὐτῷ τῷ σκέλει καὶ τῇ ἰθυωρίῃ τοῦ ἄχθεος τοῦ κατὰ τὸ σκέλος, τὸ πλεῖον ἔχει τοῦ πόνου τὸ ἔσω ὀστέον· τοῦ γὰρ μηροῦ ἡ κεφαλὴ ὑπεροχέει τὸ ὕπερθεν τοῦ σώματος, αὐτή τε εἴσωθεν πέφυκε τοῦ σκέλεος καὶ οὐκ ἔξωθεν, ἀλλὰ κατὰ τὴν τοῦ ἀντικνημίου ἴξιν· ἅμα δὲ τὸ ἄλλο ἥμισυ τοῦ σώματος γειτονεύεται ταύτῃ μᾶλλον τῇ ἴξει, ἀλλ' οὐχὶ τῇ ἔξωθεν. ἅμα δὲ ὅτι παχύτερον τὸ ἔσω τοῦ ἔξωθεν, ὥσπερ καὶ ἐν τῷ πήχει[71] τὸ κατὰ τὴν τοῦ μικροῦ δακτύλου ἴξιν λεπτότερον καὶ μακρότερον. ἐν μέντοι τῷ κάτω ἄρθρῳ τούτῳ οὐχ ὁμοίη ἡ ὑπόστασις τοῦ ὀστέου τοῦ μακροτέρου· ἀνομοίως γὰρ ὁ ἀγκὼν τε καὶ ἡ ἰγνύη κάμπτεται. διὰ οὖν ταύτας τὰς προφάσιας τοῦ μὲν ἔξωθεν ὀστέου κατεηγότος ταχεῖαι αἱ ἐπιβάσιες, τοῦ δὲ ἔσωθεν ὀστέου κατεηγότος, βραδεῖαι αἱ ἐπιβάσιες. |

482 19. Ἢν δὲ τὸ τοῦ μηροῦ ὀστέον καταγῇ, τὴν κατάτασιν χρὴ περὶ παντὸς ποιέεσθαι, ὅπως μὴ ἐνδεεστέρως σχήσει. πλεονασθεῖσα μὲν γὰρ οὐδὲν ἂν σίνοιτο· οὐδὲ γὰρ οὐδ'[72] εἰ διεστεῶτα τὰ ὀστέα ὑπὸ τῆς ἰσχύος τῆς κατατάσιος ἐπιδέοι τις, οὐκ ἂν δύναιτο κρατέειν ἡ ἐπίδεσις ὥστε διεστάναι, ἀλλὰ συνέλθοι ἂν πρὸς ἄλληλα τὰ ὀστέα ὅτι τάχιστα ἀφείησαν οἱ τείνοντες. παχεῖαι γὰρ καὶ ἰσχυραὶ αἱ σάρκες ἐοῦσαι κρατή-

fractured, he has much less inconvenience to bear, and, even if the fracture has not been well set, this is much more readily concealed, since the bone is well covered with flesh, and he can soon stand. In fact, inner bone of the shin bears the greatest part of the body's weight, since due both to the disposition of the leg itself and to the direct orientation of the weight down the leg, the inner bone does most of the work. Further, the head of the thigh supports the body from below, and is naturally directed toward the inner side of the leg and not the outer side, *viz.* it is in the line of the shinbone. So, too, the opposite half of the body is nearer the line of this bone than of the outer one; and besides, the inner bone is thicker than the outer one, just as in the forearm the bone on the side of the little finger is longer and slenderer (ulna); but in this lower articulation, the longer bone does not lie underneath in the same way, for flexion at the elbow and knee are dissimilar. For these reasons, when the outer bone is fractured, patients can soon get about; but when the inner one is broken their return to mobility is slower.

19. If the bone of the thigh is fractured, it is most important that there should be no deficiency in the extension that is made, while any overextension will do no harm. In fact, even if someone applies the bandage while the bones are separated by the force of the extension, the dressing will have no power to keep them apart, but they will come together at once when the assistants relaxed their tension: because the muscles here are thick and powerful, they will

---

[71] Jouanna suggests there is a loss in the text here, e.g., ‹*τὸ μὲν ἄνω παχύτερον*›. [72] *οὐδ'* B: om. MV

σουσι τῆς ἐπιδέσιος, ἀλλ' οὐ κρατηθήσονται. περὶ οὗ
οὖν ὁ λόγος, διατείνειν εὖ καὶ ἀδιαστρέπτως χρή, μη-
δὲν ἐπιλείποντα. μεγάλη γὰρ ἡ αἰσχύνη καὶ βλάβη
βραχύτερον τὸν μηρὸν ἀποδεῖξαι. χεὶρ μὲν γάρ, βρα-
χυτέρη γενομένη, καὶ ξυγκρυφθείη ἂν καὶ οὐ μέγα τὸ
σφάλμα· σκέλος δὲ βραχύτερον γενόμενον χωλὸν
ἀποδείξει τὸν ἄνθρωπον· τὸ γὰρ ὑγιὲς ἐλέγχει παρα-
τιθέμενον μακρότερον ἐόν. ὥστε λυσιτελέει τὸν μέλ-
λοντα κακῶς ἰητρεύεσθαι, ἀμφότερα καταγῆναι τὰ
σκέλεα μᾶλλον ἢ τὸ ἕτερον· ἰσόρροπος γ' οὖν ἂν εἴη
αὐτὸς ἑωυτῷ. ἐπὴν μέντοι ἱκανῶς καταστανύσῃς, κατ-
ορθωσάμενον χρὴ τοῖσι θέναρσι τῶν χειρῶν ἐπιδεῖν
τὸν αὐτὸν τρόπον, ὥσπερ καὶ πρόσθεν γέγραπται, καὶ
τὰς ἀρχὰς βαλλόμενον, ὥσπερ εἴρηται, καὶ νεμόμε-
νον ἐς τὸ ἄνω τῇ ἐπιδέσει. καὶ ὑποκρινέσθω ταὐτὰ
ὥσπερ καὶ πρόσθεν, καὶ πονείτω κατὰ ταὐτὰ καὶ ῥη-
484 ϊζέτω· καὶ μετεπιδείσθω | ὡσαύτως· καὶ ναρθήκων
πρόσθεσις ἡ αὐτή. κρατύνεται δὲ ὁ μηρὸς ἐν πεντή-
κοντα ἡμέρῃσι.

20. Προσξυνιέναι δὲ χρὴ καὶ τόδε, ὅτι μηρὸς γαῦ-
σός ἐστιν ἐς τὸ ἔξω μέρος μᾶλλον ἢ ἐς τὸ ἔσω, καὶ
ἐς τὸ ἔμπροσθεν μᾶλλον ἢ ἐς τοὔπισθεν· ἐς ταῦτα
τοίνυν τὰ μέρεα καὶ διαστρέφεται, ἐπὴν μὴ καλῶς
ἰητρεύηται. καὶ δὴ καὶ κατὰ ταῦτα ἀσαρκότερος αὐτὸς
ἑωυτοῦ ἐστιν, ὥστε οὐδὲ συγκρύπτειν δύνανται ἐν τῇ
διαστροφῇ. ἢν οὖν τι τοιοῦτον ὑποπτεύῃς, μηχανο-
ποιέεσθαι χρὴ ἅπερ ἐν τῷ βραχίονι τῷ διεστρεφο-
μένῳ παρῄνηται. προσπεριβάλλειν δὲ χρὴ ὀλίγα τῶν

prevail over the bandaging, and not be overcome by it. To come to the point, you must extend very strongly and without deviation, leaving no deficiency; for the disgrace and harm are great if the thigh is revealed to be shortened. An arm, indeed, when shortened may be concealed, and the fault is not great, but when a leg is shortened, it will give the patient an obvious lameness, since when the sound leg is longer than the injured one, the mistake is immediately exposed. Thus, a patient profits, if he is going to have unskillful treatment, for both of his legs to be broken rather than just one of them, since then he will at least be able to maintain his equilibrium. When, therefore, you have made sufficient extension, you should adjust the parts with the palms of your hands and apply a bandage in the same way as described above, starting the bandage as directed, and carrying it upward. Also, the patient should give the same answers as before, and have the same pains and relief. Change the dressings and apply the splints in the same way. Consolidation of the thigh occurs in fifty days.

20. You must also bear in mind that the thigh is naturally curved outward rather than inward, and to the front rather than to the back, so that it tends to be distorted in these directions, if it is not skillfully treated. Furthermore it is less covered with flesh in these parts, so that the distortions cannot be hidden. Now if you suspect anything of this kind, you should have recourse to the mechanical devices recommended for distortion of the upper arm. Some additional turns of bandage must be made around

ὀθονίων κύκλῳ ἀμφὶ τὸ ἰσχίον καὶ τὰς ἰξύας, ὅπως
ἂν οἱ βουβῶνές τε καὶ τὸ ἄρθρον τὸ κατὰ τὴν πλι-
486 χάδα καλεομένην | προσεπιδέηται. καὶ γὰρ ἄλλως
ξυμφέρει καὶ ὅπως μὴ τὰ ἄκρεα τῶν ναρθήκων σίνη-
ται πρὸς τὰ ἀνεπίδετα προσβαλλόμενα. ἀπολείπειν δὲ
χρὴ ἀπὸ τοῦ γυμνοῦ αἰεὶ τοὺς νάρθηκας καὶ ἔνθεν καὶ
ἔνθεν ἱκανόν· καὶ τὴν θέσιν αἰεὶ τῶν ναρθήκων προ-
μηθέεσθαι χρή, ὅκως μήτε κατὰ τὸ ὀστέον τῶν ἐξ-
εχόντων παρὰ τὰ ἄρθρα φύσει πεφυκότων μήτε κατὰ
τὸ[73] νεῦρον ἔσται.

21. Τὰ δὲ οἰδήματα τὰ κατ᾽ ἰγνύην, ἢ κατὰ πόδα,
ἢ κατά τι ἄλλο ἐξαειρεύμενα ὑπὸ τῆς πιέξιος, εἰρίοισι
πολλοῖσι ῥυπαροῖς, εὖ κατεργασμένοις, οἴνῳ καὶ
ἐλαίῳ ῥήνας, κηρωτῇ ὑποχρίων, καταδεῖν· καὶ ἢν πι-
έζωσιν οἱ νάρθηκες χαλᾶν. θᾶσσον ἰσχναίνοις[74] δ᾽ ἄν,
εἰ ἐπάνεις τοὺς νάρθηκας ὀθονίοις ἰσχνοῖσι συ-
488 χνοῖσι[75] | ἐπιδέοις τὰ οἰδήματα, ἀρξάμενος ἀπὸ τοῦ
κατωτάτω ἐπὶ τὸ ἄνω νεμόμενος· οὕτω γὰρ ἂν τάχι-
στα ἰσχνὸν τὸ οἴδημα γένοιτο, καὶ ὑπέλθοι[76] ἂν ὑπὸ
490 τὰ ἀρχαῖα ἐπιδέσματα· ἀλλ᾽ οὐ χρὴ | τούτῳ τῷ τρόπῳ
χρῆσθαι τῆς ἐπιδέσιος, ἢν μὴ κίνδυνος ᾖ ἐν τῷ οἰ-
δήματι φλυκταινώσιος ἢ μελασμοῦ· γίνεται δὲ οὐδὲν
τοιοῦτο, ἢν μὴ ἄγαν τις πιέζῃ τὸ κάτηγμα, ἢ κατα-
κρεμάμενον ἔχῃ, ἢ κνᾶται τῇ χειρί, ἢ ἄλλο τι προσ-
πίπτῃ ἐρεθιστικὸν πρὸς τὸν χρῶτα.

22. Σωλῆνα δὲ εἰ μέν τις ὑπ᾽ αὐτὸν τὸν μηρὸν ὑπο-
θείη μὴ ὑπερβάλλοντα τὴν ἰγνύην, βλάπτοι ἂν μᾶλ-
λον ἢ ὠφελέοι· οὔτε γὰρ ἂν τὸ σῶμα κωλύοι οὔτε τὴν

the hip and loins, so that the groins and the joint at the so-called fork (perineum) can be included; for besides other benefits, this prevents the ends of the splints from doing damage by contact with the uncovered parts. The splints should always come considerably short of the bare part of the bone at either end, and care should always be taken as to their position, so that they rest neither on the bone where there are natural projections in the area of the joint, nor on the cord.

21. As to the swellings which arise from pressure behind the knee, or at the foot, or elsewhere, dress them with plenty of crude wool, well pulled out, sprinkling it with oil and wine after anointing with cerate; and if the splints produce pressure loosen them. You will promptly reduce the swellings if you apply many narrow bandages after removing the splints, beginning from the lowest part and moving upward; for in this way the swelling will be most rapidly reduced, and the fluid will flow back down under the original dressing. But you should not use this method of bandaging unless there is a danger of blisters forming or necrosis at the swelling. In fact, nothing of this kind happens unless excessive pressure is applied on the fracture, or the part is left hanging down, or the wound is abraded with the hand, or some other irritant affects the skin.

22. As to a hollow splint, if someone places it under the thigh itself without going below the bend of the knee, it will do more harm than good; for it will prevent neither

---

73 ἄρθρον add. MV 74 -νοις Littré: -νειαις B: -νοιαις B²: -νειε MV 75 συχνοῖσι B: om. MV

76 ὑπέλθοι B: ὑπερθείη MV

κνήμην ἄνευ τοῦ μηροῦ κινεῖσθαι, ἀσηρόν τ' ἂν εἴη πρὸς τὴν ἰγνύην προσβαλλόμενον· καὶ ὃ ἥκιστα δεῖ, τοῦτ' ἂν ἐποτρύνοι ποιέειν,[77] κατὰ τὸ γόνυ κάμπτειν· πᾶσαν γὰρ ἂν τύρβην παρέχοι τῇσιν ἐπιδέσεσιν, καὶ μηροῦ ἐπιδεδεμένου καὶ κνήμης, ὅστις κατὰ τὸ γόνυ κάμπτοι. ἀνάγκη γὰρ ἂν εἴη[78] τοὺς μύας ἄλλοτε ἄλλο σχῆμα ἴσχειν· ἀνάγκη δ' ἂν εἴη καὶ τὰ ὀστέα τὰ κατεηγότα κίνησιν ἔχειν. περὶ παντὸς οὖν ποιητέον τὴν ἰγνύην ἐντετάσθαι. δοκέοι ἂν οὖν μοι[79] ὁ σωλὴν ὑπερέχων[80] πρὸς τὸν πόδα ἀπὸ τοῦ ἰσχίου, ὠφελέειν ὑποτιθέμενος· καὶ ἄλλως κατ' ἰγνύην ταινίην χαλα-
492 ρῶς περιβάλλειν | σὺν τῷ σωλῆνι, ὥσπερ τὰ παιδία ἐν τῇσι κοίτῃσι σπαργανοῦται· εἶτα ἐπὴν ὁ μηρὸς ἐς τὸ ἄνω διαστρέφηται ἢ ἐς τὸ πλάγιον, εὐκατασχετώτερον εἴη ἂν σὺν τῷ σωλῆνι οὕτως. ἢ οὖν διαμπερής σοι ποιητέος ὁ σωλήν, ἢ οὐ ποιητέος.

23. Πτέρνης δὲ ἄκρης κάρτα χρὴ ἐπιμελέεσθαι ὡς εὐθέτως ἔχει, καὶ ἐν τοῖσι κατὰ κνήμην καὶ ἐν τοῖσι κατὰ μηρὸν κατήγμασιν. ἢν μὲν γὰρ ἀπαιωρῆται ὁ ποὺς τῆς ἄλλης κνήμης ἡρματισμένης, ἀνάγκη κατὰ τὸ ἀντικνήμιον τὰ ὀστέα κυρτὰ φαίνεσθαι· ἢν δὲ ἡ μὲν πτέρνη ὑψηλοτέρη τοῦ μετρίου ἠρτισμένη ᾖ, ἡ δὲ ἄλλη κνήμη ὑπομετέωρος ᾖ, ἀνάγκη τὸ ὀστέον τούτῳ κατὰ τὸ ἀντικνήμιον κοιλότερον φανῆναι τοῦ μετρίου, προσέτι καὶ ἢν ἡ πτέρνη τυγχάνῃ ἐοῦσα τοῦ ἀνθρώπου φύσει μεγάλη. ἀτὰρ καὶ κρατύνεται πάντα τὰ ὀστέα βραδύτερον, ἢν μὴ κατὰ φύσιν κείμενα <ᾖ>,[81]

the body nor the leg from moving independently from the thigh, but it will cause discomfort by pressing against the flexure of the knee, and furthermore it will incite the patient to bend the knee, which is the last thing he should do. For when the thigh and lower leg are bandaged, anyone who bends the knee causes all sorts of disturbance to the dressings: the muscles will be forced continually to change their relative positions, and the fractured bones will be compelled to move. Special care, then, should be taken to keep the knee extended. Thus, I imagine that a hollow splint reaching from the hip to the foot might be useful, especially with a band passed loosely round at the knee to hold the splint in, as babies are swaddled in their cots. Then if the thigh is distorted upward (i.e., forward) or sideways, it will be more easily controlled by the hollow splint in this way. You must, then, use the hollow splint either for the whole limb, or not at all.

23. In fractures both of the leg and of the thigh great care should be taken that the point of the heel is in a good position. For if the foot hangs down in the air but the leg is supported, the bones at the shin will necessarily present a convex appearance, while if the foot is held higher than it should be, and the rest of the leg is rather too low, this bone at the shin must necessarily have a hollower appearance than is normal, especially if the heel happens to be overly large in the person. Here, too, all bones consolidate more slowly if they are not returned to their natural posi-

---

77 Add. ἥκιστα γὰρ δεῖ MVB[2]
78 Add. τοῦτο MV
79 μοι B: ὁμοίως MV
80 ὑπερέχων B: ὁ περιέχων MV
81 ᾗ add. Aldina: om. BMV

καὶ τὰ μὴ[82] *ἀτρεμέοντα ἐν τῷ αὐτῷ σχήματι, καὶ αἱ πωρώσιες ἀσθενέστεραι.*

24. Ταῦτα μὲν οὖν δὴ ὅσοισι τὰ μὲν ὀστέα κατέηγεν, ἐξέχει δὲ μή, μηδὲ ἄλλως ἕλκος ἐγένετο. οἷσι δὲ
494 τὰ ὀστέα | κατέηγεν ἁπλῷ τρόπῳ καὶ μὴ πολυσχιδεῖ, ‹ἐξέσχε δέ›,[83] αὐθήμερα ἐμβληθέντα ἢ τῇ ὑστεραίῃ κατὰ χώρην ἱζόμενα, καὶ μὴ ἐπίδοξος ἡ ἀπόστασις παρασχίδων ὀστέων ἀπιέναι, ἢ καὶ οἷσιν ἕλκος μὲν ἐγένετο, τὰ δὲ ὀστέα τὰ κατεηγότα οὐκ ἐξίσχει, οὐδ' ὁ τρόπος τῆς κατήξιος τοιοῦτος οἷος παρασχίδας ὀστέων ἐούσας ἐπιδόξους εἶναι ἀναπλῶσαι· τοὺς τοι-
496 ούτους οἱ μὲν μήτε μέγα | ἀγαθὸν μήτε μέγα κακὸν ποιέοντες, ἰητρεύουσι τὰ μὲν ἕλκεα καθαρτικῷ τινι, ἢ πισσηρὴν ἐπιθέντες, ἢ ἔναιμον ἢ ἄλλο τι ὧν εἰώθασι ποιέειν· ἐπάνω δὲ τοὺς οἰνηροὺς σπλῆνας ἢ εἴρια ῥυπαρὰ [ὁκόσοι][84] ἐπιδέουσιν ἢ ἄλλο τι τοιοῦτον. ἐπὴν δὲ τὰ ἕλκεα καθαρὰ γένηται καὶ ἤδη ξυμφύηται, τότε τοῖς ὀθονίοισι συχνοῖσι πειρῶνται ἐπιδεῖν καὶ νάρθηξι κατορθοῦν. αὕτη μὲν ἡ ἴησις ἀγαθόν τι ποιέει, κακὸν δὲ οὐ μέγα. τὰ μέντοι ὀστέα οὐχ ὁμοίως δύναται ἱδρύεσθαι ἐς τὴν ἑωυτῶν χώρην, ἀλλά τινι καὶ ὀγκηρότερα ὀστέα[85] τοῦ καιροῦ ταύτῃ γίνεται. γένοιτο δ' ἂν καὶ βραχύτερα, ὧν ἀμφότερα τὰ ὀστέα κατέηγεν ἢ πήχεος ἢ κνήμης.

25. Ἄλλοι δ' αὖ τινές εἰσιν οἳ ὀθονίοισι τὰ τοιαῦτα ἰητρεύουσιν εὐθέως· καὶ ἔνθεν μὲν καὶ ἔνθεν ἐπιδέουσι τοῖσιν ὀθονίοισι, κατὰ δὲ τὸ ἕλκος αὐτὸ διαλείπουσι,

tion and are not kept at rest in the same posture, and if the callus is weaker.

24. The above remarks apply to cases where the bones are fractured, but do not protrude or form a lesion in some other way. In cases where the bones are fractured in the simple way and not splintered, but they do protrude, if they are reduced and returned to their original position on the same or the following day, and if there is no reason to expect the production and expulsion of splinters, or even cases in which a lesion arises, but the fractured bones do not project, nor is the fracture by nature such that any splinters are likely to come to the surface: in such cases practitioners will do neither much good nor much harm if they treat the wound with a cleaning plaster, either pitch cerate, or an application suitable for fresh wounds, or whatever else they commonly use, and bind compresses soaked in wine over it, or uncleaned wool or anything of that kind. Similarly, if after the wounds have been cleaned and are already united, the practitioners attempt to apply many bandages and make an adjustment with splints, such a treatment does some good and no great harm. The bones, however, cannot be so well settled in their proper place, but become somewhat bulkier than normal in that place. If both bones are broken, of either the forearm or the leg, there may also be shortening.

25. Then there are others who treat such cases at once with bandages, applying them on either side, while leaving a vacancy at the lesion itself where it is exposed to the air.

---

82 μὴ MV: μὲν B see Jouanna, p. 237 vol. 2, 496n3)

83 ἐξέσχε δέ› add. Littré Kühlewein;

84 ὁκόσοι BMV: del. Bosquillon (Littré,

85 ὀστέα B: σώματα MV

καὶ ἐῶσιν ἀνεψῦχθαι· ἔπειτα ἐπιτιθέασιν ἐπὶ τὸ ἕλκος τῶν καθαρτικῶν τι, καὶ σπλήνεσιν οἰνηροῖς ἢ εἰρίοις
498 *ῥυπαροῖσι θεραπεύουσιν. αὕτη ἡ ἴησις κακή·* | *καὶ εἰκὸς τοὺς οὕτως ἰητρεύοντας τὰ μέγιστα ἀσυνετέειν, καὶ ἐν τοῖσιν ἄλλοισι κατήγμασι καὶ ἐν τοῖσι τοιούτοισι. μέγιστον γάρ ἐστι τὸ γινώσκειν καθ' ὁποῖον τρόπον χρὴ τὴν ἀρχὴν βάλλεσθαι τοῦ ὀθονίου, καὶ καθ' ὁποῖον μάλιστα πεπίεχθαι, καὶ οἷά τε ὠφελέονται ἢν ὀρθῶς τις βάλληται τὴν ἀρχὴν καὶ πιέζῃ ᾗ μάλιστα χρή, καὶ οἷα βλάπτονται ἢν μὴ ὀρθῶς τις βάλληται*[86] *μηδὲ πιέζῃ ᾗ μάλιστα χρή, ἀλλὰ ἔνθεν καὶ ἔνθεν. εἴρηται μὲν οὖν καὶ ἐν τοῖσι πρόσθεν γεγραμμένοισιν, ὁποῖα ἀφ' ἑκατέρων ἀποβαίνει· μαρτυρεῖ δὲ καὶ αὐτὴ ἡ ἰητρείη· ἀνάγκη γὰρ τῷ οὕτως ἐπιδεομένῳ τὸ οἶδος ἐξαείρεσθαι ἐς αὐτὸ τὸ ἕλκος· καὶ γὰρ εἰ ὑγιὴς χρὼς ἔνθεν καὶ ἔνθεν ἐπιδεθείη, ἐν μέσῳ δὲ διαλειφθείη, μάλιστα κατὰ τὴν διάλειψιν οἰδήσειεν ἂν καὶ ἀχροιήσειεν· πῶς οὖν οὐχὶ ἕλκος γε ταῦτα ἂν πάθοι; ἀναγκαίως οὖν ἔχει ἄχροον μὲν καὶ ἐκπεπλιγ-*
500 *μένον*[87] | *τὸ ἕλκος εἶναι, δακρυῶδες δὲ καὶ ἀνεκπύητον*[88] *εἶναι· ὀστέα δέ καὶ μὴ μέλλοντα ἀποστῆναι, ἀποστατικὰ γενέσθαι· σφυγμῶδές τε καὶ πυρετῶδες τὸ ἕλκος ἂν εἴη. ἀναγκάζονται δὲ διὰ τὸ οἶδος ἐπικαταπλάσσειν· ἀσύμφορον δὲ καὶ τοῦτο τοῖσιν ἔνθεν καὶ ἔνθεν ἐπιδεομένοισιν· ἄχθος γὰρ ἀνωφελὲς πρὸς τῷ ἄλλῳ σφυγμῷ ἐπιγίνεται. τελευτῶντες δὲ ἀπο-*

[86] Add. *ἢν μὴ ὀρθῶς ἐπιδεθῇ* B

Afterward, they put one of the cleaning applications on the wound, and treat it with compresses steeped in wine, or with crude wool. This treatment is bad, and those who use it probably show very great folly in their treatment of other fractures as well as these. For the most important thing is to know the proper way of applying the beginning of a bandage, and how the chief pressure should be exerted; also, what the benefits of a proper application would be, and of getting the chief pressure in the proper place; and what harm would result from not placing the bandage correctly, and of not producing pressure where it should chiefly be, but to one side or the other of that. Now, it has been explained in the text above what the result of each of these will be; and the treatment itself, too, is evidence, for in a patient bandaged this way swelling necessarily arises in the lesion itself. Indeed, if even healthy tissue is bandaged on both sides with a vacancy left in the middle, it will be especially at the vacant part that swelling and discoloration would occur: how then could a lesion fail to be affected in the same way? The lesion would, of course, certainly become discolored with everted edges, and have a watery discharge devoid of pus; and the bones, even those which were not destined to come away, would exfoliate. The wound will become heated and throb, and the surgeon will be obliged to apply an additional plaster on account of the swelling. This too, however, will be harmful for patients bandaged on either side of the lesion, for a worthless burden is being added to the throbbing itself. In

87 *ἐκπεπλιγμ.* Foes (n. 23): *ἐκπεπετιγμ.* B: *ἐκπεπλιεχμ.* MV
88 *ἀνεκπύητον* MV: *-ποίητον*

λύουσι τὰ ἐπιδέσματα, ὁπόταν σφιν παλιγκοτέῃ, καὶ ἰητρεύουσι τὸ λοιπὸν ἄνευ ἐπιδέσιος. οὐδὲν δὲ ἧσσον καὶ ἤν τι ἄλλο τρῶμα τοιοῦτο λάβωσι, τῷ αὐτῷ τρόπῳ ἰητρεύουσιν· οὐ γὰρ οἴονται τὴν ἐπίδεσιν τὴν ἔνθεν καὶ ἔνθεν, καὶ τὴν ἀνάψυξιν τοῦ ἕλκεος αἰτίην εἶναι,[89] ἀλλ' ἄλλην τινὰ ἀτυχίην· οὐ μέντοι γε ἂν ἔγραφον περὶ τούτου τοσαῦτα, εἰ μὴ εὖ μὲν ᾔδειν ἀσύμφορον ἐοῦσαν τὴν ἐπίδεσιν, συχνοὺς δὲ οὕτως ἰητρεύοντας· ἐπίκαιρον δὲ τὸ ἀπομάθημα· μαρτύριον δὲ τοῦ ὀρθῶς γεγράφθαι τὰ πρόσθεν γεγραμμένα ᾗ τε μάλιστα πιεστέα τὰ κατήγματα ᾗ τε[90] ἥκιστα.

26. Χρὴ δέ, ὡς ἐν κεφαλαίῳ εἰρῆσθαι, οἷσιν ἂν μὴ ἐπίδοξος ᾖ ἡ τῶν ὀστέων ἀπόστασις ἔσεσθαι, τὴν αὐτὴν ἰητρείην ἰητρεύειν, ὥσπερ οἷσιν ὀστέα μὲν κατ-
502 εηγότα εἴη, ἕλκος δὲ μὴ ἔχοντα· τάς | τε γὰρ κατατάσιας καὶ κατορθώσιας τῶν ὀστέων τὸν αὐτὸν τρόπον χρὴ[91] ποιέεσθαι τήν τε ἐπίδεσιν παραπλήσιον τρόπον. ἐπὶ μὲν γὰρ αὐτὸ τὸ ἕλκος πισσηρὴν κηρωτὴν χρίσαντα, σπλῆνα λεπτὸν διπλόον ἐπιθεῖναι,[92] τὰ δὲ πέριξ κηρωτῇ λευκῇ[93] χρίειν. τὰ δὲ ὀθόνια καὶ τὰ ἄλλα πλατύτερά τινι ἐσχισμένα[94] ἔστω, ἢ εἰ μὴ ἕλκος εἶχεν· καὶ ᾧ ἂν πρώτῳ ἐπιδέηται, συχνῷ ἔστω τοῦ ἕλκεος πλατύτερον. τὰ γὰρ στενότερα τοῦ ἕλκεος ζώσαντα ἔχει τὸ ἕλκος· τὸ δὲ οὐ χρή. ἀλλ' ἡ πρώτη περιβολὴ ὅλον κατεχέτω τὸ ἕλκος, καὶ ὑπερεχέτω τὸ

89 εἶναι B: om. MV

90 ᾗ τε . . . ᾗ τε Ermerins: ητε . . . ητε MV: εἴτε . . . εἴτε V

91 χρὴ B: om. MV

the end they take the dressings off, when they see that there are complications, and treat from then on without bandaging. But in spite of this, if they later meet another wound of the same sort, they again employ the same treatment, since they do not grasp that it is the bandaging on both sides and the exposure of the wound itself to the air that are to blame, but (sc. mistakenly) assume that some other mischance is the cause. Indeed, I would not have written so much on this subject had I not known well the harmfulness of this mode of dressing, and that many practitioners use it; it is of vital importance that this habit be unlearned. Besides, it is evidence of the truth of what has been written before on the question of where the greatest or the least pressure should be directed against a fracture.

26. To speak summarily, when there is no likelihood of the expulsion of bone fragments, the same treatment must be used as in cases of fracture without an external lesion: the extensions and adjustments of the bones should be made in the same way, and so too with the bandaging. The wound itself is to be anointed with pitch cerate, bound with a thin doubled-folded compress over it, and the surrounding parts anointed with a thin layer of cerate. The bandages and other dressings should be torn in somewhat broader strips than if there was no wound, and the one first one used should be considerably wider than the wound; for bandages narrower than the wound confine it like a belt, which must be avoided. Rather the first turn of the bandage should take in the whole lesion, and extend

---

92 ἐπιθεῖναι Kühlewein: -θῆναι B: -δεθῆναι MV

93 λευκῇ Foes (n. 23): λεπτῇ BMV, see Jouanna (pp. 250f.)

94 ἐσχισμένα MV: om. B

ὀθόνιον ἔνθεν τε καὶ ἔνθεν. βάλλεσθαι μὲν οὖν χρὴ
τὸ ὀθόνιον κατ᾽ αὐτὴν τὴν ἴξιν τοῦ ἕλκεος, πιέζειν δὲ
ὀλίγῳ ἧσσον ἢ εἰ μὴ ἕλκος εἶχεν, ἐπινέμεσθαι δὲ τῇ
504 | ἐπιδέσει ὥσπερ καὶ πρόσθεν εἴρηται. τὰ δὲ ὀθόνια
αἰεὶ μὲν τοῦ τρόπου τοῦ μαλθακοῦ ἔστωσαν, μᾶλλον
δέ τινι[95] ἐν τοῖσι τοιούτοισιν, ἢ εἰ μὴ ἕλκος εἶχεν.
πλῆθος δὲ τῶν ὀθονίων μὴ ἐλάσσω ἔστω τῶν προειρημένων, ἀλλά τινι καὶ πλείω. ἢν δὲ ἐπιδεθῇ, δοκείτω
τῷ ἐπιδεδεμένῳ ἡρμάσθαι μέν, πεπιέχθαι δὲ μή· φάτω
δὲ κατὰ τὸ ἕλκος μάλιστα ἡρμάσθαι. τοὺς δὲ χρόνους
τοὺς αὐτοὺς μὲν χρὴ εἶναι ἐπὶ τὸ μᾶλλον δοκέειν ἡρμάσθαι, τοὺς αὐτοὺς δὲ ἐπὶ τὸ μᾶλλον δοκέειν χαλᾶν,
ὥσπερ καὶ ἐν τοῖσι πρόσθεν εἴρηται. μετεπιδεῖν δὲ
διὰ τρίτης, πάντα μεταποιέοντα ἐς τοὺς τρόπους τοὺς
παραπλησίους, ὥσπερ καὶ πρόσθεν εἴρηται, πλὴν ἐς
τὸ σύμπαν ἧσσόν τινι πιέζειν ταῦτα ἢ ἐκεῖνα. καὶ ἢν
κατὰ λόγον τὰ εἰκότα γένηται, ἰσχνότερον μὲν αἰεὶ
εὑρεθήσεται τὸ κατὰ τὸ ἕλκος, ἰσχνὸν δὲ καὶ τὸ ἄλλο
πᾶν τὸ ὑπὸ τῆς ἐπιδέσιος κατεχόμενον· καὶ αἵ τε ἐκπυήσιες ἔσονται θάσσους ἢ τῶν ἄλλως ἰητρευμένων
ἑλκέων, ὅσα τε σαρκία ἐν τῷ τρώματι ἐμελάνθη καὶ
ἐθανατώθη, θᾶσσον περιρρήγνυται καὶ ἐκπίπτει ἐπὶ
ταύτῃ τῇ ἰητρείῃ ἢ ἐν τῇσι ἄλλῃσιν, ἐς ὠτειλάς τε
θᾶσσον ὁρμᾶται τὸ ἕλκος οὕτως ἢ ἄλλως ἰητρευμένον. πάντων δὲ τούτων αἴτιον, ὅτι ἰσχνὸν μὲν τὸ κατὰ
τὸ ἕλκος χωρίον γίνεται, ἰσχνὰ δὲ τὰ περιέχοντα. τὰ
μὲν οὖν ἄλλα πάντα παραπλησίως[96] ἰητρεύειν, ὡς τὰ
506 | ἄνευ ἑλκώσιος ὀστέα κατηγνύμενα· τοὺς δὲ νάρθη-

beyond it on both sides. The bandage must be placed just in the line of the lesion, exert rather less pressure than in cases without a lesion, and be distributed as directed above. The bandages should always have a pliant texture, and more so in these cases than if there was no wound. Number of bandages: not less than the ones mentioned above, but even a few more. When the bandage is completed, it should seem to the patient to be secure without exerting any pressure, and he should say that the greatest firmness is over the lesion. The periods of time in which there is a sensation of greater firmness, and those in which there is greater relaxation should be the same as were described in the former cases. Change the dressings every other day, making the changes in a similar fashion except that, on the whole, the pressure should be less in these cases. If the case takes its natural course according to rule, the part around the lesion will become progressively less swollen, as will all the rest of the limb which is included in the bandage. Putrefaction will take place more rapidly than in lesions treated in other ways, and all fragments of blackened or necrotic tissue in the wound will be more rapidly expelled and fall off with this treatment than with other methods. The wound, too, advances more quickly toward cicatrization in this way than when treated otherwise. The cause of all this is that the area of the lesion and the surrounding parts becomes free of swelling. In all other respects, then, these cases must be treated like fractured bones without a lesion, except that splints should not

[95] τινι B: τε δεῖ M: τι δεῖ V
[96] Add. χρὴ MV

κας οὐ[97] *χρὴ προστιθέναι. διὰ τοῦτο καὶ τὰ ὀθόνια χρὴ τούτοισι πλέω εἶναι ἢ τοῖσιν ἑτέροισιν, ὅτι τε ἧσσον πιέζεται, ὅτι τε οἱ νάρθηκες βραδύτερον προστίθενται· ἢν μέντοι τοὺς νάρθηκας προστιθῇς, μὴ κατὰ τὴν ἴξιν τοῦ ἕλκεος προστιθέναι, ἄλλως τε καὶ χαλαρῶς προστιθέναι, προμηθεύμενος*[98] *ὅπως μηδεμία σφίγξις μεγάλη ἔσται ἀπὸ τῶν ναρθήκων. εἴρηται δὲ τοῦτο καὶ ἐν τοῖσι πρότερον γεγραμμένοισι. τὴν μέντοι δίαιταν ἀκριβεστέρην καὶ πλείω χρόνον χρὴ ποιέεσθαι οἷσιν ἐξ ἀρχῆς ἕλκεα γίνεται καὶ οἷσιν ὀστέα ἐξίσχει· καὶ τὸ ξύμπαν δὲ εἰρῆσθαι, ἐπὶ τοῖς ἰσχυροτάτοισι τρώμασιν ἀκριβεστέρην καὶ πολυχρονιωτέρην εἶναι χρὴ τὴν δίαιταν.*

27. Ἡ *αὐτὴ δὲ ἰητρείη τῶν ἑλκέων καὶ οἷσιν ὀστέα μὲν κατέηγεν, ἕλκος δὲ ἐξ ἀρχῆς μηδὲν ᾖ, ἢν δὲ ἐν τῇ ἰητρείῃ ἕλκος γένηται, ἢ τοῖσιν ὀθονίοισι μᾶλλον πιεχθέντος, ἢ ὑπὸ νάρθηκος ἐνέδρης, ἢ ὑπὸ ἄλλης τινὸς προφάσιος. γινώσκεται μὲν οὖν τὰ τοιαῦτα, ἢν ἕλκος ὑπῇ, τῇ τε ὀδύνῃ καὶ τοῖσι σφυγμοῖσι· καὶ τὸ οἴδημα τὸ ἐν τοῖσιν ἄκροισι σκληρότερον γίνεται τῶν τοιούτων, καὶ εἰ τὸν δάκτυλον ἐπάγοις, τὸ ἔρευθος †ἐξαρ(ε)*
508 *ίαται†, ἀτὰρ καὶ αὖτις* | *ὑποτρέχει ταχέως. ἢν οὖν τι τοιοῦτον ὑποπτεύῃς, λύσαντα χρή, ἢν μὲν ᾖ κνησμὸς κατὰ τὰς ὑποδεσμίδας καὶ τὸ ἄλλο τὸ ἐπιδεδεμένον, πισσηρῇ κηρωτῇ ἀντὶ τῆς ἑτέρης χρῆσθαι. ἢν δὲ τούτων μὲν μηδὲν ᾖ, αὐτὸ δὲ τὸ ἕλκος ἠρεθισμένον εὑρίσκηται ἢ*[99] *ἐπὶ πολὺ*[100] *ἀκάθαρτον· καὶ τῶν μὲν σαρκίων ἐκπυησομένων, τῶν δὲ νεύρων προσεκπε-*

be used. This is the reason why the bandages should be more numerous than in the other cases, both because they should provide less pressure, and because the splints are applied later. If you do in fact apply splints, do not place them in the line of the lesion; also apply them very loosely, taking care that there will be no great compression from the splints: this same direction was also given above. Diet, however, should be stricter and kept up longer in cases where there are lesions from the first, and where the bones protrude. On the whole, the greater the injury, the stricter and more prolonged the dieting should be.

27. The same treatment of lesions also applies to cases of bone fractures which are at first without a lesion, but one arises during treatment, from either too great compression by the bandages, or the position of a splint, or some other cause. Such cases are recognized when the ulceration is facing downward, by the pain and throbbing; the swelling in the extremities becomes harder, and if you apply your finger the redness disappears, but then quickly runs back in. Now, if you suspect anything of this kind, you should undo the dressings, and if there is irritation below the underbandages, or in the rest of the bandaged part, use pitch cerate instead of the other plaster. If none of these symptoms are present, but the sore itself is found to be irritated or very putrid, and the tissues are about to suppurate, and cords are on the way to be sloughed off, it

---

97 *οὐ* B: om. MV
98 *προμηθεύμενος* Chartier: -*μένοις* BMV
99 *ἢ* B: *μέλαν* MV
100 Add. *ἢ* MV

σουμένων, τούτους οὐδὲν δεῖ ἀναψύχειν παντάπασιν, οὐδέ τι φοβεῖσθαι τὰς ἐκπυήσιας ταύτας· ἀλλ᾽ ἰητρεύειν τὰ μὲν ἄλλα παραπλήσιον τρόπον, ὥσπερ καὶ οἷσιν ἐξ ἀρχῆς ἕλκος ἐγένετο. τοῖσι δὲ ὀθονίοισιν ἄρχεσθαι χρὴ ἐπιδέοντα ἀπὸ τοῦ οἰδήματος τοῦ ἐν τοῖς ἀκραίοις πάνυ χαλαροῖσι| καὶ ἔπειτα ἐπινέμεσθαι τῇ ἐπιδέσει αἰεὶ ἐς τὸ ἄνω· καὶ πεπιέχθαι μὲν μηδαμῇ, ἡρμάσθαι δὲ μάλιστα κατὰ τὸ ἕλκος, τὰ δὲ ἄλλα ἐπὶ ἧσσον. τὰ δὲ ὀθόνια τὰ πρῶτα, ταῦτα μὲν καθαρὰ ἔστω καὶ μὴ στενά· τὸ δὲ πλῆθος τῶν ὀθονίων ἔστω ὅσον περ καὶ ἐν τοῖσι νάρθηξιν, εἰ ἐπιδέοιτο, ἢ ὀλίγῳ ἔλασσον. ἐπὶ δὲ αὐτὸ τὸ ἕλκος ἱκανὸν σπληνίον τῇ λευκῇ κηρωτῇ κεχρισμένον· ἤν τε γὰρ σὰρξ ἤν τε νεῦρον μελανθῇ, προσεκπεσεῖται. τὰ γὰρ τοιαῦτα οὐ χρὴ δριμέσιν ἰητρεύειν, ἀλλὰ μαλθακοῖσιν ὥσπερ τὰ πυρίκαυστα. μετεπιδεῖν δὲ διὰ τρίτης, νάρθηκας δὲ μὴ προστιθέναι· ἀτρεμέειν δὲ ἐπὶ μᾶλλον ἢ τὸ πρόσθεν, καὶ ὀλιγοσιτέειν. εἰδέναι δὲ χρὴ εἴτε σάρξ, εἴτε νεῦρον τὸ ἐκπεσούμενόν ἐστι, ὅτι οὕτω πολλῷ μὲν ἧσσον νέμεται ἐπὶ πλεῖον, πολλῷ δὲ θᾶσσον ἐκπεσεῖται, πολλῷ δὲ ἰσχνότερα τὰ περιέχοντα ἔσται, ἢ εἴ τις ἀπολύσας τὰ ὀθόνια ἐπιθείη τι τῶν καθαρτικῶν φαρμάκων ἐπὶ τὸ ἕλκος. καίτοι καὶ ἢν ἐκπέσῃ τὸ ἐκπυησόμενον, θᾶσσόν τε σαρκοῦται ἐκείνως ἢ ἑτέρως ἰητρευόμενον, καὶ θᾶσσον ὠτειλοῦται. πάντα μήν ἐστι ταῦτα ὀρθῶς ἐπιδεῖν καὶ μετρίως ἐπίστασθαι. προσσυμβάλλεται δὲ καὶ τὰ σχήματα, ἢν is by no means appropriate to leave them exposed to the air, or to be alarmed in any way by the suppurations. Rather, you should treat them from then on in the same manner as cases in which there has been a wound from the first. The bandages must start from the swelling at the extremities and be left quite slack; then they should be carried right on upward; no pressure should be applied in any place, and special support should be provided at the wound, but less elsewhere. The first bandages must be clean and not narrow, their number as many as when splints are applied, or a little fewer. On the wound itself a compress anointed with white cerate is sufficient; for if flesh or a cord turns black, it too will fall off. You should treat such cases not with irritants, but with mild applications, as you do burns. Change the dressing every other day, but do not apply splints. Keep the patient more strictly at rest than in the former case, and on a mild diet. You must be aware that if either flesh or a cord is going to come away, in this case the loss will be much less extensive and occur much quicker, and the surrounding parts will be much less swollen (sc. with this treatment), than if someone removes the bandages and applies some detersive plaster to the wound. Besides, when the part that is going to putrefy does come away, new flesh will be built up more rapidly with the former treatment than with any other, and the cicatrization will occur sooner. The whole point is to know the correct method and due measure in bandaging these cases. Correctness of position also con-

<ᾗ>[101] οἵα χρὴ εἶναι, καὶ ἡ ἄλλη δίαιτα, καὶ τῶν ὀθονίων ἡ ἐπιτηδειότης.

28. Ἢν δὲ ἄρα ἐξαπατηθῇς ἐν τοῖς νεοτρώτοισι, μὴ οἰόμενος ὀστέων ἀπόστασιν ἔσεσθαι, τὰ δ' ἐπίδοξα ᾖ
512 ἀναπλῶσαι, οὐ | χρὴ ὀρρωδέειν τὸν τρόπον τῆς ἰητρείης. οὐδὲν γὰρ ἂν μέγα φλαῦρον γένοιτο, ἢν μοῦνον οἷός τε ᾖς τῇ χειρὶ τὰς ἐπιδέσιας ἀγαθὰς καὶ ἀσινέας ποιέεσθαι. σημεῖον δὲ τόδε, ἢν μέλλῃ ὀστέων ἀπόστασις ἔσεσθαι ἐν τῷ τρόπῳ τούτῳ τῆς ἰητρείης· πῦον γὰρ συχνὸν ῥεῖ ἐκ τοῦ ἕλκεος καὶ ὀργᾶν φαίνεται. πυκνότερον οὖν μετεπιδεῖν[102] διὰ τὸ πλάδον· ἐπεὶ ἄλλως τε καὶ ἀπύρετοι[103] γίνονται, καὶ ἢν μὴ[104] κάρτα πιέζωνται ὑπὸ τῆς ἐπιδέσιος, καὶ τὸ ἕλκος καὶ τὰ περιέχοντα ἰσχνά. ὅσαι μὲν οὖν λεπτῶν πάνυ ὀστέων ἀποστάσιες, οὐδεμίης μεγάλης μεταβολῆς δέονται, ἀλλ' ἢ χαλαρώτερα ἐπιδεῖν, ὡς μὴ ἀπολαμβάνηται τὸ
514 | πύον, ἀλλ' εὐαπόρρυτον ᾖ· καὶ πυκνότερον μετεπιδεῖν ἔστ' ἂν ἀποστῇ τὸ ὀστέον· καὶ νάρθηκας μὴ προστιθέναι.

29. Ὁκόσοισι δὲ μέζονος ὀστέου ἀπόστασις ἐπίδοξος γίνεται, ἤν τε ἐξ ἀρχῆς προγνῷς, ἤν τε καὶ ἔπειτα μεταγνῷς, οὐκ ἔτι τῆς αὐτῆς ἰητρείης δεῖ, ἀλλὰ τὰς μὲν κατατάσιας καὶ τὰς διορθώσιας οὕτω ποιέεσθαι ὥσπερ εἴρηται· σπλῆνας δὲ χρὴ διπλοῦς, πλάτος μὲν ἡμισπιθαμιαίους, μὴ ἐλάσσους (ὁκοῖον δὲ ἄν τι καὶ τὸ τρῶμα ᾖ, πρὸς τοῦτο τεκμαίρεσθαι), μῆκος δὲ

tributes to the result if it is as it should be, as well as diet and the suitability of the bandages.

28. If you happen to be deceived in a new case, and think there will be no exfoliation of bones, but then signs of bone coming to the surface appear, you have no need to be apprehensive about this mode of treatment, since no great damage will be done, as long as you possess sufficient manual skill to apply the dressings properly and in a way that will do no harm. The following is a sign of approaching exfoliation of bone in a case treated this way: a large amount of pus flows from the lesion, which takes on a turgid appearance. Therefore, the dressing should be changed more often in accordance with the discharge, for, in this way in particular, patients become free of fever, and if there is no great compression by the bandages, the lesion and its surrounding parts will not be engorged. The expulsion of very small bone fragments requires no great alteration of treatment beyond looser bandaging, so as not to intercept the pus, but to allow it freely to escape; also, change the dressing more frequently until the bone separates, and avoid the application of splints.

29. But in cases where exfoliation of a rather large bone is expected, whether you foresee this from the start, or recognize it only later, the treatment should not be identical, but, while the extensions and adjustments should be done as was directed, the compresses should be double, half a span in breadth at least—take the form of the wound as your standard for this—and in length a little shorter

101 <ᾗ> add. Kühlewein 102 μετεπιδεῖν B: -δέεσθαι MV
103 ἀπύρετοι B: πυρετοὶ MV 104 μὴ B: μὲν MV

βραχυτέρους μὲν ὀλίγῳ ἢ ὥστε δὶς περικνέεσθαι περὶ τὸ σῶμα τὸ τετρωμένον, μακροτέρους δὲ συχνῷ ἢ ὥστε ἅπαξ περικνέεσθαι· πλῆθος δὲ ὁπόσους ἂν ξυμφέρῃ, ποιησάμενον, τούτους ἐν οἴνῳ μέλανι αὐστηρῷ βρέχοντα χρὴ ἐκ μέσου ἀρχόμενον, ὡς ἀπὸ δύο ἀρχέων ὑποδεσμὶς ἐπιδεῖται, περιελίσσειν, κἄπειτα σκεπαρνηδὸν παραλλάσσοντα τὰς ἀρχὰς ἀφιέναι. ταῦτα κατά τε αὐτὸ τὸ ἕλκος ποιέειν καὶ κατὰ τὸ ἔνθεν καὶ ἔνθεν τοῦ ἕλκεος· καὶ πεπιέχθω μὲν μή, ἀλλ'
516 ὅσον ἑρμασμοῦ ἕνεκεν τοῦ ἕλκεος | προσκείσθω. ἐπὶ δὲ αὐτὸ τὸ ἕλκος ἐπιτιθέναι χρὴ πισσηρήν, ἤ τι τῶν ἐναίμων ἤ τι τῶν ἄλλων φαρμάκων, ὅ τι ξύντροφόν ἐστιν ἐπιτέγξει. καὶ ἢν μὲν ἡ ὥρη θερινὴ ᾖ, ἐπιτέγγειν τῷ οἴνῳ τοὺς σπλῆνας πυκνά· ἢν δὲ χειμερινὴ ἡ ὥρη ᾖ, καὶ[105] εἴρια πολλὰ ῥυπαρὰ νενοτισμένα οἴνῳ καὶ ἐλαίῳ ἐπικείσθω. ἰξαλῆν δὲ χρὴ [αἰγὸς][106] ὑποτετάσθαι καὶ εὐαπόρρυτα ποιέειν, φυλάσσοντα τοὺς ὑπορρόους, μεμνημένον ὅτι οἱ τόποι οὗτοι, ἐν τοῖσι αὐτοῖσι σχήμασι πολλὸν χρόνον κειμένοισι ἐκτρίμματα δυσάκεστα ποιέουσι.

30. Ὅσους δὲ μὴ οἷόν τε ἐπιδεσει ἰήσασθαι διά[107] τούτων τῶν εἰρημένων τρόπων ἢ τῶν εἰρησομένων, τούτους περὶ πλέονος χρὴ ποιέεσθαι ὅκως εὐθέτως σχήσουσι τὸ κατεηγὸς τοῦ σώματος κατ' ἰθυωρίην, προσέχοντα τὸν νόον καὶ τῷ ἀνωτέρω δὲ μᾶλλον ἢ τῷ κατωτέρω. εἰ δέ τις μέλλοι καλῶς καὶ εὐχερῶς ἐργάζεσθαι, ἄξιον καὶ μηχανοποιήσασθαι, ὅπως κατάτασιν δικαίην καὶ μὴ βιαίην σχήσει τὸ κατεηγὸς τοῦ

than would go around the wounded part twice, but a good deal more than would go around it once. Provide as many of these as should suffice, and after soaking them in dark astringent wine, apply them starting from their middle, as is done with a two-headed underbandage, enveloping the part and then leaving the ends crossed obliquely, as with the adze-shaped bandage. Put them both over the lesion itself and on either side of it; although there should be no compression, the bandages should be applied firmly so as to support the wound. On the lesion itself, pitch cerate should be anointed, or one of the applications for fresh injuries, or any other appropriate remedy that will serve as an embrocation. If it is summer soak the compresses frequently with wine, but in winter also apply in addition plenty of crude wool moistened with wine and oil. Spread a goat's skin underneath to make a free course for the discharges, and giving heed to drainage bear in mind that these regions—when patients lie a long time in the same position—develop sores difficult to heal.

30. As to cases that cannot be treated by bandaging in one of the ways that have been or will be described, all the more should be done in order that the fractured part of the body will be well positioned according to its normal orientation, paying attention that it slopes upward rather than downward. If someone intends to do this work well and skillfully, it is worthwhile in addition to have recourse to a mechanism, in order that the fractured part of the

105 καὶ B: om. MV
106 αἰγὸς BMV: del. Littré
107 τινα add. MV

518 σώματος· μάλιστα ἐν κνήμῃ ἐνδέχεται μηχανοποιέ-
ειν. εἰσὶ μὲν οὖν τινες οἳ ἐπὶ πᾶσι τοῖς τῆς κνήμης
κατήγμασι, καὶ τοῖσι ἐπιδεομένοισι καὶ τοῖσι μὴ ἐπι-
δεομένοισι, τὸν πόδα ἄκρον προσδέουσι πρὸς τὴν
κλίνην ἢ πρὸς ἄλλο τι ξύλον παρὰ τὴν κλίνην κατ-
ορύξαντες. οὗτοι μὲν οὖν πάντα κακὰ ποιέουσιν, ἀγα-
θὸν δὲ οὐδέν· οὔτε γὰρ τοῦ κατατείνεσθαι ἄκος ἐστὶ
τὸ προσδεδέσθαι τὸν πόδα, οὐδὲν γὰρ ἧσσον τὸ ἄλλο
σῶμα προσχωρήσει πρὸς τὸν πόδα καὶ οὕτως οὐκ ἂν
ἔτι τείνοιτο. οὔτ᾽ αὖ ἐς τὴν ἰθυωρίην οὐδὲν ὠφελέει,
ἀλλὰ καὶ βλάπτει· στρεφομένου γὰρ τοῦ ἄλλου σώ-
ματος ἢ τῇ ἢ τῇ, οὐδὲν[108] κωλύσει ὁ δεσμὸς τὸν πόδα
καὶ τὰ ὀστέα τὰ τῷ ποδὶ προσηρτημένα ἐπακολουθεῖν
τῷ ἄλλῳ σώματι· εἰ δὲ μὴ προσεδέδετο, ἧσσον ἂν
διεστρέφετο· ἧσσον γὰρ ἂν ἐγκατελείπετο ἐν τῇ κινή-
σει τοῦ ἄλλου σώματος. εἰ δέ τις σφαίρας δύο ῥά-
ψαιτο ἐκ σκύτεος Αἰγυπτίου τοιαύτας οἵας φορέουσιν
οἱ ἐν τῇσι μεγάλῃσι πέδῃσι πολλὸν χρόνον πεπεδη-
μένοι· αἱ δὲ σφαῖραι ἔχοιεν ἔνθεν καὶ ἔνθεν χιτῶνας
τὰ μὲν πρὸς τοῦ τρώματος βαθυτέρους, τὰ δὲ πρὸς
τῶν ἄρθρων βραχυτέρους· εἶεν δὲ ὀγκηραὶ μὲν καὶ
μαλθακαί, ἁρμόζουσαι δέ, ἡ μὲν τῷ ἄνωθεν τῶν σφυ-
520 ρῶν, ἡ δὲ τῷ κάτωθεν τοῦ γούνατος· | ἐκ δὲ πλαγίης
ἑκατέρη δισσὰ ἑκατέρωθεν ἔχοι προσηρτημένα ἢ
ἁπλόου ἱμάντος ἢ διπλόου, βραχέα[109] ὥσπερ ἀγκύλας,
τὰ μέν τι τοῦ σφυροῦ ἑκατέρωθεν, τὰ δέ τι τοῦ γού-
522 νατος· καὶ ἡ ἄνωθεν σφαῖρα | ἕτερα τοιαῦτα ἔχοι κατὰ
τὴν ἰθυωρίην τὴν αὐτήν. κἄπειτα κρανάϊνας ῥάβδους

body will have adequate but not excessive extension. This method is particularly applicable for the leg. Now, there are some practitioners who in all cases of leg fractures, whether they are bandaged or not, fasten the sole of the foot to the bed, or to a wooden post fixed in the ground by the bed: in truth this does all sorts of mischief and no good; for extension is not achieved by fastening the foot, since the rest of the body will still move toward the foot, and thus the extension will not be maintained. Nor is it of any use for preserving the normal alignment, but even harmful: for when the rest of the body turns one way or another, the ligature is completely ineffective in preventing the foot and the bones connected with it from following the body's movement. Besides, if the leg had not been tied up, there would be less distortion, since it would not be left behind so much in the movement of the rest of the body. But suppose someone sewed two rounded circlets of Egyptian leather like the ones worn by prisoners who are kept shackled for a long time in large fetters: the circlets should have coverings on both sides, deeper on the side facing the injury and shallower on the side facing the joints; they should be large and soft, one fitting above the ankles, and the other below the knee, and on each side they should have two attachments in the form of leather thongs, single or double, short like loops, one set placed at the ankle on either side, the other on either side of the knee; and the upper circlet should have attachments like them in the same straight line, i.e., just opposite the ones

108 *οὐδὲν* MV: om. B
109 *βραχέα* B: *βραχύτερα* MV

τέσσαρας[110] λαβών, ἴσας μέγεθος ἀλλήλῃσιν ἐχούσας, πάχος μὲν ὡς δακτυλιαίας, μῆκος δέ, ὡς κεκαμμέναι ἐναρμόσουσιν ἐς τὰ ἀπαιωρήματα, ἐπιμελόμενος ὅπως τὰ ἄκρα τῶν ῥάβδων μὴ ἐς τὸν χρῶτα, ἀλλ' ἐς τὰ ἄκρα τῶν σφαιρέων ἐγκέλσῃ. εἶναι δὲ χρὴ ζεύγεα τρία τῶν ῥάβδων, καὶ πλείω, καί τινι μακροτέρας τὰς ἑτέρας τῶν ἑτέρων καί τινι καὶ βραχυτέρας[111] καὶ σμικροτέρας, ὡς καὶ μᾶλλον διατείνῃς, ἢν βούλῃ,[112] καὶ ἧσσον. ἔστωσαν δὲ αἱ ῥάβδοι ἑκάτεραι ἔνθεν καὶ ἔνθεν τῶν σφυρῶν. ταῦτα δ' εἰ καλῶς μηχανοποιηθείη, τήν τε κατάτασιν καὶ δικαίην ἂν παρέχοι καὶ ὁμαλὴν κατὰ τὴν ἰθυωρίην, καὶ τῷ τρώματι πόνος οὐδεὶς ἂν εἴη· τὰ γὰρ ἀποπιέσματα, εἴ τι καὶ ἀποπιέζοιτο, τὰ μὲν ἂν ἐς τὸν πόδα ἀπάγοιτο, τὰ δὲ ἐς τὸν μηρόν· αἵ τε ῥάβδοι εὐθετώτεραι, αἱ μὲν ἔνθεν, αἱ δὲ ἔνθεν τῶν σφυρῶν, ὥστε μὴ κωλύεσθαι τὴν θέσιν τῆς κνήμης· τό τε τρῶμα εὐκατάσκεπτον καὶ εὐβάστακτον. οὐδὲν γὰρ ἐμποδών, εἴ τις ἐθέλοι τὰς δύο τῶν
524 ῥάβδων τὰς ἀνωτέρω αὐτὰς πρὸς ἀλλήλας| ζεῦξαι· καὶ ἤν τις κούφως βούλοιτο ἐπιβάλλειν, ὥστε τὸ ἐπιβαλλόμενον μετέωρον ἀπὸ τοῦ τρώματος εἶναι. εἰ μὲν οὖν αἵ τε σφαῖραι προσηνέες καὶ καλαὶ καὶ μαλθακαὶ καὶ καιναὶ ῥαφεῖεν, καὶ ἡ ἔντασις τῶν ῥάβδων χρηστῶς ἐνταθείη, ὥσπερ ἤδη εἴρηται, εὔχρηστον τὸ μηχάνημα· εἰ δέ τι τούτων μὴ καλῶς ἕξει, βλάπτοι ἂν μᾶλλον ἢ ὠφελέοι. χρὴ δὲ καὶ τὰς ἄλλας μηχανὰς ἢ καλῶς μηχανᾶσθαι, ἢ μὴ μηχανᾶσθαι·[113] αἰσχρὸν γὰρ καὶ ἄτεχνον μηχανοποιέοντα ἀμηχανοποιέεσθαι.

below. Then take four rods of cornel wood of equal size, one inch thick, and of such a length that when they are bent they fit into the attachments, taking care that the ends of the rods do not press on the skin, but on the projecting edges of the circlets. There should be three pairs of rods, or even more, some longer than others and some shorter and slenderer, so as to exert greater or less tension as desired; place the rods separately on either side of the ankles. This mechanism, if well arranged, will produce an extension, both proper and even, in a straight line, and cause no pain to the wound; for the outward pressure of humors, if there is any, will be diverted partly to the foot and partly to the thigh. The rods must be correctly positioned, two on one side of the ankles and two on the other side, so as not to interfere with the position of the leg; this makes the wound both easy to examine and easy to handle. For, if you care to, there is nothing to prevent the two upper rods from being tied together, or something being lightly spread over them, in such a way that the covering will be kept up away from the wound. If then the circlets are supple, of good quality, soft and newly sewn, and the extension by the bent rods suitably regulated as just described, the mechanism will be effective; but if any of these things is not well executed, it will cause more harm than good. Other mechanisms too should either be employed well or not used at all, for it is shameful and unprofessional to use a machine but achieve no mechanical effect.

---

110 τέσσαρες B: om. MV   111 βραχυτέρας MV: παχυτέρας B   112 διατείνῃς . . . βούλῃ B: διατείνειν . . . βούλεται MV   113 ἢ μὴ μηχανᾶσθαι MV: om. B

31A. Τοῦτο δέ, οἱ πλεῖστοι τῶν ἰητρῶν τὰ κατήγματα καὶ τὰ σὺν ἕλκεσι καὶ τὰ ἄνευ ἑλκέων, τὰς πρώτας τῶν ἡμερέων ἰητρεύουσιν εἰρίοισι ῥυπαροῖσιν· καὶ οὐδέν τι ἄτεχνον δοκέει τοῦτο εἶναι. ὅσοι μὲν ἀναγκάζονται ὑπὸ τῶν αὐτίκα νεοτρώτων ἐόντων, μὴ ἔχοντες ὀθόνια εἰρίοισι παρασκευάσασθαι, τούτοισι πλείστη συγγνώμη· οὐδὲ γὰρ ἄν τις ἔχοι ἄνευ ὀθονίων ἄλλο τι πολλῷ βέλτιον εἰρίου ἐπιδῆσαι ἐπὶ τὰ τοιαῦτα. εἶναι δὲ χρὴ πάμπολλα καὶ πάνυ καλῶς εἰργασμένα καὶ μὴ τρηχέα· τῶν γὰρ ὀλίγων καὶ φλαύρων ὀλίγη καὶ ἡ δύναμις. ὅσοι δὲ[114] μίαν ἢ δύο ἡμέρας εἴρια ἐπιδέειν δικαιοῦσι, τρίτῃ δὲ καὶ τετάρτῃ ὀθονίοις ἐπιδέοντες πιέζουσι, καὶ κατατείνουσι τότε μάλιστα, οὗτοι πολύ τῆς[115] ἰητρικῆς καὶ κάρτα ἐπίκαιρον ἀσυνετέουσι· ἥκιστα γὰρ χρὴ τῇ τρίτῃ καὶ τετάρτῃ
526 στυφελίζειν πάντα τὰ τρώματα, ὡς ἐν | κεφαλαίῳ εἰρῆσθαι· καὶ μηλώσιας χρὴ[116] πάσας φυλάσσεσθαι[117] ἐν ταύτῃσι τῇσιν ἡμέρῃσι, καὶ ὁπόσοις ἄλλοισι τρώματα[118] ἠρέθισται. τὸ ἐπίπαν γὰρ ἡ τρίτη καὶ ἡ τετάρτη ἡμέρη ἐπὶ τοῖσι πλείστοισι τῶν τρωμάτων τίκτει τὰς παλιγκοτίας καὶ ὅσα ἐς φλεγμονὴν καὶ ἀκαθαρσίην ὁρμᾷ, καὶ ὅσα ἂν ἐς πυρετοὺς ἴῃ. καὶ μάλα πολλοῦ ἄξιον τοῦτο τὸ μάθημα, εἴπέρ τι καὶ ἄλλο. τίνι γὰρ οὐκ ἐπικοινωνέει τῶν ἐπικαιροτάτων ἐν ἰητρικῇ, οὐ κατὰ τὰ ἕλκεα μοῦνον, ἀλλὰ καὶ κατὰ ἄλλα πολλὰ νοσήματα—εἰ μή τις φήσειε καὶ τἆλλα νοσήματα ἕλκεα εἶναι; ἔχει γάρ τινα καὶ οὗτος ὁ λόγος ἐπιείκειαν· πολλαχῇ γὰρ ἠδέλφισται τὰ ἕτερα

31A. Furthermore, most practitioners treat fractures, whether with or without lesions, by applying crude wool during the first days, and this appears in no way contrary to the art. Those who are compelled, because they have no bandages available, to take wool for first-aid treatment are altogether excusable, for in the absence of bandages nothing is much better than wool to dress such cases. But it must be plentiful, well-carded and fine; if it is small in amount and of poor quality, its effect will also be small. Now, those who think it is right to dress with wool for one or two days, and then on the third or fourth day, use bandages with compression, and apply extension mainly at this period, are very ignorant of the healing art, and on a most vital point. For, in short, the third or fourth day is the very last one on which any lesion should be actively interfered with; and all probings as well as everything else by which wounds are irritated should be avoided on these days. For, as a rule, the third or fourth day sees the arrival of the complications in the majority of injuries, both where the tendency is to inflammation and putrefaction, and in those in which fever supervenes. And indeed, if any rule is of special value, this one is, for to which of medicine's most important fields does it not apply, not only as regards lesions, but also in many other maladies—unless a person calls the other maladies lesions too? And this view too is not unreasonable, seeing that they have an af-

---

114 μὴ add. V 115 τῆς B: τι MV
116 χρὴ B: δὲ MV
117 χρὴ add. MV
118 τρώματα B: τρώμασι MV

τοῖσι ἑτέροισι. ὁκόσοι μέντοι δικαιοῦσιν εἰρίοισι χρῆσθαι, ἔστ' ἂν ἑπτὰ ἡμέραι παρέλθωσιν, ἔπειτα κατατείνειν τε καὶ κατορθοῦν καὶ ὀθονίοισιν ἐπιδεῖν, οὗτοι οὐκ ἂν ἀσύνετοι ὁμοίως φανεῖεν· καὶ γὰρ τῆς φλεγμονῆς τὸ ἐπικαιρότατον παρελήλυθε, καὶ τὰ ὀστέα χαλαρὰ καὶ εὔθετα μετὰ ταύτας τὰς ἡμέρας ἂν εἴη. πολλῷ μέντοι ἥσσηται καὶ αὕτη ἡ μελέτη τῆς ἐξ ἀρ-
528 χῆς τοῖσιν ὀθονίοισιν ἐπιδέσιος· κεῖνος | μὲν γὰρ ὁ τρόπος ἑβδομαίους ἐόντας ἀποδείκνυσιν ἀφλεγμάντους,[119] καὶ παρασκευάζει νάρθηξι τελέως ἐπιδεῖν· οὗτος δὲ ὁ τρόπος πολὺ ὑστερεῖ, βλάβας δέ τινας καὶ ἄλλας ἔχει· ἀλλὰ μακρὸν ἂν εἴη πάντα γράφειν.

31B. Οἷσι δὲ τὰ ὀστέα κατεηγότα καὶ ἐξίσχοντα μὴ δύνηται ἐς[120] χώρην καθιδρύεσθαι, ἥδε ἡ κατάστασις·[121] σιδήρια χρὴ ποιέεσθαι ἐς τοῦτον τὸν τρόπον ὅνπερ οἱ μοχλοὶ ἔχουσιν, οἷς οἱ λατύποι χρέονται, τὸ μέν τι πλατύτερον, τὸ δέ τι στενότερον· εἶναι δὲ χρὴ καὶ τρία καὶ ἔτι πλείω, ὡς τοῖσι μάλιστα ἁρμόζουσί τις χρήσεται. ἔπειτα τούτοισι χρὴ ἅμα τῇ κατατάσει μοχλεύειν ὑπερβάλλοντα,[122] πρὸς μὲν τὸ κατωτέρω τοῦ ὀστέου τὸ κατώτερον ἐρείδοντα, πρὸς δὲ τὸ ἀνωτέρω τὸ ἀνώτερον τοῦ σιδηρίου, ἁπλῷ δὲ λόγῳ, ὥσπερ εἰ λίθον τις ἢ ξύλον μοχλεύοι ἰσχυρῶς. ἔστω δὲ σθεναρὰ τὰ σιδήρια ὡς οἷόν τε, ὡς μὴ κάμπτηται. αὕτη μεγάλη τιμωρίη, ἤν τά τε σιδήρια ἐπιτήδεια ᾖ καὶ μοχλεύηταί τις ὡς χρή· ὁκόσα γὰρ ἀνθρώποις ἀρfinity to one another in many ways. But those who think it correct to use wool until seven days have passed, and then proceed to extension, coaptation and bandaging, might not appear so unintelligent either, since by then the most dangerous time for inflammation is past, and the bones will be found loose and easy to put into place. Still, even this latter treatment is much inferior to the use of bandages from the beginning, since that method results in the patients being without inflammation on the seventh day, and ready for complete dressing with splints, while the former one is much slower, and has some other disadvantages; but it would take a long time to go through everything.

31B. In cases where fractured and projecting bones cannot be set back into their proper place, the method of reduction is the following. Have iron rods made like the levers stonemasons use, broader at one end and narrower at the other; you should acquire three, or even more of these, so that the most suitable size can be chosen. Employ the iron by inserting it while extension is being exerted, and pressing its underside on the lower bone to generate leverage, and its upper side against the upper bone, in short, exactly as someone would do who was violently levering up some stone or log. The irons should be as strong as possible so as not to bend. This is a great service, if the irons are suitable and someone carries out the leverage properly; for of all the apparatus contrived by humans,

119 ἀφλεγμάντους B^2 above the line: om. BMV
120 Add. τὴν ἑωυτῶν MVB^2 above the line
121 κατάστασις B: κατάτασις MV
122 ὑπερβ. H^2 Petrequin Withington: ὑποβ. BMV

μενα μεμηχάνηται, πάντων ἰσχυρότατά ἐστι τρία ταῦτα, ὄνου τε περιαγωγὴ καὶ μόχλευσις καὶ σφήνωσις. ἄνευ δὲ τούτων ἑνός τε τινος ἢ πάντων, οὐδὲν τῶν
530 ἔργων τῶν ἰσχυροτάτων οἱ | ἄνθρωποι ἐπιτελέουσιν. οὔκουν ἀτιμαστέη ἡ μόχλευσις αὕτη· ἢ γὰρ οὕτως ἐμπεσεῖται τὰ ὀστέα, ἢ οὐκ ἄλλως. ἢν δ' ἄρα τοῦ ὀστέου τὸ ἄνω παρηλλαγμένον μὴ ἐπιτήδειον ἔχῃ ἐνέδρην τῷ μοχλῷ, ἀλλὰ πάροξυ ἐὸν[123] παραφέρῃ, παραγλύψαντα χρὴ τοῦ ὀστέου ἐνέδρην τῷ μοχλῷ ἀσφαλέα ποιῆσαι. μοχλεύειν δὲ χρὴ καὶ τείνειν αὐθήμερα ἢ δευτεραῖα, τριταῖα δὲ μή, τεταρταῖα δὲ ὡς ἥκιστα καὶ πεμπταῖα. καὶ γὰρ μὴ ἐμβάλλοντι, ὀχλήσαντι δὲ ἐν ταύτῃσι τῇσιν ἡμέρῃσι, φλεγμονὴν ἂν ποιήσειε, καὶ ἐμβάλλοντι οὐδὲν ἧσσον· σπασμὸν μέντοι ἐμβάλλοντι πολὺ ἂν μᾶλλον ποιήσειεν ἢ ἀπορήσαντι ἐμβαλεῖν. ταῦτα εὖ χρὴ εἰδέναι· καὶ γὰρ εἰ ἐπιγένοιτο σπασμὸς ἐμβάλλοντι, ἐλπίδες μὲν οὐ πολλαὶ σωτηρίης. λυσιτελέει δὲ ὀπίσω ἐκβάλλειν τὸ ὀστέον, εἰ οἷόν τε εἴη ἀόχλως. οὐ γὰρ ἐπὶ τοῖσι χαλαρωτέροισι τοῦ καιροῦ σπασμοὶ καὶ τέτανοι ἐπιγίνονται, ἀλλὰ ἐπὶ τοῖς ἐντεταμένοισι μᾶλλον. περὶ οὗ οὖν ὁ λόγος, οὐ χρὴ ἐνοχλέειν ἐν τῇσι προειρημένῃσιν ἡμέρῃσι ταύτῃσι, ἀλλὰ μελετᾶν ὅπως ἥκιστα φλεγμανεῖ τὸ ἕλκος καὶ μάλιστα ἐκπυήσει. ἐπὴν δὲ ἑπτὰ ἡμέραι παρέλθωσιν ἢ ὀλίγῳ πλείους, ἢν ἀπύρετος ᾖ καὶ μὴ φλεγμαίνῃ τὸ ἕλκος, τότε ἧσσον
532 κωλύει πειρᾶσθαι ἐμβάλλειν, ἢν | ἐλπίζῃς κρατήσειν· εἰ δὲ μή, οὐδὲν δεῖ μάτην ὀχλέειν καὶ ὀχλέεσθαι.

these three are the most powerful: the wheel and axle, the lever, and the wedge. Without some one, or all of these, in fact, people would not be able to accomplish any work requiring very great force. This lever method, then, is not to be despised, for displaced bones will be reduced either in this way, or not at all. If it happens that the upper part of the bone overriding the rest affords no suitable hold for the lever, but because of its pointed shape the lever slips past, it is necessary to cut a notch in the bone to form a secure lodgment for the lever. The leverage and extension should be done on the first or second day, but not on the third, and least of all on the fourth or fifth. For to cause a disturbance on these days without achieving a successful reduction would provoke inflammation, and no less so, even if there was a reduction; convulsions, indeed, would much more likely be caused if reduction succeeded than if it failed. This must be well understood, for if a convulsion supervenes after reduction there is not much hope of survival. It can also be advantageous to dislocate the joint again, if this can be done without causing a disturbance: for it is not when the parts are more relaxed than usual that spasms and tetanus supervene, but when they are more contracted. As regards our subject, then, you should not disturb the parts on the days indicated above, but take care how best to oppose inflammation in the lesion, and favor the production of pus. At the end of seven days, or a few more, if the patient is free of fever and the lesion is not inflamed, there is less objection to an attempt at reduction, if you expect to succeed; otherwise you should not subject the patient and yourself to trouble that brings no advantage.

123 ἐὸν Littré: -ν B: ὃ MV

32. Ἢν μὲν οὖν ἐμβάλῃς τὰ ὀστέα ἐς τὴν ἑωυτῶν χώρην, γεγράφαται ἤδη οἱ τρόποι οἵως χρὴ ἰητρεύειν, ἤν τε ἐλπίζῃς ὀστέα ἀποστήσεσθαι ἤν τε μή. χρὴ δὲ καὶ ἢν μὲν ἐλπίζῃς ὀστέα ἀποστήσεσθαι, τῷ τρόπῳ τῶν ὀθονίων ἐπὶ πᾶσι τοῖσι τούτοισι τὴν ἐπίδεσιν ποιέεσθαι ἐκ μέσου τοῦ ὀθονίου ἀρχόμενον, ὡς ἐπὶ τὸ πολύ, ὡς ἀπὸ δύο ἀρχέων ὑποδεσμὶς ἐπιδεῖται· τεκμαίρεσθαι δὲ χρὴ πρὸς τὴν μορφὴν τοῦ ἕλκεος, ὅπως ἥκιστα σεσηρὸς καὶ ἐκπεπλιγμένον ἔσται παρὰ τὴν ἐπίδεσιν· τοῖσι μὲν γὰρ ἐπὶ δεξιὰ ἐπιδεῖν συντρόφως ἔχει, τοῖσι δὲ ἐπ᾽ ἀριστερά, τοῖσι δὲ ἀπὸ δύο ἀρχέων.

33. Ὁπόσα δὲ κατηπορήθη ὀστέα ἐμπεσεῖν, ταῦτα αὐτὰ εἰδέναι χρὴ ὅτι ἀποστήσεται καὶ ὅσα τελέως ἐψιλώθη τῶν σαρκῶν. ψιλοῦται δὲ ἐνίων μὲν τὸ ἄνω μέρος, μετεξετέρων δὲ κυκλόθεν περιθνῄσκουσιν[124] αἱ σάρκες· καὶ τῶν μὲν ἀπὸ τοῦ ἀρχαίου τρώματος σεσάπρισται ἔνια τῶν ὀστέων, τῶν δὲ οὔ· καὶ τῶν μὲν μᾶλλον, τῶν δὲ ἧσσον, καὶ τὰ μὲν σμικρά, τὰ δὲ μεγάλα. διὰ οὖν ταῦτα τὰ εἰρημένα οὐκ ἔστιν ἑνὶ ὀνόματι εἰπεῖν, ὁπότε τὰ ὀστέα ἀποστήσεται· τὰ μὲν γὰρ διὰ μικρότητα, τὰ δὲ διὰ τὸ ἐπ᾽ ἄκρου ἔχεσθαι θᾶσσον ἀφίσταται· τὰ δέ, διὰ τὸ μὴ ἀφίστασθαι, ἀλλὰ λεπιδοῦσθαι, καταξηρανθέντα καὶ σαπρὰ γενό-
534 μενα· πρὸς δὲ τούτοις, διαφέρει τι καὶ ἰητρείη | ἰητρείης. ὡς μὲν οὖν τὸ ἐπίπαν τάχιστα τούτων ὀστέα ἀφίσταται, ὧν τάχιστα μὲν αἱ ἐκπυήσιες, τάχισται δὲ

[124] περιθνῄσκουσιν B: ἀμφιθνῄ- MV

32. Now for cases where you have already reduced the bones to their place, the proper modes of treatment have been described above, whether you expect the bones to exfoliate or not. And even when you expect the bones to exfoliate, you should use in all such cases the method of bandaging with linen cloths, generally beginning with the middle of the bandage, the way an underbandage is applied from two beginnings. Regulate the process according to the shape of the lesion, so that it will be drawn aside or everted as little as possible by the bandaging: for in some cases it is appropriate to bandage toward the right, in others toward the left, and in others from two beginnings.

33. About bones that cannot be reduced, it must be understood that these are bound to exfoliate, as also are bones completely denuded of flesh. In some cases, the upper part of the bones is denuded, while in others all the flesh surrounding them perishes. The starting point of the necrosis in some bones is the old wound, but not in others; it is more extensive in some and less so in others, and some of the bones affected are small, but others large. It follows from what I have written that it is impossible to explain in a single account when bones will exfoliate, for some separate sooner, owing to their small size, and others because they are situated at the end (of a fracture); others again, because they fail to come away (when they are whole), exfoliate after desiccation and decomposition. Furthermore, the effects of various treatments are different. As a general rule, bones are most quickly eliminated in cases

καὶ κάλλισται αἱ σαρκοφυΐαι· καὶ γὰρ αἱ ὑποφυόμεναι σάρκες κατὰ τὸ σιναρὸν αὗται μετεωρίζουσι τὰ ὀστέα ὡς ἐπὶ τὸ πολύ. ὅλος μὴν κύκλος ὀστέου, ἢν ἐν τεσσαράκοντα ἡμέρῃσιν ἀποστῇ, καλῶς ἀποστήσεται· ἔνια γὰρ ἐς ἑξήκοντα ἡμέρας ἀφικνεῖται. τὰ μὲν γὰρ ἀραιότερα τῶν ὀστέων θᾶσσον ἀφίσταται, τὰ δὲ στερεώτερα βραδύτερον· τὰ δὲ ἄλλα τὰ μείω πολλὸν ἐνδοτέρω, ἄλλα δ' ἄλλως.

Ἀποπρίειν δὲ ὀστέον ἐξέχον ἐπὶ τῶνδε τῶν προφασίων χρή· ἢν μὴ δύνηται ἐμβάλλειν, μικροῦ δέ τινος αὐτῷ δοκέῃ δεῖν παρελθεῖν, καὶ οἷόν τε ᾖ παραιρεθῆναι· ἤν τε ἀσηρὸν ᾖ καὶ θραῦόν τι τῶν σαρκίων, καὶ δυσθεσίην παρέχῃ, ψιλόν τε τυγχάνῃ ἐόν, καὶ τὸ τοιοῦτο ἀφαιρέειν χρή. τὰ δὲ ἄλλα οὐδὲν μέγα διαφέρει, οὔτε ἀποπρῖσαι οὔτε μὴ ἀποπρῖσαι. σαφέως γὰρ εἰδέναι χρὴ ὅτι ὀστέα, ὅσα τελέως στερέεται τῶν
536 | *σαρκῶν καὶ ἐπιξηραίνεται, ὅτι πάντα τελέως ἀποστήσεται. ὅσα δὲ ἀπολεπιδοῦσθαι μέλλει, ταῦτα οὐ χρὴ ἀποπρίειν· τεκμαίρεσθαι δὲ χρὴ ἀπὸ τῶν τεταγμένων σημείων τὰ τελέως ἀποστησόμενα.*

34. *Ἰητρεύειν δὲ τοὺς τοιούτους σπλήνεσι καὶ τῇ οἰνηρῇ ἰητρείῃ, ὥσπερ καὶ πρόσθεν γέγραπται ἐπὶ τῶν ἀποστησομένων ὀστέων. φυλάσσεσθαι δὲ χρὴ μὴ καταψύχροισι τέγγειν τὸν πρῶτον χρόνον· ῥιγέων γὰρ πυρετωδέων κίνδυνος· κίνδυνος δὲ καὶ σπασμῶν· προκαλέεται γὰρ σπασμὸν τὰ ψυχρά, ποτὶ δὲ καὶ ἕλκεα. εἰδέναι δὲ χρὴ ὅτι ἀνάγκη βραχύτερα τὰ σώματα*

where suppuration is quickest, and the growth of new flesh most rapid and successful; for it is the new flesh proliferating in the wound that generally lifts the bony fragments up. If the whole circle of bone comes away in forty days, it will be a good separation, for some cases go on to sixty days or even more. More porous bones come away more quickly, more solid ones more slowly; otherwise, smaller splinters take much less time, etc.

The following are the indications for resection of a protruding piece of a bone: if a bone cannot be reduced, but only some small portion of it seems to be in the way, and it is possible to remove this by sawing; also, if the projecting bone is troublesome and is irritating some of the tissues, and causing the part to be held in the wrong position, and if it is denuded, it should be resected. In other cases, it makes no great difference whether there is a resection or not. For it must be kept clearly in mind that when bones are entirely deprived of soft parts and dried out they will all come completely away, and that bones about to be naturally exfoliated in this way have no need of being resected. Base your decision about which bones are likely to come away completely on the symptoms I have set forth.

34. Treat such cases with compresses and vinous applications as described above in the case of bones about to be eliminated. Take care not to moisten with cold fluids at first, for there is a risk of feverish rigors, and a further risk of convulsions, since cold substances provoke convulsions, and sometimes also ulcerations. Bear in mind that there must be shortening of the parts in cases where, when both

ταύτῃ γίνεσθαι, ὧν ἀμφότερα τὰ ὀστέα κατεηγότα καὶ παρηλλαγμένα ἰητρεύηται, καὶ οἷς ὅλος ὁ κύκλος τοῦ ὀστέου ἀπέστη.

35. Ὅσων δὲ μηροῦ ὀστέον ἢ βραχίονος ἐξέσχεν, οὗτοι οὐ μάλα περιγίνονται· τὰ γὰρ ὀστέα μεγάλα καὶ πολυμύελα, καὶ πολλὰ καὶ ἐπίκαιρα τὰ συντιτρωσκόμενα καὶ νεῦρα καὶ μύες καὶ φλέβες. καὶ ἢν μὲν ἐμβάλλοις, σπασμοὶ φιλέουσιν ἐπιγίνεσθαι, μὴ ἐμβληθεῖσι δὲ πυρετοὶ ὀξέες καὶ ἐπίχολοι καὶ λυγγώδεες, καὶ ἐπιμελαίνονται. περιγίνονται δὲ οὐχ ἧσσον, οἷσι μὴ ἐμβληθῇ, μηδὲ πειρηθῇ ἐμβάλλεσθαι· ἔτι δὲ μᾶλλον περιγίνονται, οἷσι τὸ κάτω μέρος τοῦ ὀστέου ἐξέσχεν, ἢ οἷσι τὸ ἄνω· περιγένοιντο δ' ἂν καὶ οἷς ἐμβληθείη, σπανίως γε μήν. μελέται γὰρ μελετέων
538 μέγα διαφέρουσι, | καὶ φύσιες φυσίων τῶν σωμάτων ἐς εὐφορίην. διαφέρει δὲ μέγα, καὶ ἢν ἔσω τοῦ βραχίονος καὶ τοῦ μηροῦ τὰ ὀστέα ἐξέχῃ· πολλαὶ γὰρ καὶ ἐπίκαιροι κατατάσιες φλεβῶν ἐν τῷ ἔσω μέρει, ὧν ἔνιαι τιτρωσκόμεναι σφάγιαί εἰσιν· εἰσὶ δὲ καὶ ἐν τῷ ἔξω μέρει, ἧσσον δέ. ἐν τοῖσιν οὖν τοιούτοισι τρώμασι τοὺς μὲν κινδύνους οὐ χρὴ λήθειν, ὁποῖοί τινές εἰσι, καὶ προλέγειν χρὴ πρὸς τοὺς καιρούς. εἰ δὲ ἀναγκάζοιο μὲν ἐμβάλλειν, ἐλπίζοις δὲ ἐμβάλλειν, καὶ μὴ πολλὴ ἡ παράλλαξις εἴη τοῦ ὀστέου, καὶ μὴ ξυνδεδραμήκοιεν οἱ μύες—φιλέουσι γὰρ συνθεῖν—ἡ μόχλευσις καὶ τούτοισι μετὰ τῆς κατατάσιος εὖ ἂν συλλαμβάνοιτο.

36. Ἐμβάλλοντα δέ ἐλλέβορον μαλθακὸν πῖσαι

bones are broken, they are treated while overlapping; also, in cases where a circle of bone is eliminated entirely.

35. Cases where the bone of the thigh or upper arm protrudes rarely recover; for the bones are large and contain much marrow, while the cords, muscles, and blood vessels which share in the injury are numerous and important. Besides, if you reduce the fracture, convulsions are likely to supervene, while in cases not reduced there are acute bilious fevers with hiccups and necrosis. Cases where reduction has not been made or even attempted are no less likely to recover, and recovery is more frequent when the lower part of a bone projects than when the upper part does. There may be survival even in cases where a reduction is made, but it is very rare. One mode of treatment differs greatly from others, and one bodily constitution from others in their effect on a patient's endurance. It also makes a great difference whether a bone protrudes on the inner side of the arm or thigh, for many important blood vessels stretch along the inner side, and lesions of some of these are fatal; there are also vessels on the outside, but fewer. In such injuries, then, the dangers must not remain unreported, nor their nature, but be foretold in good time. If you are compelled to attempt reduction and expect to succeed, and no great overriding of the bone is present, and the muscles are not retracted (for they often do tend to retract), leverage combined with extension would be well employed even in these cases.

36. After reduction give a mild dose of hellebore to

χρὴ αὐθήμερον, ἢν αὐθημερὸν ἐμβληθῇ· εἰ δὲ μή, οὐδ' ἐγχειρέειν χρή. τὸ δὲ ἕλκος ἰητρεύειν χρή, οἷσί περ κεφαλῆς ὀστέα κατεηγυίης· καὶ ψυχρὸν μηθὲν προσφέρειν, σιτίου δὲ στερῆσαι τελέως· καὶ ἢν μὲν[125] *πικρόχολος φύσει ᾖ, ὀξύγλυκυ εὐῶδες ὀλίγον ἐπὶ ὕδωρ ἐπιστάζοντα τούτῳ διαιτᾶν· ἢν δὲ μὴ*[126] *πικρό-*
540 *χολος ᾖ, ὕδατι | πόματι χρῆσθαι· καὶ ἢν μὲν πυρεταίνῃ συνεχῶς, τεσσαρεσκαίδεκα ἡμέρας τὸ ἐλάχιστον οὕτω διαιτᾶν, ἢν δὲ ἀπύρετος ᾖ, ἑπτὰ ἡμέρῃσιν. ἔπειτα ἐκ προσαγωγῆς κατὰ λόγον ἐς φαύλην δίαιταν ἄγειν. καὶ οἷσιν ἂν μὴ ἐμβληθῇ τὰ ὀστέα, καὶ τὴν φαρμακείην χρὴ τοιαύτην ποιέεσθαι, καὶ τῶν ἑλκέων τὴν μελέτην καὶ τὴν δίαιταν ὡσαύτως· καὶ τὸ ἀπαιωρεύμενον τοῦ σώματος μὴ κατατείνειν, ἀλλὰ προσάγειν μᾶλλον ὥστε χαλαρώτερον εἶναι τὸ κατὰ τὸ ἕλκος. τῶν δὲ ὀστέων ἡ ἀπόστασις χρονίη, ὥσπερ καὶ πρόσθεν εἴρηται. μάλιστα δὲ χρὴ τὰ τοιαῦτα διαφυγεῖν, ἅμα*[127] *ἤν τις καλὴν φυγὴν ἔχῃ. αἵ τε γὰρ ἐλπίδες ὀλίγαι καὶ οἱ κίνδυνοι πολλοί· καὶ μὴ ἐμβαλὼν ἄτεχνος ἂν δοκέοι εἶναι, καὶ ἐμβαλὼν ἂν ἐγγυτέρω θανάτου ἀγάγοι ἢ σωτηρίης.*

37. *Τὰ δὲ ὀλισθήματα τὰ κατὰ τὰ γούνατα καὶ τὰ διακινήματα τῶν ὀστέων εὐηθέστερα πολὺ τῶν κατ' ἀγκῶνα κινημάτων καὶ ὀλισθημάτων. τό τε γὰρ ἄρθρον τοῦ μηροῦ εὐσταλέστερον ὡς ἐπὶ τῷ μεγέθει ἢ τὸ τοῦ βραχίονος, καὶ δικαίην φύσιν μοῦνον ἔχον καὶ ταύτην περιφερέα· τὸ δὲ τοῦ βραχίονος ἄρθρον μέγα τε καὶ βαθμίδας πλείονας ἔχον. πρὸς δὲ τούτοις τὰ*

drink on the first day, if the reduction has succeeded on that day, but otherwise do not even attempt treatment. The lesion should be treated with the remedies used for the bones of a fractured skull: apply nothing cold and prescribe strict abstinence from solid food. If the patient is of a bilious nature, give him a little aromatic oxymel sprinkled in water, but if not, use water as beverage. If there is continuous fever, maintain this regimen for at least fourteen days, but if there is no fever, for seven days; then return by regular steps to an ordinary diet. In cases where the bones are not reduced, a similar use of medication should be made, and likewise with the management of the lesions and the regimen. Furthermore, do not stretch the unreduced part of the body, but on the contrary bring it more together so that the seat of the lesion will be more relaxed. The elimination of the bones takes time, as was said before. Generally, the acceptance of such cases should be avoided if a creditable excuse can be found, for the favorable chances are few, and the risks many. Besides, if a man does not reduce the fracture, he will be thought unskillful, while if he does reduce it, he will bring the patient nearer to death than to recovery.

37. Dislocations and displacements of the bones at the knee are much less trouble than displacements and dislocations at the elbow, since the articular end of the thigh is simpler in relationship to its size than the joint of the arm, and it alone has a regular conformation, in fact a rounded one, whereas the joint of the arm is large and has several facets. Besides this, the two bones of the lower leg are

---

125 μὲν B: μὴ MV 126 μὴ B: καὶ MV
127 ἅμα B: ἀλλ' MV

μὲν τῆς κνήμης ὀστέα παραπλήσια μῆκός ἐστι καὶ σμικρόν τι οὐκ ἄξιον λόγου τὸ ἔξω ὀστέον ὑπερέχει,
542 *οὐδενὸς μεγάλου κώλυμα ἐόν,* | *ἀφ᾽ οὗ πέφυκεν ὁ ἔξω τένων ὁ παρὰ τὴν ἰγνύην. τὰ δὲ τοῦ πήχεος ὀστέα ἄνισά ἐστι, καὶ τὸ βραχύτερον παχύτερον συχνῷ, τὸ δὲ λεπτότερον πολλὸν ὑπερβάλλει καὶ ὑπερέχει τὸ ἄρθρον· ἐξήρτηται μέντοι καὶ τοῦτον τῶν νεύρων*
544 *κατὰ τὴν κοινὴν ξύμφυσιν* | *τῶν ὀστέων· πλέον δὲ μέρος ἔχει τῆς ἐξαρτήσιος τῶν νεύρων ἐν τῷ βραχίονι τὸ λεπτὸν ὀστέον ἤπερ τὸ παχύ. ἡ μὲν οὖν φύσις τοιουτότροπος τῶν ἄρθρων τούτων καὶ τῶν ὀστέων τοῦ ἀγκῶνος. καὶ διὰ τὸν τρόπον τῆς φύσιος τὰ κατὰ γόνυ ὀστέα πολλάκις μὲν ὀλισθάνει, ῥηϊδίως δὲ ἐμπίπτει· φλεγμονὴ δὲ οὐ μεγάλη προσγίνεται, οὐδὲ δεσμὸς τοῦ ἄρθρου. ὀλισθάνει δὲ τὰ πλεῖστα ἐς τὸ εἴσω μέρος, ἔστι δὲ ὅτε καὶ ἐς τὸ ἔξω, ποτὲ δὲ καὶ ἐς τὴν ἰγνύην. τούτων ἁπάντων ἐμβολαὶ οὐ χαλεπαί, ἀλλὰ τὰ μὲν ἔσω καὶ ἔξω ὀλισθάνοντα καθῆσθαι μὲν χρὴ τὸν ἄνθρωπον χαμαὶ ἢ*[128] *ἐπὶ χαμαιζήλου τινός, τὸ δὲ σκέλος ἀνωτέρω ἔχειν, μὴ μὴν πολλῷ. κατάτασις δὲ ὡς ἐπὶ τὸ πολὺ μετρίη ἀρκέει· τῇ μὲν κατατείνειν τὴν κνήμην, τῇ δὲ ἀντιτείνειν τὸν μηρόν.*

38. Τὰ *δὲ κατὰ τὸν ἀγκῶνα ὀχλωδέστερά ἐστι τῶν κατὰ τὸ γόνυ, καὶ δυσεμβολώτερα καὶ διὰ τὴν φύσιν καὶ διὰ τὴν φλεγμονήν, ἢν μή τις αὐτίκα ἐμβάλοι· ὀλισθάνει μὲν γὰρ ἧσσον ἢ ἐκεῖνα, δυσεμβολώτερα δὲ καὶ δυσθετώτερα, καὶ ἐπιφλεγμαίνει μᾶλλον καὶ ἐπιπωροῦται.*

about the same length, and the outer one (fibula) hardly extends beyond the other one, and so presents no hindrance to any great displacement, where the external tendon of the ham arises from it. But the bones of the forearm are unequal, and the shorter one (radius) is much thicker, whereas the slenderer one (ulna) extends far beyond and overreaches the joint, but this one too is attached to the cords at the common junction of the bones. The slenderer bone has a larger share of the attachments of the cords in the arm than the thicker one does. Such then is the arrangement of these articulations and bones of the elbow. Owing to the way the bones at the knee are disposed, they are often dislocated, but easily reduced, and no very great inflammation or fixation of the joint supervenes. Most dislocations are inward, although some outward, and some are into the ham. Reduction is not difficult in any of these cases, but external and internal dislocations require the patient to be seated on the ground or some low object, and hold his leg raised, although not very much. Moderate extension as a rule suffices: exert extension on the leg and counterextension on the thigh.

38. Dislocations at the elbow are more troublesome than those at the knee, and harder to reduce, both because of the conformation of the joint and because of inflammation, unless someone reduces them immediately. It is true that the bones of the elbow are more rarely dislocated than the cases described, but they are harder to reduce and keep in position, and inflammation and excessive formation of callus are more likely to supervene in them.

---

[128] χαμαὶ ἢ B: om. MV

39. Ἔστι δὲ καὶ τούτων τὰ μὲν πλεῖστα σμικραὶ
ἐγκλίσιες, ἄλλοτε ἐς τὸ πρὸς τῶν πλευρέων μέρος,
ἄλλοτε ἐς τὸ ἔξω· οὐ πᾶν δὲ τὸ ἄρθρον μεταβεβηκός,
546 ἀλλὰ μένον τι[129] τὸ κατὰ τὸ κοῖλον | τοῦ ὀστέου τοῦ
βραχίονος, ᾗ τὸ τοῦ πήχεος ὀστέον τὸ ὑπερέχον ἐξ-
έσχεν·[130] τὰ μὲν οὖν τοιαῦτα, κἂν τῇ ἢ τῇ ὀλίσθῃ,
ῥηΐδιον ἐμβάλλειν, καὶ ἀπόχρη ἡ κατάτασις ἡ ἐς τὸ
ἰθὺ γινομένη κατ' ἰθυωρίην τοῦ βραχίονος, τὸν μὲν
κατὰ τὸν καρπὸν τῆς χειρὸς τείνειν, τὸν δὲ κατὰ τὴν
μασχάλην περιβάλλοντα, τὸν δὲ τῇ ἑτέρῃ πρὸς τὸ
ἐξεστεὸς ἄρθρον τὸ θέναρ προσβάλλοντα ὠθέειν, τῇ
δὲ ἑτέρῃ ἀντωθέειν προσβάλλοντα πρὸς τοῦ πήχεος[131]
ἐγγὺς τῷ ἄρθρῳ.

40. Ἐνακούει δὲ οὐ βραδέως ἐμβαλλόμενα τὰ τοιαῦτα ὀλισθήματα, ἢν πρὶν φλεγμήνῃ ἐμβάλλῃ τις. ὀλισθάνει δὲ ὡς ἐπὶ τὸ πολὺ μᾶλλον ἐς τὸ ἔσω μέρος, ὀλισθάνει δὲ καὶ ἐς τὸ ἔξω· εὔδηλα δὲ τῷ σχήματι. καὶ πολλάκις ἐμπίπτει τὰ τοιαῦτα καὶ ἄνευ ἰσχυρῆς κατατάσιος. χρὴ δὲ τῶν ἔσω ὀλισθανόντων, τὸ μὲν ἄρθρον ἀπωθέειν ἐς τὴν φύσιν, τὸν δὲ πῆχυν ἐς τὸ καταπρηνὲς μᾶλλον ῥέποντα περιάγειν. τὰ μὲν πλεῖστα ἀγκῶνος τοιαῦτα ὀλισθήματα.

41. Ἢν δὲ ὑπερβῇ τὸ ἄρθρον ἢ ἔνθα ἢ ἔνθα ὑπὲρ
τὸ ὀστέον τὸ τοῦ πήχεος τὸ ἐξέχον ἐς τὸ κοῖλον τοῦ
βραχίονος—γίνεται μὲν[132] ὀλιγάκις τοῦτο—ἢν δὲ γέ-
νηται, οὐκ ἔτι ὁμοίως ἡ κατάτασις ἡ ἐς τὴν ἰθυωρίην
548 γινομένη ἐπιτηδείη τῶν τοιούτων ὀλισθημάτων· | κω-
λύει γὰρ ἐν τῇ τοιαύτῃ κατατάσει τὸ ἀπὸ τοῦ πήχεος

39. The majority of these (sc. dislocations of the elbow) are small displacements, sometimes inward toward the ribs, and sometimes outward (our "forward" and "backward"). The joint is not dislocated as a whole, but maintains some of its connection with the cavity of the arm bone, where the projecting part of the ulna sticks out. Such cases, then, whether dislocation is to one side or the other, are easy to reduce, and direct extension in the line of the upper arm is quite enough: one person provides traction on the wrist, another does so by clasping the arm at the axilla, while a third presses with the palm of one hand on the projecting part, and with the other hand exerts counterpressure toward the forearm near the joint.

40. Such dislocations yield readily to reduction if someone reduces them before they become inflamed; the dislocation is usually rather inward (forward), but may also be outward; it is easily recognized by the position of the limb, and can often be reduced even without strong extension. In the case of internal dislocations, you should press the joint back into its natural place, and bring the forearm around to the prone position. Most dislocations of the elbow are of this kind.

41. If the articular end (sc. of the humerus) slips in one direction or the other over the part of the bone of the forearm (ulna) that projects into the cavity of the arm—this does rarely occur—extension in the line of the limb is no longer equally suitable for these dislocations, since in this extension the projecting part of the forearm prevents

129 *μένον τι* MV: *μονον* B
130 *ἐξέσχεν* B: *ἔχει* MV
131 *πρ.τ. π.* B: om. MV
132 Add. *οὖν* MV

ὑπερέχον ὀστέον τὴν ὑπέρβασιν τοῦ βραχίονος. χρὴ τοίνυν τοῖσιν οὕτως ἐκβεβληκόσι τὴν κατάτασιν ποιέεσθαι τοιαύτην, οἵη περ πρόσθεν γέγραπται, ἐπήν τις ὀστέα βραχίονος κατεηγότα ἐπιδέῃ· ἀπὸ μὲν τῆς μασχάλης ἐς τὸ ἄνω τείνεσθαι, ἀπὸ δὲ τοῦ ἀγκῶνος αὐτοῦ ἐς τὸ κάτω ἀναγκάζειν· οὕτω γὰρ ἂν μάλιστα ὁ βραχίων ὑπεραιωρηθείη ὑπὲρ τῆς ἑωυτοῦ βαθμίδος· ἢν δὲ ὑπεραιωρηθῇ, ῥηϊδίη ἡ κατάστασις,[133] τοῖσι θέναρσι τῶν χειρῶν· τὸ μὲν ἐς τὸ[134] ἐξεστεὸς τοῦ βραχίονος ἐμβάλλοντα ὠθέειν, τὸ δὲ ἐς τὸ τοῦ πήχεος ὀστέον τὸ παρὰ τὸ ἄρθρον ἐμβάλλοντα ἀντωθέ-
550 ειν, τὸν | αὐτὸν τρόπον ἄμφω. ἧσσον μέντοι ἡ τοιαύτη κατάτασις τοῦ τοιούτου ὀλισθήματος δικαιοτάτη· ἐμβληθείη δ᾽ ἂν καὶ ἀπὸ τῆς ἐς ἰθὺ κατατάσιος, ἧσσον δὲ ἢ οὕτως.

42. Ἢν δὲ ἐς τοὔμπροσθεν ὀλίσθῃ ὁ βραχίων, ἐλαχιστάκις μὲν τοῦτο γίνεται—ἀλλὰ τί ἂν ἐξαπίνης ἐκπάλησις οὐκ ἐκβάλλοι; πολλὰ γὰρ καὶ παρὰ τὴν ἐοικυῖαν φύσιν ἐκπίπτει, καὶ ἢν μέγα τι ᾖ τὸ κωλῦον.
552 ταύτῃ δὲ τῇ ἐκπαλήσει μέγα τι τὸ | ὑπερβαινόμενον τὸ ὑπὲρ τὸ παχύτερον τῶν ὀστέων, καὶ τῶν νεύρων συχνὴ κατάτασις· ὅμως δὲ δή τισιν ἐξεπάλησεν. σημεῖον δὲ τοῖς οὕτως ἐκπαλήσασιν· οὐδὲν γὰρ χρῆμα τοῦ ἀγκῶνος κάμψαι δύνανται. ἔνδηλον δὲ καὶ τὸ ἄρθρον ψαυόμενον. ἢν μὲν οὖν μὴ αὐτίκα ἐμβληθῇ, ἰσχυραὶ καὶ βίαιαι φλεγμοναὶ καὶ πυρετώδεες γίνονται· ἢν δὲ δὴ αὐτίκα τις παρατύχῃ, εὐέμβολον. ὀθόνιον γὰρ σκληρὸν εἱλιγμένον ἀρκέει μὴ μέγα ἐν-

the arm (sc. bone) from passing over it. In patients with such dislocations, extension should be made in the manner described above for bandaging a fractured bone of the upper arm: provide traction upward from the axilla, and exert force downward from the elbow itself. This is the most likely way to get the arm lifted above its own socket, and if it is raised this way, replacement by the palms of hands is easy: exert pressure with one hand to reduce the projecting part of the arm, and exert counterpressure on the bone of the forearm (ulna) at the joint to reduce it, using the same method for both. This method of extension has, admittedly, less claim to be called the regular treatment for such a dislocation, since reduction could also be in a straight line, but that less easily.

42. If the arm dislocates forward, this happens very rarely—but what (sc. bone) might a sudden shock not dislocate? Indeed, many other bones are displaced from their natural position, even though the opposing obstacle is great. Now, there is a great obstacle to this violent dislocation, namely passage over the thicker bone (radius), and an extensive stretching of the cords, but nevertheless in some cases it does happen. Signs indicating the dislocation: patients cannot bend the elbow at all, and besides palpation of the joint makes the situation clear. If, then, it is not reduced at once, violent and severe inflammation supervenes together with fever: but if a competent person happens to be present, the dislocation is easily reduced. Take a hard-rolled bandage of no great size, place this

---

133 *κατάστασις* Foes (n. 32): *κατάτασις* BMV
134 *ἐς τὸ* B: om. MV

θέντα πλάγιον ἐς τὴν καμπὴν τοῦ ἀγκῶνος, ἐξαπίνης ξυγκάμψαι τὸν ἀγκῶνα καὶ προσαγαγεῖν ὡς μάλιστα τὴν χεῖρα πρὸς τὸν ὦμον. ἱκανὴ μὲν αὕτη ἡ ἐμβολὴ τῷ τοιούτῳ.[135] ἀτὰρ καὶ ἡ ἐς τὸ ἰθὺ κατάτασις δύναται εὐθετίζειν τοῦτον τὸν τρόπον τῆς ἐμβολῆς· τοῖσι μέντοι θέναρσι τῆς χειρὸς χρὴ τὸν μὲν ἐμβάλλοντα ἐς τὸ τοῦ βραχίονος ἐξέχον τὸ παρὰ τὴν καμπὴν ὀπίσω ἀπωθέειν, τὸν δέ τινα κάτωθεν ἐς τὸ τοῦ ἀγκῶνος
554 ὀξὺ[136] ἀντωθέειν ἐς τὴν ἰθυωρίην τοῦ πήχεος | ῥέποντα. δύναται δὲ ἐν τούτῳ τῷ τρόπῳ τῆς ὀλισθήσιος κἀκείνη ἡ κατάτασις ἡ πρόσθεν γεγραμμένη, ὡς χρὴ κατατείνειν τὰ ὀστέα τοῦ βραχίονος κατεηγότα, ἐπὴν μέλλωσιν ἐπιδεῖσθαι. ἐπὴν δὲ καταταθῇ, οὕτω χρὴ τοῖσι θέναρσι τὰς προσβολὰς ποιέεσθαι, ὥσπερ καὶ πρόσθεν γέγραπται.

43. Ἢν δὲ ἐς τοὐπίσω βραχίων ἐκπέσῃ—ὀλιγάκις δὲ τοῦτο γίνεται, ἐπωδυνώτατόν τε τοῦτο πάντων καὶ πυρετωδέστατον· συνεχέων πυρετῶν καὶ ἀκρητοχόλων, θανατωδέων καὶ ὀλιγημέρων—οἱ τοιοῦτοι ἐκτανύειν οὐ δύνανται. ἢν μὲν οὖν αὐτίκα παρατύχῃς, βιάζεσθαι χρὴ ἐκτανύσαντα τὸν ἀγκῶνα, καὶ αὐτομάτως ἐμπίπτει. ἢν δέ σε φθάσῃ πυρετήνας, οὐκ ἔτι χρὴ ἐμβάλλειν· κατακτείνειε γὰρ ἂν ἡ ὀδύνη ἀναγκαζομένου. ὡς δ᾽ ἐν κεφαλαίῳ εἰρῆσθαι, οὐδὲ ἄλλο χρὴ ἄρθρον πυρεταίνοντι ἐμβάλλειν, ἥκιστα δὲ ἀγκῶνα.

44. Ἔστι δὲ καὶ ἄλλα σίνεα κατ᾽ ἀγκῶνα ὀχλώδεα. τοῦτο μὲν γάρ, τὸ παχύτερον ὀστέον ἔστιν ὅτε ἐκινήθη ἀπὸ τοῦ ἑτέρου, καὶ οὔτε συγκάμπτειν οὔτ᾽

crosswise in the bend of the elbow, suddenly flex the elbow, and bring the hand as close as possible to the shoulder. This mode of reduction is sufficient for such dislocations. Direct extension, too, can accomplish the same reduction. You must, however, use the palms of the hands, putting one on the projecting part of the arm at the elbow and pushing backward (our inward), and with the other making counterpressure downward on the point of the elbow, inclining it into line with the forearm. In this form of dislocation, the mode of extension described above as necessary for stretching a fractured arm, when it is going to be bandaged, is also effective. But when extension is made, application of the palms should be done as described above.

43. If the arm is dislocated backward (our "inward"), this occurs rarely and is the most painful and febrile of all: there are continuous fevers and the vomiting of pure bile, and the outcome is fatal in a few days. Such patients cannot extend the arm: if you happen to be quickly on the spot, you must extend the elbow forcibly, and it will go back into place of its own accord. But if the person is already feverish when you arrive, do not perform any reduction, since the pain of such a violent operation will cause death. In short, it is a general rule not to reduce any joint in a patient with fever, least of all the elbow.

44. There are also other troublesome injuries of the elbow. For example, the thicker bone (radius) is sometimes separated from the other one (ulna), and such pa-

---

135 *τῷ τοιούτῳ* B: *τοῖσι(ν) οὕτω ἐκπαλ(λ)ήσασι(ν)* MV
136 Add. *ἐμβάλλοντα* MV

ἐκτανύειν[137] ὁμοίως δύνανται. δῆλον δὲ γίνεται ψαυόμενον κατὰ τὴν σύγκαμψιν τοῦ ἀγκῶνος παρὰ τὴν διασχίδα τῆς φλεβὸς τὴν ἄνωθεν τοῦ μυὸς τείνουσαν. οἷσι δὲ τὸ τοιοῦτον, οὐκ ἔτι ῥηΐδιον ἐς τὴν ἑωυτοῦ
556 φύσιν ἀγαγεῖν· οὐδὲ γὰρ | ἄλλην οὐδεμίαν ῥηΐδιον συμφυάδα κοινὴν δύο ὀστέων κινηθεῖσαν ἐς τὴν ἀρχαίην φύσιν ἱδρυθῆναι, ἀλλ' ἀνάγκη ὄγκον ἴσχειν τὴν διάστασιν. ὡς δὲ ἐπιδεῖν χρὴ ἐν ἄρθρῳ, ἐν τῇ κατὰ σφυρὸν ἐπιδέσει εἴρηται.

45. Ἔστι δ' οἷσι ἀπάγνυται[138] τοῦ πήχεος τὸ ὀστέον τὸ ὑποτεταγμένον τῷ βραχίονι—ὁτὲ μὲν τὸ χονδρῶδες αὐτοῦ, ἀφ' οὗ πέφυκεν ὁ τένων ὁ ὄπισθεν τοῦ βραχίονος <ὁτὲ δὲ τὰ πρόσω κατὰ τὴν ἀρχὴν τῆς ἐκφύσιος τοῦ προσθίου κορωνοῦ>[139]—καί ἐπὴν τοῦτο κινηθῇ, πυρετῶδες καὶ κακόηθες γίνεται· τὸ μέντοι ἄρθρον μένει ἐν τῇ ἑωυτοῦ χώρῃ· πᾶσα[140] γὰρ ἡ βάσις αὐτοῦ ταύτῃ ὑπερέχει.[141] ἢν[142] δὲ ἀπαγῇ ταύτῃ ᾗ ὑπερέχει ἡ κεφαλὴ τοῦ βραχίονος, πλανωδέστερον τὸ ἄρθρον γίνεται, ἢν παντάπασιν ἀποκαυλισθῇ. ἀσινέστερα δέ, ὡς ἐν κεφαλαίῳ εἰρῆσθαι, πάντα τὰ καταγνύμενα τῶν ὀστέων ἐστὶν ἢ οἷσι τὰ μὲν ὀστέα οὐ κατάγνυται, φλέβες δὲ καὶ νεῦρα ἐπίκαιρα ἀμφιφλᾶται ἐν τούτοισι τοῖσι χωρίοισιν. ἐγγυτέρω γὰρ θανάτῳ πελάζει ταῦτα ἢ ἐκεῖνα, ἢν ἐκπυρωθῇ ξυνεχεῖ πυρέτῳ. ὀλίγα γε μὴν τὰ τοιαῦτα κατήγματα γίνεται.

137 οὔτ' ἐκτανύειν B: οὔτε κατατανύειν MV
138 ἀπάγνυται B: κατήγν. MV

tients can neither flex nor extend the joint as before. The lesion is made clear by palpation at the bend of the elbow around the bifurcation of the blood vessel which passes upward along the muscle (biceps). In such cases it is not easy to restore the natural state; nor can any other shared symphysis of two bones once displaced be easily restored to its original configuration, but the diastasis (separation) necessarily remains as a swelling. How the joint ought to be bandaged was described in the case of the ankle.

45. There are cases in which the bone of the forearm (ulna) is fractured where it is subjacent to the arm—sometimes the cartilaginous part of it from which the tendon at the back of the arm arises, ‹and sometimes the part in front at the origin of the anterior coronoid process›—and when this separation occurs, it is complicated by fever and becomes dangerous. The joint (articular end of humerus), however, remains in its place, since its entire base protrudes at that point. But when the fracture is in the place on which the articular head of the arm rests, the joint becomes more mobile if it is a complete cabbage-stalk fracture (i.e., straight across). Speaking generally, fractures are always less troublesome than cases where no bones are broken but blood vessels and important cords in the regions have been contused. The latter lesions, in fact, involve greater risk of death than do the former, if continuous fevers supervene. Still, fractures of this kind rarely occur.

---

139 *ὅτε δὲ . . . κορωνοῦ* add. Littré Petrequin Kühlewein Withington from Oribasius' account (CMG VI 2,1, p. 216)

140 *πᾶσα* B: *ἴσως* MV   141 *ὑπέχει* B: *ὑπερέχει* B²MV

142 *ἣν* B: *ὅταν* MV

46. Ἔστι δὲ ὅτε αὐτὴ ἡ κεφαλὴ τοῦ βραχίονος
558 κατὰ τὴν ἐπίφυσιν | ἀπάγνυται· τοῦτο δὲ δοκέον κακοσινώτατον εἶναι πολλῷ τῳ εὐηθέστερον τῶν κατ' ἀγκῶνα σινέων ἐστίν.

47. Ὡς μὲν οὖν ἕκαστα τῶν ὀλισθημάτων ἁρμόσει[143] μάλιστα ἰητρεύειν, γέγραπται, καὶ ὅτι παραχρῆμα ἐμβάλλειν μάλιστα ἄρθρον συμφέρει διὰ τὸ τάχος τῆς φλεγμονῆς τῶν νεύρων. καὶ γὰρ ἢν ἐκπεσόντα αὐτίκα ἐμπέσῃ, ὅμως φιλέει τὰ νεῦρα σύντασιν ποιέεσθαι, καὶ κωλύειν ἐπὶ ποσὸν χρόνον τήν τε ἔκτασιν, ὅσην περ πέφυκε[144] ποιέεσθαι, τήν τε σύγκαμψιν. ἰητρεύειν δὲ πάντα παραπλησίως τὰ τοιαῦτα συμφέρει καὶ ὅσα ἀπάγνυται, καὶ ὅσα διίσταται, καὶ ὅσα ὀλισθάνει· πάντα γὰρ χρὴ ὀθονίοισι πολλοῖσι καὶ σπλήνεσι καὶ κηρωτῇ ἰητρεύειν, ὥσπερ καὶ τὰ ἄλλα κατήγματα. τὸ δὲ σχῆμα τοῦ ἀγκῶνος ἐν τούτοισι δὴ καὶ παντάπασι χρὴ[145] τοιοῦτον ποιέεσθαι, οἷόν περ οἷσι βραχίων ἐπεδεῖτο καταγείς, καὶ πῆχυς· κοινότατον μὲν γὰρ πᾶσι τοῖσιν ὀλισθήμασι καὶ τοῖσι κινήμασι καὶ τοῖσι κατήγμασι τοῦτο τὸ σχῆμά ἐστι· κοινότατον δὲ πρὸς τὴν ἔπειτα διάστασιν, καὶ τὸ ἐκτανύειν ἕκαστα καὶ ξυγκάμπτειν· ἐντεῦθεν γὰρ ὁδοὶ ἐς ἀμφότερα παραπλήσιοι. εὐοχώτατον δὲ καὶ εὐαναληπτότατον[146] αὐτῷ τῷ κάμνοντι τοῦτο τὸ σχῆμα. ἔτι δὲ πρὸς τούτοισι, εἰ ἄρα κρατηθείη ὑπὸ τοῦ πωρώματος, εἰ μὲν ἐκτεταμένη ἡ χεὶρ κρατηθείη, κρέσσων ἂν εἴη μὴ προσεοῦσα· πολλῶν μὲν γὰρ ἂν κώλυμα εἴη,
560 ὠφελίη δὲ ὀλίγων,[147] εἰ δ' αὖ | ξυγκεκαμμένη, μᾶλλον

46. Sometimes the actual head of the arm is fractured at the epiphysis (medial epicondyle of the humerus), but this, although apparently a very grave lesion, is much less serious than injuries of the elbow joint.

47. How, then, each dislocation is most appropriately treated has been described; especially the value of immediate reduction owing to the rapid inflammation of the cords. For, even when dislocated parts are reduced at once, the cords are apt to become contracted and to hinder the natural amount of extension and flexion for a considerable time. All such injuries, whether avulsions, separations, or dislocations, require similar treatment, and should be treated with plenty of bandages, compresses, and cerate, like other fractures. The position of the elbow should in these cases, too, be the same in all respects as in the bandaging of patients with a fractured arm or forearm; for this position is the one most generally used for all the luxations, displacements, and fractures; it is also most favorable as regards the limb's future position, in respect of extension and flexion in the various cases, since from it the path in both directions is similar. This position is also most easily kept up or returned to by the patient himself. And besides this, if the joint comes to be dominated by callus, an arm ankylosed in the extended position would be better removed, since it would be a hindrance in many cases and of use in few. If flexed, on the other hand, it would be more

143 Add. ἐμβάλλειν MV 144 πέφυκε B: φιλέει MV
145 χρὴ B: δεῖ M: om. V
146 -ληπτότατον B: -ληπτον MV
147 πολλῶν . . . ὀλίγων B: πολλῷ . . . ὀλίγον MV

εὔχρηστος ἂν εἴη· πολλῷ δὲ εὐχρηστοτέρη, εἰ τὸ διὰ μέσον σχῆμα ἔχουσα πωρωθείη [κρέσσον].[148] τὰ μὲν περὶ τοῦ σχήματος τοιαῦτα.

48. Ἐπιδεῖν δὲ χρὴ τήν τε ἀρχὴν τοῦ πρώτου ὀθονίου βαλλόμενον κατὰ τὸ βλαφθέν, ἤν τε καταγῇ, ἤν τε ἐκστῇ, ἤν τε διαστῇ· καὶ τὰς περιβολὰς τὰς πρώτας κατὰ τοῦτο ποιέεσθαι, καὶ ἐρηρείσθω μάλιστα ταύτῃ, ἔνθεν δὲ καὶ ἔνθεν ἐπὶ ἧσσον. τὴν δὲ ἐπίδεσιν κοινὴν ποιέεσθαι χρὴ τοῦ τε πήχεος καὶ τοῦ βραχίονος, καὶ ἐπὶ πολὺ πλέον ἑκάτερον ἢ ὡς οἱ πλεῖστοι ποιέουσιν, ὅπως †ἐξαριαται† ὡς μάλιστα ἀπὸ τοῦ σίνεος τὸ οἴδημα ἔνθεν καὶ ἔνθεν. προσπεριβαλλέσθω δὲ καὶ τὸ ὀξὺ τοῦ πήχεος, ἢν τὸ σίνος κατὰ τοῦτο ᾖ, ἤν τε μή, ἵνα μὴ τὸ οἴδημα ἐνταῦθα περὶ αὐτὸ ξυλλέγηται. περιφεύγειν δὲ χρὴ ἐν τῇ ἐπίδεσει, ὅπως μὴ κατὰ τὴν καμπὴν πολλὸν τοῦ ὀθονίου ἠθροισμένον ἔσται ἐκ τῶν δυνατῶν. πεπιέχθαι δὲ κατὰ τὸ σίνος ὡς μάλιστα· καὶ τὰ ἄλλα καταλαβέτω αὐτὸν περὶ τῆς πιέξιος καὶ τῆς χαλάσιος ταὐτά, καὶ κατὰ τοὺς αὐτοὺς χρόνους ἕκαστα, ὥσπερ τῶν ὀστέων τῶν κατεηγότων ἐν τῇ ἰητρείῃ πρόσθε γέγραπται. καὶ αἱ μετεπιδέσιες διὰ τρίτης ἔστωσαν· χαλᾶν δὲ δοκείτω τῇ τρίτῃ, ὥσπερ καὶ τότε. καὶ νάρθηκας προσπερι-
562 βάλλειν | ἐν τῷ ἱκνεομένῳ χρόνῳ—οὐδὲν γὰρ ἀπὸ τρόπου, καὶ τοῖσι τὰ ὀστέα κατεηγόσι, καὶ τοῖσι μή, ἢν μὴ καὶ[149] πυρεταίνῃ—ὡς χαλαρωτάτους δέ, τοὺς μὲν ἀπὸ βραχίονος κατατεταμένους, τοὺς δὲ ἀπὸ τοῦ πήχεος ἀνειμένους. ἔστωσαν δὲ μὴ παχέες οἱ νάρθη-

useful, and even more useful if the ankylosis occurred in an attitude of semiflexion. So much concerning the position (of the elbow).

48. You should bandage by applying the start of the first roll to the place injured, whether it is fractured, dislocated, or separated. The first turns should be made there and the pressure should be firmest, slackening off toward each side. The bandaging should include both the fore- and upper arm, and be carried much further each way than most practitioners normally do, so that the edema will be held off as far as possible from the lesion in both directions. Include the point of the elbow, too, in the bandage, whether the lesion is there or not, so that edema will not collect around this part. You must take special care in the dressing in order that, so far as possible, there will be no great accumulation of bandage in the bend of the elbow, and that the firmest pressure will be made at the lesion. For the rest, the surgeon must deal with the case, as regards pressure and slackness, in the same way and at the same respective periods as were previously described in the treatment of fractured bones. Change the dressings every other day, and the patient should feel them become relaxed on the third day, as in the former case. Apply splints at the proper time—for their use is not unsuitable whether there is a fracture or not, as long as there is no fever—but they should be applied as loosely as possible, those of the arm being underneath and those of the forearm on the top. The splints should not be thick, and must

---

148 *κρέσσων* BM: *κρέσσον* V: del. Bosquillon (Littré)
149 *καὶ* B: om. MV

κες· ἀναγκαῖον δὲ καὶ ἀνίσους αὐτοὺς εἶναι ἀλλήλοισι, παραλλάσσειν δὲ παρ' ἀλλήλους ᾗ ἂν ξυμφέρῃ, τεκμαιρόμενον πρὸς τὴν ξύγκαμψιν. ἀτὰρ καὶ τῶν σπληνῶν τὴν πρόσθεσιν τοιαύτην χρὴ ποιέεσθαι, ὥσπερ καὶ τῶν ναρθήκων εἴρηται, ὀγκηροτέρους δὲ ὀλίγῳ κατὰ τὸ σίνος προστίθεσθαι.[150]

[150] B: add. τοὺς δὲ χρόνους τοὺς ἀπὸ τῆς φλεγμονῆς τεκμαίρεσθαι χρὴ καὶ ἀπὸ τῶν πρόσθεν γεγραμμένων. MV

be unequal in length in order to overlap one another where it is convenient, judging this by the degree of flexion. Regarding the application of compresses, you should follow the directions for splints. They should be somewhat thicker at the site of the lesion.

ΠΕΡΙ ΑΡΘΡΩΝ[1]

IV 78 Littré 1. Ὤμου δὲ ἄρθρον ἕνα τρόπον οἶδα ὀλίσθανον, τὸν ἐς τὴν μασχάλην· ἄνω δὲ οὐδέπω εἶδον, οὐδὲ ἐς τὸ ἔξω· οὐ μέντοι διισχυριείω γε εἰ ὀλίσθοι ἂν ἢ οὔ, καίπερ ἔχων περὶ αὐτοῦ ὅ τι λέγω. ἀτὰρ οὐδὲ ἐς τοὔμπροσθεν οὐδέπω ὄπωπα, ὅ τι ἔδοξέν μοι ὠλισθηκέναι. τοῖσι μέντοι ἰητροῖσι δοκέει κάρτα ἐς τοὔμπροσθεν ὀλισθάνειν, καὶ μάλιστα ἐξαπατῶνται ἐν τούτοις, ὧν ἂν φθίσις καταλάβῃ τὰς σάρκας τὰς περὶ τὸ ἄρθρον τε καὶ τὸν βραχίονα· φαίνεται γὰρ τοῖσι τοιούτοισι παντάπασιν ἡ κεφαλὴ τοῦ βραχίονος ἐξέχουσα ἐς τοὔμπροσθεν. καὶ ἔγωγέ[2] ποτε τὸ τοιοῦτον οὐ φὰς ἐκπεπτωκέναι ἤκουσα φλαύρως ὑπό τε τῶν ἰητρῶν, ὑπό τε τῶν δημοτέων διὰ τοῦτο τὸ πρῆγμα· ἐδόκεον γὰρ αὐτοῖσιν ἠγνοηκέναι μοῦνος, οἱ δὲ ἄλλοι ἐγνωκέναι,
80 καὶ οὐκ ἠδυνάμην αὐτοὺς ἀναγνῶσαι, | εἰ μὴ μόλις, ὅτι τόδ᾽ ἐστὶ τοιόνδε· εἴ τις τοῦ βραχίονος ψιλώσειε μὲν τῶν σαρκῶν τὴν ἐπωμίδα, ψιλώσειε δὲ ᾗ ὁ μῦς ἀνατείνει, ψιλώσειε δὲ τὸν τένοντα τὸν κατὰ τὴν μασχάλην καὶ τὴν κληῖδα πρὸς τὸ στῆθος ἔχοντα, φαί-

[1] ΕΜΒΟΛΗΣ add. BV [2] καὶ ἔγωγε B: ἐγὼ δέ MV

JOINTS[1]

1. Of the shoulder joint, I know only one kind of dislocation, the one into the axilla. I have never observed a dislocation upward, nor to the outside. Still, I do not mean to assert positively whether or not such dislocations occur, although I do have something to say on the subject. Nor have I ever seen anything that seemed to me to be a dislocation forward. Practitioners do in fact believe that there is dislocation forward, being especially deceived in cases where there is wasting of the flesh in the area of the joint and arm; for in all such instances the head of the arm has an obvious projection forward. Indeed, I myself, on denying that there was a dislocation in one such instance, came to be disparaged by both practitioners and laymen on this account. They thought that I was the only person ignorant of what the others knew, and I was unable, or practically so, to make them see that the matter stood as follows: suppose someone were to lay the arm at the point of the shoulder bare of its fleshy parts, and also denude it where the muscle is attached, and expose the tendon running along the axilla and collarbone toward the chest. Then the

[1] This is the title in the manuscript M: B and V add "Reduction of the."

νοιτο ἂν ἡ κεφαλὴ τοῦ βραχίονος ἐς τοὔμπροσθεν ἐξέχουσα ἰσχυρῶς, καίπερ οὐκ ἐκπεπτωκυῖα· πέφυκε γὰρ ἐς τοὔμπροσθεν προπετὴς ἡ κεφαλὴ τοῦ βραχίονος· τὸ δὲ ἄλλο ὀστέον τοῦ βραχίονος ἐς τὸ ἔξω καμπύλον. ὁμιλέει δὲ ὁ βραχίων τῷ κοίλῳ[3] τῆς ὠμοπλάτης πλάγιος, ὅταν παρὰ τὰς πλευρὰς παρατεταμένος ἔῃ. ὅταν μέντοι ἐς τοὔμπροσθεν ἐκτανυσθῇ ἡ σύμπασα χείρ, τότε ἡ κεφαλὴ τοῦ βραχίονος κατὰ τὴν ἴξιν τῆς ὠμοπλάτης τῷ κοίλῳ[4] γίνεται, καὶ οὐκ ἔτι ἐξέχειν ἐς τοὔμπροσθεν φαίνεται. περὶ οὗ οὖν ὁ λόγος, οὐδέπω εἶδον οὐδὲ ἐς τοὔμροσθεν ἐκπεσόν· οὐ μὴν ἰσχυριείω γε οὐδὲ περὶ τούτου, εἰ μὴ ἐκπέσοι ἂν οὕτως ἢ οὔ. ὅταν οὖν ἐκπέσῃ ὁ βραχίων ἐς τὴν μασχάλην, ἅτε πολλοῖς ἐκπίπτοντος, πολλοὶ ἐπίστανται ἐμβάλλειν· εὐπαίδευτον δέ ἐστι τὸ εἰδέναι πάντας τοὺς τρόπους, ὅσοισιν οἱ ἰητροὶ ἐμβάλλουσι, καὶ ὡς ἄν τις αὐτοῖσι τοῖσι τρόποισι τούτοισι κάλλιστα χρῷτο· χρῆσθαι δὲ χρὴ τῷ κρατίστῳ τῶν τρόπων, ἢν τὴν ἰσχυροτάτην ἀνάγκην ὁρᾷς· κράτιστος δὲ ὁ ὕστατος γεγραψόμενος.

2. Οἷσι μὲν οὖν πυκνὰ ἐκπίπτει ὦμος, ἱκανοὶ ὡς ἐπὶ
82 τὸ | πολὺ[5] αὐτοὶ σφίσιν αὐτοῖσιν ἐμβάλλειν εἰσίν· ἐνθέντες γὰρ τῆς ἑτέρης χειρὸς τοὺς κονδύλους ἐς τὴν μασχάλην, ἀναγκάζουσιν ἄνω τὸ ἄρθρον, τὸν δὲ ἀγκῶνα παράγουσι παρὰ τὸ στῆθος. τὸν αὐτὸν δὲ τρόπον τοῦτον καὶ ἰητρὸς ἂν ἐμβάλλοι, εἰ αὐτὸς μὲν ὑπὸ τὴν μασχάλην ἐσωτέρω τοῦ ἄρθρου τοῦ ἐκπεπτωκότος ὑποτείνας τοὺς δακτύλους ἀπαναγκάζοι[6] ἀπὸ

head of the arm would be revealed as having a strongly marked projection to the front although the joint was not dislocated, for the head of the arm inclines forward in its very structure, while the rest of the bone is curved toward the outside. And the arm meets the (glenoid) cavity of the shoulder blade obliquely when it is extended along the ribs, but when the whole arm is raised toward the front, the head of the arm comes into line with the cavity of the shoulder blade, and no longer appears to project forward. To return to my subject, I have never seen a dislocation forward, although I would hesitate to make a rule about the possibility of this dislocation either. When, then, the arm is displaced into the axilla, since this is a common accident, many practitioners know how to reduce it, for it is easy to teach all the methods by which practitioners effect reduction, and the best way for someone to use these methods. You should use the most effective of the methods where you see the most pressing need: the most effective method is the one that will be described last.

2. Those who have frequent dislocations of the shoulder are usually capable of reducing it for themselves; for by inserting the fist of the opposite hand into the axilla, they forcibly move the joint upward, while simultaneously drawing the elbow toward their chest. A practitioner would also be reducing it in the same way if, placing his fingers under the axilla on the inner side of the dislocated

---

3 τῷ κοίλῳ M: τῷ κῦλω B: τὸ κοῖλον V

4 τῷ κοίλῳ Galen: τοῦ κύλου B: τῷ κύλῷ B²: τοῦ κοίλου MV 5 πολυ B: πλεῖστον MV

6 ἀπαναγκάζοι BM: ἀναγκάζοι V

τῶν πλευρέων ἐμβαλὼν τὴν ἑωυτοῦ κεφαλὴν ἐς τὸ ἀκρώμιον ἀντερείσιος εἵνεκα, τοῖσι δὲ γούνασι παρὰ τὸν ἀγκῶνα ἐς τὸν βραχίονα ἐμβαλών, ἀντωθέοι πρὸς τὰς πλευράς—συμφέρει δὲ καρτερὰς τὰς χεῖρας ἔχειν τὸν ἐμβάλλοντα—ἢ εἰ αὐτὸς μὲν τῇσι χερσὶ καὶ τῇ κεφαλῇ οὕτω ποιοίη, ἕτερος[7] δέ τις τὸν ἀγκῶνα παράγοι παρὰ τὸ στῆθος.

Ἔστι δὲ ἐμβολὴ ὤμου καὶ ἐς τοὐπίσω ὑπερβάλλοντα τὸν πῆχυν ἐπὶ τὴν ῥάχιν, ἔπειτα τῇ μὲν ἑτέρῃ χειρὶ ἀνακλᾶν ἐς τὸ ἄνω τοῦ ἀγκῶνος ἐχόμενον, τῇ δὲ ἑτέρῃ παρὰ τὸ ἄρθρον ὄπισθεν ἐρείδειν. αὕτη ἡ ἐμβολὴ καὶ ἡ πρόσθεν εἰρημένη, οὐ κατὰ φύσιν ἐοῦσαι, ὅμως ἀμφισφάλλουσαι τὸ ἄρθρον ἀναγκάζουσιν ἐμπίπτειν.

3. Οἱ δὲ τῇ πτέρνῃ πειρώμενοι ἐμβάλλειν, ἐγγύς τι τοῦ κατὰ φύσιν ἀναγκάζουσιν· χρὴ δὲ τὸν μὲν ἄνθρωπον χαμαὶ κατακλῖναι ὕπτιον, τὸν δὲ ἐμβάλλοντα χαμαὶ ἵζεσθαι ἐφ᾽ ὁπότερα ἂν τὸ ἄρθρον ἐκπεπτώκῃ. ἔπειτα λαβόμενον τῇσι χερσὶ τῇσιν ἑωυτοῦ τῆς χειρὸς τῆς σιναρῆς, κατατείνειν αὐτήν, τὴν δὲ πτέρνην
84 ἐς τὴν μασχάλην | ἐμβάλλοντα ἀντωθέειν, τῇ μὲν δεξιῇ ἐς τὴν δεξιήν, τῇ δὲ ἀριστερῇ ἐς τὴν ἀριστερήν. δεῖ δὲ ἐς τὸ κοῖλον τῆς μασχάλης ἐνθεῖναι στρογγύλον τι ἐνάρμοττον· ἐπιτηδειόταται δὲ αἱ πάνυ σμικραὶ σφαῖραι καὶ σκληραί, οἷαι ἐκ τῶν πολλῶν σκυτέων ῥάπτονται· ἢν γὰρ μή τι τοιοῦτον ἐγκέηται,[8] οὐ δύναται ἡ πτέρνη ἱκνέεσθαι πρὸς τὴν κεφαλὴν τοῦ βραχίονος· κατατεινομένης γὰρ τῆς χειρὸς κοιλαίνε-

joint, he forced it away from the ribs, pressing his own head against the top of the shoulder to make counterpressure, and by applying his knees against the arm at the elbow he pressed it against the ribs—it is advantageous for the operator to have strong hands—or, while he used his hands and head in this way, an assistant drew the elbow toward the chest.

There is also a way of reducing the shoulder by bringing the forearm backward to the spine, then, taking hold of the arm at the elbow with one hand to turn it upward, and with the other hand to support it from behind at the joint. This method, like the one described before, although not in conformity with nature, nevertheless by rotating the joint forces it to fall into place.

3. Those who attempt to reduce the shoulder with their heel, operate in a way nearly conformable with nature. The patient should lie on his back on the ground, and the operator should sit on the ground on the side where the joint is dislocated. Then grasping the injured arm with both his hands, he should pull it, while at the same time exerting counterpressure by putting his heel in the axilla, using his right heel for the right axilla, and left for the left. Into the hollow of the axilla should be placed some rounded object fitted to it: most suitable are very small, hard balls sewed together out of many scraps of leather. For unless something of this kind is inserted, the heel cannot reach the head of the arm, for when the arm is

---

[7] ἕτερος B: ἄλλος MV
[8] ἐγκέηται Froben: ἐκκέηται BMV

ται ἡ μασχάλη· οἱ γὰρ τένοντες οἱ ἔνθεν καὶ ἔνθεν τῆς μασχάλης ἀντισφίγγοντες ἐναντίοι εἰσίν. χρὴ δέ τινα ἐπὶ θάτερα τοῦ κατατεινομένου καθήμενον κατέχειν κατὰ τὸν ὑγιέα ὦμον, ὡς μὴ περιέλκηται τὸ σῶμα, τῆς χειρὸς τῆς σιναρῆς ἐπὶ θάτερα τεινομένης· ἔπειτα ἱμάντος μαλθακοῦ πλάτος ἔχοντος ἱκανόν, ὅταν ἡ σφαῖρα ἐντεθῇ ἐς τὴν μασχάλην, περὶ τὴν σφαίραν περιβεβλημένου τοῦ ἱμάντος, καὶ κατέχοντος, λαβόμενον ἀμφοτέρων τῶν ἀρχέων τοῦ ἱμάντος, ἀντικατατείνειν τινά, ὑπὲρ κεφαλῆς τοῦ κατατεινομένου καθήμενον, τῷ ποδὶ προσβάντα πρὸς τοῦ ἀκρωμίου τὸ ὀστέον. ἡ δὲ σφαῖρα ὡς ἐσωτάτω καὶ ὡς μάλιστα πρὸς τῶν πλευρέων κείσθω, καὶ μὴ ὑπὸ[9] τῇ κεφαλῇ τοῦ βραχίονος.

4. Ἔστι δὲ καὶ ἄλλη ἐμβολή, ὡς κατωμίζουσιν ἐς ὀρθόν· μείζω μέντοι χρὴ εἶναι τὸν κατωμίζοντα, διαλαβόντα δὲ τὴν χεῖρα ὑποθεῖναι τὸν ὦμον τὸν ἑωυτοῦ
86 ὑπὸ τὴν μασχάλην ὀξύν· κἄπειτα | ἐπιστρέψαι ὡς ἂν ἐνίζηται ἕδρῃ, οὕτω στοχασάμενον ὅπως ἀμφὶ τὸν ὦμον τὸν ἑωυτοῦ κρεμάσαι τὸν ἄνθρωπον κατὰ τὴν μασχάλην· αὐτὸς δὲ ἑωυτὸν ὑψηλότερον ἐπὶ τοῦτον τὸν ὦμον ποιέετω ἢ ἐπὶ τὸν ἕτερον· τοῦ δὲ κρεμαμένου τὸν βραχίονα πρὸς τὸ ἑωυτοῦ στῆθος προσαναγκαζέτω ὡς μάλιστα· ἐν τούτῳ δὲ τῷ σχήματι προσανασειέτω, ὅταν μετεωρίσῃ τὸν ἄνθρωπον, ὡς ἀντιρρέπῃ τὸ ἄλλο σῶμα αὐτῷ, ἀντίον τοῦ βραχίονος τοῦ κατεχομένου· ἢν δὲ ἄγαν κοῦφος ᾖ ὤνθρωπος, καὶ προσεκκρεμασθήτω[10] ὄπισθεν κοῦφος παῖς. αὗται δὲ αἱ

stretched, the axilla becomes hollow and the tendons on either side of it, by their contraction, form an obstacle to the reduction. Someone else should be seated on the other side of the person undergoing extension to hold the sound shoulder where it is, so that the patient's body is not dragged along when the injured arm on the other side is pulled. Then, when the ball has been inserted into the axilla, a fairly broad strap of soft leather is put round it to fix it in place, and someone seated at the head of the patient undergoing traction makes counterextension by holding the ends of the strap, and pressing with his foot against the top of the bone at the shoulder. The ball should be put as far as possible into the axilla and as near to the ribs, and not under the head of the arm.

4. There is also another mode of reduction in which they apply force from below with their shoulder in a standing patient. The operator, who must be taller (sc. than the patient), after taking hold of the patient's arm, should insert the tip of his own shoulder under the patient's axilla, and then make a twist to get it seated there, the aim of the maneuver being to suspend the patient by his axilla from the operator's shoulder, which should be held higher than his other shoulder. He should force the arm of the suspended patient in as far as possible toward his own chest. In this attitude, let him proceed to shake the patient as he lifts him up, so that the rest of the body may act as a counterpoise to the arm that is being held down. If the patient is too light, a child of no great weight should be hung on to him from behind. All these methods are very useful in

[9] ὑπὸ B: ἐπὶ MV [10] τις add. MV

ἐμβολαὶ πᾶσαι κατὰ παλαίστρην εὔχρηστοί εἰσιν, ὅτι οὐδὲν ἀλλοίων ἀρμένων δέονται ἐπεσενεχθῆναι· χρήσαιτο δ᾽ ἄν τις καὶ ἄλλοθι.

5. Ἀτὰρ καὶ οἱ περὶ τὰ ὕπερα ἀναγκάζοντες ἐγγύς τι τοῦ κατὰ φύσιν ἐμβάλλουσι. χρὴ δὲ τὸ μὲν ὕπερον κατειλίχθαι ταινίῃ τινὶ μαλθακῇ—ἧσσον γὰρ ἂν ὑπολισθάνοι—ὑπηναγκάσθαι δὲ μεσηγὺ τῶν πλευρέων καὶ τῆς κεφαλῆς τοῦ βραχίονος· καὶ ἢν μὲν βραχὺ ᾖ τὸ ὕπερον, καθῆσθαι χρὴ τὸν ἄνθρωπον ἐπί
88 τινος ὡς μόλις | τὸν βραχίονα περιβάλλειν δύνηται περὶ τὸ ὕπερον· μάλιστα δὲ ἔστω μακρότερον τὸ ὑπεροειδές,[11] ὡς ἂν ὤνθρωπος ἑστεὼς μικροῦ δέῃ κρέμασθαι ἀμφὶ τῷ ξύλῳ. κἄπειτα ὁ μὲν βραχίων καὶ ὁ πῆχυς παρατεταμένος παρὰ τὸ ὕπερον ἔστω, τὸ δὲ ἐπὶ θάτερα τοῦ σώματος καταναγκαζέτω τις, περιβάλλων κατὰ τὸν αὐχένα παρὰ τὴν κληῖδα τὰς χεῖρας. αὕτη ἡ ἐμβολὴ κατὰ φύσιν ἐπιεικέως ἐστὶ καὶ ἐμβάλλειν δύναται, ἢν χρηστῶς σκευάσωνται αὐτήν.

6. Ἀτὰρ καὶ ἡ διὰ τοῦ κλιμακίου ἑτέρη τις τοιαύτη, καὶ ἔτι βελτίων, ὅτι ἀσφαλεστέρως ἂν τὸ σῶμα, τὸ μὲν τῇ, τὸ δὲ τῇ ἀντισηκωθείη μετεωρισθέν· περὶ γὰρ τὸ ὑπεροειδὲς ὁ ὦμος ἢν καὶ καταπεπήγῃ, περισφάλλεσθαι κίνδυνος τὸ σῶμα ἢ τῇ ἢ τῇ· χρὴ μέντοι καὶ ἐπὶ τῷ κλιμακτῆρι ἐπιδεδέσθαι τι ἄνωθεν στρογγύλον ἐνάρμοσσον ἐς τὸ κοῖλον τῆς μασχάλης, ὃ προσδιαναγκάσει τὴν κεφαλὴν τοῦ βραχίονος ἐς τὴν φύσιν ἀπιέναι.

7. Κρατίστη μέντοι πασέων τῶν ἐμβολέων ἡ τοιήδε·

the wrestling school, since they do not require any other kinds of apparatus to be brought in, and you can also use them elsewhere.

5. Then again, those who reduce (sc. this dislocation) by forcibly bending (sc. the arm) around a pestle come fairly near to the natural method. The pestle should have a soft band wrapped round it, making it less slippery, and be pressed in between the ribs and the head of the arm. If the pestle is short, the patient should be seated on something such that he can just get his arm over the pestle, but as a rule the wooden pestle should be rather long, so that the patient, when standing, is almost suspended on the post. Then let the arm and forearm be pulled down beside the pestle, while on the other side an assistant, by putting his arms around the patient's neck at the collarbone, forces the body down. This method is reasonably natural, and usually successful if people employ it as they should.

6. Then again, there is another similar method with the ladder, which is better still, since the body when lifted up is more safely kept in equilibrium on either side. For with the pestle, although the shoulder may be fixed, there is a danger of the body slipping around to one side or the other. Something rounded should also be fastened on to the upper side of the ladder step, which by fitting into the hollow of the axilla, helps to force the head of the arm back to its natural place.

7. But the most effective of all methods of reduction is

---

11 ὑπεροειδές B: ὕπερον MV

ξύλον χρὴ εἶναι πλάτος μὲν ὡς πενταδάκτυλον, ἢ τετραδάκτυλον τὸ ἐπίπαν, πάχος δὲ ὡς διδάκτυλον ἢ καὶ λεπτότερον, μῆκος δὲ δίπηχυ, ἢ ὀλίγῳ ἔλασσον. ἔστω δὲ ἐπὶ θάτερα τὸ ἄκρον περιφερὲς καὶ στενότατον ταύτῃ καὶ λεπτότατον· ἄμβην δὲ ἐχέτω σμικρὸν | ὑπερέχουσαν ἐπὶ τῷ ὑστάτῳ τοῦ περιφερέος, ἐπὶ τῷ μέρει μὴ τῷ πρὸς τὰς πλευράς, ἀλλὰ τῷ πρὸς τὴν κεφαλὴν τοῦ βραχίονος ἔχοντι, ὡς ὑπαρμόσειεν τῇ μασχάλῃ παρὰ τὰς πλευρὰς ὑπὸ τὴν κεφαλὴν τοῦ βραχίονος ὑποτιθέμενον. ὀθονίῳ δὲ ἢ ταινίῃ μαλθακῇ κατακεκολλήσθω ἄκρον τὸ ξύλον, ὅπως προσηνέστερον ᾖ. ἔπειτα χρὴ ὑπώσαντα τὴν κεφαλὴν τοῦ ξύλου ὑπὸ τὴν μασχάλην ὡς ἐσωτάτω μεσηγὺ τῶν πλευρέων καὶ τῆς κεφαλῆς τοῦ βραχίονος, τὴν δὲ ὅλην χεῖρα πρὸς τὸ ξύλον κατατείναντα προσκαταδῆσαι κατά τε τὸν βραχίονα, κατά τε τὸν πῆχυν, κατά τε τὸν καρπὸν τῆς χειρός, ὡς ἀτρεμῇ ὅτι μάλιστα. περὶ παντὸς δὲ χρὴ ποιέεσθαι, ὅπως τὸ ἄκρον τοῦ ξύλου ὡς ἐσωτάτω τῆς μασχάλης ἔσται, ὑπερβεβηκὸς τὴν κεφαλὴν τοῦ βραχίονος. ἔπειτα χρὴ μεσηγὺ δύο στύλων στρωτῆρα πλάγιον εὖ προσδῆσαι, ἔπειτα ὑπερενεγκεῖν τὴν χεῖρα σὺν τῷ ξύλῳ ὑπὲρ τοῦ στρωτῆρος, ὡς ἡ μὲν χεὶρ ἐπὶ θάτερα ᾖ, ἐπὶ θάτερα δὲ τὸ σῶμα, κατὰ δὲ τὴν μασχάλην ὁ στρωτήρ· ἔπειτα ἐπὶ μὲν θάτερα τὴν χεῖρα καταναγκάζειν σὺν τῷ ξύλῳ περὶ[12] τὸν στρωτῆρα, ἐπὶ θάτερα δὲ τὸ ἄλλο σῶμα. ὕψος δὲ ἔχων ὁ στρωτὴρ προσδεδέσθω, ὥστε μετέωρον τὸ ἄλλο σῶμα εἶναι ἐπ᾽ ἄκρων τῶν ποδῶν. οὗτος the following. Prepare a piece of wood about five inches wide, or at least four, about two inches thick or even thinner, and in length two cubits or a little less; it should be rounded at one end and be thinnest and narrowest there; at the extremity of the rounded end, there should be a slightly projecting rim (*ambé*), not on the side toward the ribs, but on the side facing the head of the arm, so that it fits into the axilla along the ribs and under the head of the arm when it is inserted; the end of the wood should have a piece of linen or a soft band glued over it so that it will be more comfortable. Then, after inserting the head of the instrument as far as possible under the axilla between the ribs and the head of the arm, and stretching the whole arm down along the wood, you should fasten it at the arm, the forearm, and the wrist, so as to make the arm as immobile as possible. Beyond all else, make sure to get the end of the instrument as far into the axilla as possible, up above the head of the arm. Then a crossbar should be firmly fastened between two posts, and next you should use the instrument to bring the arm over the crossbar, so that the arm is on the one side, the body is on the other side, and the crossbar is under the axilla. Then force the arm down on one side around the crossbar with the instrument, while the rest of the body (sc. hangs) on the other side. The crossbar should be fastened high up in order that the rest of the body will be raised on the tiptoes. This is by far

---

12 *περὶ* MV: *πρὸς* B

ὁ τρόπος παρὰ πολὺ κράτιστος ἐμβολῆς ὤμου· δικαιότατα μὲν γὰρ μοχλεύει, ἢν μοῦνον ἐσωτέρω ᾖ τὸ ξύλον τῆς κεφαλῆς τοῦ βραχίονος· δικαιόταται δὲ αἱ
92 ἀντιρροπαί, | ἀσφαλέες δὲ τῷ ὀστέῳ τοῦ βραχίονος. τὰ μὲν οὖν νεαρὰ ἐμπίπτει θᾶσσον ἢ ὡς ἄν τις οἴοιτο, πρὶν ἢ καὶ κατατετάσθαι δοκεῖν· ἀτὰρ καὶ τὰ παλαιὰ μούνη αὕτη τῶν ἐμβολέων οἵη τε ἐμβιβάσαι, ἢν μὴ ἤδη ὑπὸ χρόνου σὰρξ μὲν ἐπεληλύθῃ ἐπὶ τὴν κοτύλην, ἡ δὲ κεφαλὴ τοῦ βραχίονος ἤδη τρίβον ἑωυτῇ πεποιημένη ᾖ ἐν τῷ χωρίῳ, ἵνα ἐξεκλίθη· οὐ μὴν ἀλλ' ἐμβάλλειν γάρ μοι δοκέοι καὶ οὕτω πεπαλαιωμένον ἔκπτωμα τοῦ βραχίονος· τί γὰρ ἂν δικαίη μόχλευσις οὐχὶ κινήσειεν; μένειν μέντοι οὐκ ἄν μοι δοκέοι κατὰ χώρην, ἀλλ' ὀλισθάνοι ἂν ἐς[13] τὸ ἔθος.

Τὸ αὐτὸ δὲ ποιέει καὶ περὶ κλιμακτῆρα καταναγκάζειν τοῦτον τὸν τρόπον σκευάσαντα. πάνυ μὴν ἱκανῶς ἔχει καὶ περὶ ἕδος μέγα Θεσσαλικὸν ἀναγκάζειν, ἢν νεαρὸν ᾖ τὸ ὀλίσθημα. ἐσκευάσθαι μέντοι χρὴ τὸ ξύλον οὕτως, ὥσπερ εἴρηται· ἀτὰρ τὸν ἄνθρωπον καθίσαι πλάγιον ἐπὶ τῷ δίφρῳ· ἔπειτα τὸν βραχίονα σὺν τῷ ξύλῳ ὑπερβάλλειν ὑπὲρ τοῦ ἀνακλισμοῦ, καὶ ἐπὶ μὲν θάτερα τὸ σῶμα καταναγκάζειν, ἐπὶ θάτερα δὲ τὸν βραχίονα σὺν τῷ ξύλῳ. τὸ αὐτὸ δὲ ποιεῖν καὶ
94 ὑπὲρ δίκλειδος θύρης | ἀναγκάζειν. χρῆσθαι δὲ χρὴ αἰεὶ τούτοισιν, ἃ ἂν τύχῃ παρεόντα.

8. Εἰδέναι μὲν οὖν χρὴ ὅτι φύσιες φυσίων μέγα διαφέρουσιν ἐς τὸ ῥηϊδίως ἐμπίπτειν τὰ ἐκπίπτοντα· διενέγκα μὲν γὰρ ἄν τι καὶ κοτύλη κοτύλης, ἡ μὲν

the most effective method for reducing the shoulder, since it levers in the straightest direction, provided only that the instrument is placed further in than the head of the arm; the counterpoising weights will be properly adjusted, and it is safest for the bone of the arm. Now recent cases are reduced sooner than might be thought, even before any extension is apparent, and this method alone is able to reduce even long-standing cases, unless due to the time (sc. that has passed) tissues have already invaded the (glenoid) cavity, and the head of the arm has made a friction cavity for itself in the place to which it has slipped. But I think it would reduce even a long-standing dislocation of the arm—for what would a correct leverage not move—although I do not expect that the arm will stay where it should, but rather slip back to its accustomed dislocation.

The same result is obtained by pressure over the rung of a ladder, by preparing it in the same way. It is also very effective to exert force over a large Thessalian chair, if the dislocation is recent: the wooden instrument should be prepared as has been described. First seat the patient sideways in the chair, and then lift the arm over the chair back with the wooden instrument, and press the body down on one side, and the arm with the instrument on the other side. The same result is also obtained by operating over (sc. the lower half of) a double door. You must always make use of the means that happen to be at hand.

8. Now you must be aware that there are great natural differences as to the ease with which dislocations fall back into place. For one socket differs from another, having a

13 ἐς BV: ὡς M

εὐυπέρβατος ἐοῦσα, ᾗ δὲ καὶ ἧσσον· πλεῖστον δὲ διαφέρει καὶ τῶν νεύρων ὁ σύνδεσμος, τοῖσι μὲν ἐπιδόσιας ἔχων, τοῖσι δὲ συντεταμένος. καὶ γὰρ ἡ ὑγρότης τοῖσιν ἀνθρώποισι γίνεται ἡ ἐκ τῶν ἄρθρων, διὰ τῶν νεύρων τὴν ἀπάρτησιν, ἣν χαλαρά τε ᾖ φύσει καὶ τὰς ἐπιτάσιας εὐφόρως φέρῃ. συχνοὺς γὰρ ἄν τις ἴδοι, οἳ οὕτως ὑγροί εἰσιν, ὥστε, ὅταν ἐθέλωσι, τότε αὐτοῖσι τὰ ἄρθρα ἐξίστανται ἀνωδύνως, καὶ καθίσταται.[14] *διαφέρει μέντοι τι καὶ σχέσις τοῦ σώματος· τοῖσι μὲν γὰρ εὖ ἔχουσι τὸ γυῖον καὶ σεσαρκωμένοισιν ἐκπίπτει τε ἧσσον, ἐμπίπτει τε χαλεπώτερον· ὅταν δὲ αὐτοὶ σφέων αὐτῶν λεπτότεροι καὶ ἀσαρκότεροι ἔωσι, τότε ἐκπίπτει τε μᾶλλον, ἐμπίπτει δὲ ῥᾷον.*

Σημεῖον δέ, ὅτι ταῦθ' οὕτως ἔχει, καὶ τόδε· τοῖσι γὰρ βουσὶ τότε ἐκπίπτουσι μᾶλλον οἱ μηροὶ ἐκ τῆς
96 *κοτύλης, ἡνίκα ἂν αὐτοὶ σφέων | αὐτῶν λεπτότεροι*[15]
ἔωσι· γίνονται δὲ βόες λεπτότατοι, τοῦ χειμῶνος τελευτῶντος· τότε οὖν καὶ ἐξαρθρέουσι μάλιστα, εἰ δή τι καὶ τοιοῦτο δεῖ ἐν ἰητρικῇ γράψαι· δεῖ δέ· καλῶς γὰρ Ὅμηρος καταμεμαθήκει ὅτι πάντων τῶν προβάτων[16] *βόες μάλιστα πονέουσι ταύτην τὴν ὥρην, καὶ βοῶν οἱ ἀρόται, ὅτι τὸν χειμῶνα ἐργάζονται. τούτοισι τοίνυν καὶ ἐκπίπτει μάλιστα· οὗτοι γὰρ μάλιστα λεπτύνονται· τὰ μὲν γὰρ ἄλλα πρόβατα*[17] *δύναται βραχείην τὴν ποίην βόσκεσθαι· βοῦς δὲ οὐ μάλα, πρὶν βαθεῖα γένηται· τοῖσι μὲν γὰρ ἄλλοισίν ἐστι λεπτὴ*

[14] Add. *ἀνωδύνως* M [15] *λεπτότεροι* B: *-τατοι* MV

rim that is easy to cross, while the other one is less easy to cross. But there is also a very great diversity in the attachment of the cords, some tending naturally to yield, but others tending to contract, since the humidity in the joints of individuals is connected with the disposition of the cords, which may be slack by nature and lend themselves easily to extensions. In fact, you can see many persons of so humid a temperament that they can dislocate their joints without pain whenever they choose, and also reduce them. The condition of the body makes a further difference, for in those who are muscular and whose limb is in good condition, there is less disposition to dislocation, and reduction is also more difficult, but when they are thinner and less muscular than usual, dislocation is more frequent and reduction is easier.

Here is evidence that this is so, for in cattle the thighs dislocate out of their socket, especially when they are particularly lean; now the cattle become thinnest at the end of winter, and it is then especially that they have dislocations—if indeed I might be allowed to cite such a fact in a medical work. But it should be cited, for Homer correctly observes that, of all farm animals, cattle suffer most during this season, and among cattle the plowing oxen, because these labor in the winter. It is in these, then, that dislocation is most prevalent, since at that time they are most emaciated: for while other farm animals can graze on grass while it is still short, cattle can hardly do so until it becomes deeper, due to the fact that in the other animals

---

16 *προβάτων* B: *θρεμμάτων* MV
17 *πρόβατα* B: *βοσκήματα* MV

μὲν ἡ προβολὴ τοῦ χείλεος, λεπτὴ δὲ ἡ ἄνω γνάθος· βοῒ δὲ παχείη μὲν ἡ προβολὴ τοῦ χείλεος, παχείη δὲ καὶ ἀμβλεῖα ἡ ἄνω γνάθος· διὰ ταῦτα ὑποβάλλειν ὑπὸ τὰς βραχείας ποίας οὐ δύναται. τά τε αὖ μώνυχα,[18] *ἅτε ἀμφόδοντα ἐόντα, δύναται μὲν σαρκάζειν, δύναται δὲ ὑπὸ τὴν βραχείην ποίην ὑποβάλλειν τοὺς ὀδόντας, καὶ ἥδεται τῇ τοιαύτῃ μᾶλλον ἢ τῇ βαθείῃ· καὶ γὰρ τὸ ἐπίπαν ἀμείνων καὶ στερεωτέρη ἡ βραχείη ποίη τῆς βαθείης, ποτὶ καὶ πρὶν ἐκκαρπεῖν τὴν βα-*
98 *θείην. | διὰ τοῦτο οὖν ἐποίησε τάδε τὰ ἔπη—Ὡς δ' ὁπότ' ἀσπάσιον ἔαρ ἤλυθε βουσὶν ἕλιξιν—ὅτι ἀσμενωτάτη*[19] *αὐτοῖσιν ἡ βαθείη ποίη φαίνεται. ἀτὰρ καὶ ἄλλως βοῦς χαλαρὸν φύσει τὸ ἄρθρον τοῦτο ἔχει μᾶλλον τῶν ἄλλων ζώων· καὶ διὰ τοῦτο εἰλίπουν ἐστὶ μᾶλλον τῶν ἄλλων ζώων, καὶ μάλιστα ὅταν λεπτὸν καὶ γηραλέον*[20] *ᾖ. διὰ ταῦτα πάντα καὶ ἐκπίπτει βοῒ μάλιστα. πλείω δὲ γέγραπται περὶ αὐτοῦ, ὅτι πάντων τῶν προειρημένων ταῦτα μαρτύριά ἐστιν.*

Περὶ οὗ οὖν ὁ λόγος, *ὅτι*[21] *τοῖσιν ἀσάρκοισι μᾶλλον ἐκπίπτει καὶ θᾶσσον ἐμπίπτει ἢ τοῖσι σεσκωμένοισι· καὶ ἧσσον ἐπιφλεγμαίνει τοῖσι ὑγροῖσι καὶ τοῖσιν ἀσάρκοισιν ἢ τοῖσι σκληροῖσι*[22] *καὶ σεσαρκωμένοισι, καὶ ἧσσόν γε δέδεται ἐς τὸν ἔπειτα χρόνον· ἀτὰρ καὶ εἰ μύξα πλείων ὑπείη τοῦ μετρίου μὴ σὺν φλεγμονῇ, καὶ οὕτως ἂν ὀλισθηρὸν εἴη· μυξωδέστερα*

[18] *τῶν ζῴων* add. MV
[19] *τοῖσιν* add. MV

the projection of the lip is thin, as is also the upper jaw, but in the ox the projection of their lip is thick and their upper jaw thick and blunt, so that they cannot catch hold of the short grass. Furthermore, the solid-hoofed animals, having a double row of teeth, can not only browse, but can also catch the short grass with their teeth, and they also prefer this kind of grass to the deep grass: in fact, short grass is on the whole better and lusher than deep grass, especially when the long grass is just going to seed. It is in allusion to this that Homer composed the verse—As when the spring arrives, welcome to curve-horned cattle[2]—since the long grass appears most appealing to them. Moreover, in the ox this joint (i.e., the shoulder) is by nature laxer than in other animals, and for this reason the ox has a more shambling gait than they do, especially when the ox is thin and old. For all these reasons this joint is especially liable to dislocation in the ox. I have written about this because these facts support all my preceding arguments.

To return to my subject, dislocation occurs more easily and is more quickly reduced in emaciated than in muscular persons. There is also less inflammation in people who are moist and thin than in those of a muscular and dry habit, although the joint is not so firm afterward. Further, if an excess of mucous substance is engendered without inflammation, this too will make joints liable to slip, and,

2 This verse is not present in the Homeric text as we have it.

20 γηραλέον MV: γέρων B
21 ὅτι B: om. MV
22 σκληροῖσι B: σκελίφροισι MV

δὲ τὸ ἐπίπαν τὰ ἄρθρα τοῖσι ἀσάρκοισιν ἢ τοῖσι σεσαρκωμένοισίν ἐστιν· καὶ γὰρ αὗται αἱ σάρκες τῶν μὴ ἀπὸ τέχνης ὀρθῆς λελιμαγχημένων, αἱ τῶν λεπτῶν, μυξωδέστεραί εἰσιν ἢ αἱ τῶν παχέων. οἷσι μέντοι σὺν φλεγμονῇ μύξα ὑπογίνεται, ἡ φλεγμονὴ δήσασα ἔχει τὸ ἄρθρον· διὰ τοῦτο οὐ μάλα ἐκπίπτει[23] τὰ ὑπόμυξα, ἐκπίπτοντα ἄν, εἰ μή τι ἢ πλέον ἢ ἔλασσον φλεγμονῆς ὑπεγίνετο. |

100 9. Οἷσι μὲν οὖν ἂν[24] ἐμπέσῃ τὸ ἄρθρον καὶ μὴ ἐπιφλεγμήνῃ τὰ περιέχοντα, χρῆσθαί τε ἀνωδύνως αὐτίκα δύνανται τῷ ὤμῳ, οὗτοι μὲν οὐδὲν νομίζουσι σφᾶς αὐτοὺς ἔτι δεῖν[25] ἐπιμελεῖσθαι· ἰητροῦ μήν ἐστι καταμαντεύσασθαι τῶν τοιούτων· τοῖσι τοιούτοισι γὰρ ἐκπίπτει καὶ αὖτις μᾶλλον ἢ οἷσιν ἂν ἐπιφλεγμήνῃ τὰ νεῦρα. τοῦτο κατὰ πάντα τὰ ἄρθρα οὕτως ἔχει, καὶ μάλιστα κατ᾽ ὦμον καὶ κατὰ γόνυ· μάλιστα γὰρ οὖν[26] ὀλισθάνει ταῦτα. οἷσι δ᾽ ἂν ἐπιφλεγμήνῃ,[27] οὐ δύνανται χρῆσθαι τῷ ὤμῳ· κωλύει γὰρ ἡ ὀδύνη καὶ ἡ σύντασις τῆς φλεγμονῆς. τοὺς οὖν τοιούτους χρὴ ἰῆσθαι τῇ κηρωτῇ καὶ τοῖσι σπλήνεσι καὶ ὀθονίοισι πολλοῖσι ἐπιδέοντα· ὑποτιθέναι δὲ ἐς τὴν μασχάλην εἴριον μαλθακὸν καθαρὸν συνελίσσοντα, ἐκπλήρωμα τοῦ κοίλου ποιέοντα, ἵνα ἀντιστήριγμα μὲν τῇ ἐπιδέσει ᾖ, ἀνακωχέῃ δὲ τὸ ἄρθρον· τὸν δὲ βραχίονα χρὴ ἐς τὸ ἄνω ῥέποντα ἴσχειν τὰ πλεῖστα· οὕτω γὰρ ἂν ἑκαστάτω εἴη τοῦ χωρίου, ἐς ὃ ὤλισθεν

[23] ἐκπίπτει MV: ἐμπ. B [24] ἂν Littré: ὅταν codd.

on the whole, the joints of lean persons contain more mucus than those of muscular individuals. One sees, in fact, that these tissues in emaciated persons, who have not been correctly reduced according to the principles of the art, have more mucosity than the tissues of stout people. But in those in whom mucus develops along with inflammation, the inflammation keeps the joint firm. This is why the joints do not often suffer dislocation from a slight excess of mucus, although they would, if the mucosity was not accompanied by more or less inflammation.

9. Should, however, no inflammation of the surrounding parts supervene after the reduction of the joint, patients can immediately use the shoulder without pain, and such patients think there is no further necessity for them to take care of themselves. Therefore, the practitioner must act as a prophet for such people, since it is in these cases that the joint re-dislocates, more so than in cases where inflammation of the cords has supervened. This is the case with all joints, and especially those at the shoulder and knee, for they especially are liable to dislocation. Patients in whom inflammation has supervened cannot use the shoulder, since the pain and inflammatory tension prevent it. You should treat such cases with cerate, compresses, and plenty of bandages; also put a ball of soft, clean wool into the axilla, to fill up its cavity, so that it will be a support for the bandage and hold the joint tight. The arm should be oriented mainly upward, since in this way the head of the arm will be held furthest from the place

---

25 σφᾶς αὐτοὺς . . . δεῖν B: δεῖν ἑωυτῶν MV
26 καὶ add. MV 27 τὰ νεῦρα add. MV

ἡ κεφαλὴ τοῦ ὤμου. χρὴ δὲ ὅταν ἐπιδήσῃς τὸν ὦμον,
ἔπειτα προσκαταδεῖν τὸν βραχίονα πρὸς τὰς πλευρὰς
ταινίῃ τινὶ κύκλῳ περὶ τὸ σῶμα περιβάλλοντα. χρὴ
δὲ καὶ ἀνατρίβειν τὸν ὦμον ἡσυχαίως καὶ λιπαρῶς·
πολλῶν δὲ ἔμπειρον εἶναι δεῖ τὸν ἰητρόν, ἀτὰρ καὶ
ἀνατρίψιος· ἀπὸ γὰρ τοῦ αὐτοῦ ὀνόματος οὐ τὠυτὸ |
102 ἀποβαίνει· καὶ γὰρ ἂν δήσειεν ἄρθρον ἀνάτριψις,
χαλαρώτερον ἐὸν τοῦ καιροῦ, καὶ λύσειεν ἄρθρον
σκληρότερον τοῦ καιροῦ ἐόν· ἀλλὰ διοριεῖται περὶ
ἀνατρίψιος ἐν ἄλλῳ λόγῳ. τὸν δὲ οὖν τοιοῦτον ὦμον
μαλθακῇσι συμφέρει τῇσι χερσὶν ἀνατρίβειν, καὶ
ἄλλως πρηέως· τὸ δὲ ἄρθρον διακινέειν μὴ βίῃ, ἀλλὰ
τοσοῦτον ὅσον ἀνωδύνως κινῆσαι. καθίσταται δὲ
πάντα, τὰ μὲν ἐν πλέονι χρόνῳ, τὰ δ᾽ ἐν ἐλάσσονι.

10. Γινώσκειν δὲ εἰ ἐκπέπτωκεν ὁ βραχίων τοῖσδε
χρὴ τοῖσι σημείοισι· τοῦτο μέν, ἐπειδὴ δίκαιον ἔχουσι
τὸ σῶμα οἱ ἄνθρωποι, καὶ τὰς χεῖρας καὶ τὰ σκέλεα,
παραδείγματι χρῆσθαι δεῖ τῷ ὑγιεῖ πρὸς τὸ μὴ ὑγιές,
καὶ τῷ μὴ ὑγιεῖ πρὸς τὸ ὑγιές, μὴ τὰ ἀλλότρια ἄρθρα
καθορῶντα—ἄλλοι γὰρ ἄλλων μᾶλλον ἔξαρθροι πε-
φύκασιν—ἀλλὰ τὰ αὐτοῦ τοῦ κάμνοντος, ἢν ἀνόμοιον
ᾖ τὸ ὑγιὲς τῷ κάμνοντι. καὶ τοῦτο εἴρηται μὲν ὀρθῶς,
παρασύνεσιν δὲ ἔχει πάνυ πολλὴν διὰ τὰ τοιαῦτα, καὶ
οὐκ ἀρκέει μοῦνον λόγῳ εἰδέναι ταύτην τὴν τέχνην,
ἀλλὰ καὶ ὁμιλίῃ ὁμιλέειν· πολλοὶ γάρ, ὑπὸ ὀδύνης ἢ
καὶ ὑπ᾽ ἀλλοίης προφάσιος οὐκ ἐξεστεώτων αὐτοῖσι
104 τῶν ἄρθρων ὅμως οὐ δύνανται | ἐς τὰ ὅμοια σχήματα
καθεστάναι ἐς οἷάπερ τὸ ὑγιηρὸν σῶμα σχηματίζε-

into which it was dislocated. After bandaging the shoulder, you should proceed to fasten the arm against the ribs with some sort of circular bandage passing around the body; the shoulder should be gently and softly massaged. A practitioner must be skilled in many practices, but especially in massaging. Though massage has a single name, these procedures do not have one and the same effect, for friction may make a joint firm, when it is looser than is requisite, and relax it when it is too stiff. But we shall define the rules for friction in another treatise. Now, for such a shoulder the proper friction is with soft hands, and above all should be gentle. Move the joint about, without force, but only as far as it can be moved without pain. All disorders will subside, some in a longer, others in a shorter time.

10. A dislocation of the arm you should recognize by the following signs. First, since human bodies are symmetrical in the arms and legs, you should compare the sound configuration with the unsound, and the unsound with the sound; do not observe other people's joints—for in some individuals the joints project more than in others—but those of the patient himself, to see whether the sound joint is dissimilar to the one affected. And although this is correct advice, there is a good deal of misunderstanding about it. This is why it is not enough to know the art in theory only, but necessary to be versed in its practice. For many patients owing to pain or some other cause, though their joints are not dislocated, cannot hold themselves in the attitude the healthy body assumes. There-

ται· προσσυνιέναι μὲν οὖν καὶ ἐννοεῖν καὶ τὸ τοιόνδε σχῆμα χρή. ἀτὰρ τοῦτο μὲν τῇ μασχάλῃ ἡ κεφαλὴ τοῦ βραχίονος φαίνεται ἐγκειμένη πολλῷ μᾶλλον τοῦ ἐκπεπτωκότος ἢ τοῦ ὑγιέος. τοῦτο δὲ ἄνωθεν κατὰ τὴν ἐπωμίδα κοῖλον φαίνεται τὸ χωρίον· καὶ τὸ τοῦ ἀκρωμίου ὀστέον ἔξοχον φαίνεται, ἅτε ὑποδεδυκότος τοῦ ἄρθρου ἐς τὸ κάτω χωρίον.[28] *παρασύνεσιν μὴν καὶ ἐν τούτῳ ἔχει τινά, ἀλλὰ ὕστερον περὶ αὐτοῦ γεγράψεται· ἄξιον γὰρ γραφῆς ἐστί. τοῦτο δὲ τοῦ ἐκπεπτωκότος ὁ ἀγκὼν φαίνεται ἀφεστεὼς μᾶλλον ἀπὸ τῶν πλευρέων ἢ τοῦ ἑτέρου. εἰ μέντοι τις προσαναγκάζοι, προσάγεται μέν, ἐπιπόνως δέ. τοῦτο δέ, ἄνω τὴν χεῖρα ἆραι εὐθεῖαν παρὰ τὸ οὖς, ἐκτεταμένου τοῦ ἀγκῶνος οὐ μάλα δύνανται, ὡς τὴν ὑγιέα, οὐδὲ παράγειν ἔνθα καὶ ἔνθα ὁμοίως. τά τε οὖν σημεῖα ταῦτά ἐστιν ὤμου ἐκπεπτωκότος, αἵ τε ἐμβολαὶ αἱ γεγραμμέναι αἵ τε ἰατρεῖαι αὗται.*

11. Ἐπάξιον *δὲ τὸ μάθημα ὡς χρὴ ἰητρεύειν τοὺς πυκνὰ ἐκπίπτοντας ὤμους· πολλοὶ μὲν γὰρ ἤδη ἀγωνίης ἐκωλύθησαν διὰ ταύτην τὴν ξυμφορήν, τἆλλα ἀξιόχρεοι ἐόντες, πολλοὶ δὲ ἐν πολέμοις*[29] *ἀχρεῖοι ἐγένοντο καὶ διεφθάρησαν διὰ ταύτην τὴν ξυμφορήν· ἅμα τε ἐπάξιον διὰ τόδε, ὅτι οὐδένα οἶδα ὀρθῶς ἰητρεύοντα, ἀλλὰ τοὺς μὲν οὐδ᾽ ἐγχειρέοντας, τοὺς δὲ τἀναντία τοῦ συμφέροντος φρονέοντάς τε καὶ ποιέ-*
106 *οντας. συχνοὶ γὰρ ἤδη ἰητροὶ* | *ἔκαυσαν ὤμους ἐκπίπτοντας, κατά τε τὴν ἐπωμίδα, κατά τε τοὔμπροσθεν,*

fore, you must take this too into consideration and keep the following in mind: a) in a dislocated joint, the head of the arm is much more obviously present in the axilla on the injured side than on the sound side; b) at the top of the shoulder the part appears hollow, while the bone at the shoulder point (acromion) is seen to project, since the joint has sunk down—yet there is some fallacy in this too, but it will be explained later on, since it merits special attention; c) the elbow of the dislocated limb appears to be further away from the ribs than the other one; if someone then forcibly adducts it, it yields, but with much pain; d) the patient is quite unable to raise his dislocated arm in a straight line alongside his ear with the elbow extended, as he can the sound one, or to move it in one direction or another in the same way as before. These, then, are the signs of a dislocated shoulder, and the modes of reduction and the methods of treatment have been described.

11. It is worth becoming acquainted with how you should treat shoulders that frequently dislocate, for many men have been debarred from athletic contests, although they were well fitted in all other respects, and many others have become unserviceable in war and even perished through this misfortune. Another reason for the subject's importance is the fact, that I know of no one who treats them correctly, some not even attempting to take such cases in hand, while others have theories and practices the reverse of what is correct. Often practitioners have made it their practice to cauterize shoulders liable to dislocation both at the point of the shoulder and at the front where

---

28 *χωρίον* B: *τοῦ χωρίου* MV
29 *πολέμοις* B: *πολεμικοῖσι(ν)* MV

ᾗ ἡ κεφαλὴ τοῦ βραχίονος ἐξογκεῖ, κατά τε τοὔπισθεν ὀλίγον τῆς ἐπωμίδος. αὗται οὖν αἱ καύσιες, εἰ μὲν ἐς τὸ ἄνω ἐξέπιπτεν ὁ βραχίων, ἢ ἐς τοὔμπροσθεν ἢ ἐς τοὔπισθεν, ὀρθῶς ἂν ἔκαιον· νῦν δὲ δή, ὅτε ἐς τὸ κάτω ἐκπίπτει, ἐκβάλλουσιν αὗται αἱ καύσιες μᾶλλον ἢ κωλύουσιν· ἀποκλείουσι γὰρ τῆς ἄνω εὐρυχωρίης τὴν κεφαλὴν τοῦ βραχίονος.

Χρὴ δὲ ὧδε καίειν ταῦτα· ἀπολαμβάνοντα τοῖσι δακτύλοισι κατὰ τὴν μασχάλην τὸ δέρμα ἀφελκύσαι κατ᾽ αὐτὴν τὴν ἴξιν μάλιστα, καθ᾽ ἣν ἡ κεφαλὴ τοῦ βραχίονος ἐκπίπτει· ἔπειτα οὕτως ἀφειλκυσμένον τὸ δέρμα διακαῦσαι ἐς τὸ πέρην. σιδηρίοισι δὲ χρὴ τὰ τοιαῦτα[30] μὴ παχέσι μηδὲ λίην φαλακροῖσιν, ἀλλὰ προμήκεσι—ταχυπορώτερα γάρ—καὶ τῇ χειρὶ ἐπερείδειν. χρὴ δὲ καὶ διαφανέσι καίειν, ὡς ὅτι τάχιστα περαιωθῇ κατὰ δύναμιν· τὰ γὰρ παχέα βραδέως περαιούμενα πλατυτέρας τὰς ἐκπτώσιας τῶν ἐσχαρῶν ποιέεται, καὶ κίνδυνος ἂν εἴη ξυρραγῆναι τὰς ὠτειλάς· καὶ κάκιον μὲν ἂν οὐδὲν εἴη, αἴσχιον δὲ καὶ ἀτεχνότερον. ὅταν δὲ διακαύσῃς ἐς τὸ πέρην, τῶν μὲν πλείστων ἱκανῶς ἂν ἔχοι ἐν τῷ κάτω μέρει ταύτας μόνας τὰς ἐσχάρας θεῖναι· ἢν δὲ μὴ κίνδυνος φαίνηται εἶναι[31] ξυρραγῆναι τὰς ὠτειλάς, ἀλλὰ πολὺ τὸ διὰ μέσου ᾖ, ὑπάλειπτρον χρὴ λεπτὸν διέρσαι διὰ τῶν καυ-
108 μάτων, | ἔτι ἀναλελαμμένου τοῦ δέρματος· οὐ γὰρ ἂν ἄλλως δύναιο διέρσαι· ἐπὴν δὲ διέρσῃς, ἀφεῖναι τὸ δέρμα, ἔπειτα μεσηγὺ τῶν ἐσχαρέων ἄλλην ἐσχάρην ἐμβάλλειν λεπτῷ σιδηρίῳ καὶ διακαῦσαι ἄχρι ἂν τῷ

the head of the arm forms a prominence, and also at the back a little distance away from the top of the shoulder. Now these cauterizations would be properly done if the arm had dislocated upward, forward or backward; but, as it is, since the dislocation is in fact downward, these cauterizations cause dislocation rather than prevent it, since they shut the head of the arm out from the space above it.

You should cauterize in these cases as follows: take hold of the skin at the axilla between your fingers and draw it into the line in which the head of the arm is dislocated (i.e., downward), then pass the cautery iron right through it. The cautery irons for this operation should not be wide or very rounded, but elongated—for these will pierce more quickly—and pressure should be made with your hand. You should cauterize with irons that are red-hot, so that the operation can be finished as fast as possible. Wide irons, since they pass through slowly, will leave larger eschars, which need to come away, and there is also a danger that the cicatrices will break into one another; although this would be no very great evil, it does look rather bad and reveals a want of skill. When your cautery iron pierces right through, it will be sufficient in most cases to place eschars only in the lower part. If there appears to be no risk of the cicatrices breaking into one another because there is a good interval between them, you should pass a thin spatula through the cautery holes, while still holding

---

[30] καὶ add. MV

[31] The text in M breaks off here due to the loss of a folio and recommences on page 230 below.

ὑπαλείπτρῳ ἐγκύρσῃ. ὁπόσον δέ τι χρὴ τὸ δέρμα ἀπὸ τῆς μασχάλης ἀπολαμβάνειν, τοισίδε χρὴ τεκμαίρεσθαι. ἀδένες ὕπεισιν ἢ ἐλάσσους ἢ μείζους πᾶσιν ὑπὸ τῇ μασχάλῃ, πολλαχῇ δὲ καὶ ἄλλῃ τοῦ σώματος. ἀλλὰ ἐν ἄλλῳ λόγῳ περὶ ἀδένων οὐλομελίης γεγράψεται, ὅ τι τέ εἰσι καὶ οἷα ἐν οἷσι σημαίνουσί τε καὶ δύνανται. τοὺς μὲν οὖν ἀδένας οὐ χρὴ προσαπολαμβάνειν, οὐδ᾽ ὅσα ἐσωτέρω τῶν ἀδένων· μέγας γὰρ κίνδυνος. τοῖσι γὰρ ἐπικαιροτάτοισι τόνοισι[32] γειτονεύονται· ὅσον δὲ ἐξωτέρω τῶν ἀδένων ὡς πλεῖστον ἀπολαμβάνειν· ἀσινέα γάρ. γινώσκειν δὲ καὶ τάδε[33] χρὴ ὅτι ἢν μὲν ἰσχυρῶς τὸν βραχίονα ἀνατείνῃς, οὐ δυνήσῃ τοῦ δέρματος ἀπολαβεῖν οὐδὲν τοῦ ὑπὸ τῇ μασχάλῃ, ὅ τι καὶ ἄξιον λόγου· καταναισιμοῦται[34]
110 γὰρ ἐν τῇ ἀνατάσει. οἱ δὲ αὖ τόνοι, οὓς | οὐδεμιῇ μηχανῇ δεῖ τιτρώσκειν, οὗτοι πρόχειροι γίνονται καὶ κατατεταμένοι ἐν τούτῳ τῷ σχήματι. ἢν δὲ σμικρὸν ἐπάρῃς τὸν βραχίονα, πολὺ μὲν τοῦ δέρματος ἀπολήψει, οἱ δὲ τόνοι, ὧν δεῖ προμηθέεσθαι, ἔσω καὶ πρόσω τοῦ χειρίσματος γίνονται. ἆρ᾽ οὖν οὐκ ἐν πάσῃ τῇ τέχνῃ περὶ παντὸς χρὴ ποιέεσθαι, τὰ δίκαια σχήματα ἐξευρίσκειν ἐφ᾽ ἑκάστοισι; ταῦτα μὲν τὰ κατὰ τὴν μασχάλην, καὶ ἱκαναὶ αὗται αἱ καταλήψιες, ἢν ὀρθῶς τεθῶσιν αἱ ἐσχάραι.

[32] τόνοισι B: τόποισι V [33] καὶ τάδε B: om. V

[34] καταναισιμοῦται Galen *Glossary* s.v. (K 21): κατασεῖσιμουται V: κατατεῖναι· σιμοῦται B

the skin up, since otherwise it would not be able to go through. After passing these through, release the skin and then make another eschar between the other eschars with a thin cautery iron, and burn through until you reach the spatula.[3] The amount of skin you should take up from the axilla should be estimated as follows: all people have glands, many or few, in the axilla and various other parts of the body—in another treatise about the system of glands,[4] there will be a description of what these are, what they signify, and what they are capable of doing—and these glands must not be caught up along with the skin, nor anything that is inside them, on account of the great attendant danger, since they lie close to cords of extreme vulnerability (brachial plexus). But to take up as much as you like, of what is superficial to the glands, would cause no harm. You must also be aware that if you stretch the arm strongly upward you will not be able to take up any part of the skin under the axilla worth mentioning, since it is used up for it the extension. These cords, again, which must in no case be injured, pass close to the surface and are on the stretch in this attitude; but if you raise the arm only a little, you can take up a good deal of skin, while the cords which are to be protected lie inside, and are located far from the surgical field. Ought we not then, in all our practice, consider it of the highest importance to discover the proper attitudes for each case? So much for the parts about the axilla, and these gatherings will be sufficient if the eschars are properly placed.

[3] For an explanation and illustration of this procedure, see Majno, pp. 165–66.

[4] See Loeb *Hippocrates* vol. 8 (LCL 482), 103–25.

Ἔκτοσθεν δὲ τῆς μασχάλης δισσὰ μοῦνα ἐστὶ χωρία, ἵνα ἐσχάρας ἄν τις θείη, τιμωρεούσας τῷ παθήματι, μίαν μὲν ἐν τῷ ἔμπροσθεν μεσηγὺ τῆς τε κεφαλῆς τοῦ βραχίονος καὶ τοῦ τένοντος τοῦ κατὰ τὴν μασχάλην· καὶ ταύτῃ τὸ μὲν δέρμα τελείως χρὴ διακαίειν. βαθύτερον δὲ οὐ χρή· φλέψ τε γὰρ παχείη
112 πλησίη καὶ νεῦρα, | ὧν οὐδέτερα θερμαντέα. ὄπισθέν τε αὖ ἄλλην ἐσχάρην ἐνδέχεται ἐνθεῖναι ἀνωτέρω μὲν συχνῷ τοῦ τένοντος τοῦ κατὰ τὴν μασχάλην, κατωτέρω δὲ ὀλίγῳ τῆς κεφαλῆς τοῦ βραχίονος· καὶ τὸ μὲν δέρμα τελέως χρὴ διακαίειν, βαθείην δὲ μηδὲ ταύτην κάρτα ποιέειν· πολέμιον γὰρ πῦρ νεύροις· ἰητρεύειν μὲν οὖν χρὴ διὰ πάσης τῆς ἰητρείης τὰ ἕλκεα, μηδέποτε ἰσχυρῶς ἀνατείνοντα τὸν βραχίονα, ἀλλὰ μετρίως ὅσον τῶν ἑλκέων ἐπιμελείης εἵνεκα· ἧσσον μὲν γὰρ ἂν διαψύχοιτο, συμφέρει γὰρ πάντα τὰ καύματα σκέπειν, ὡς ἐπιεικέως ἰητρεύεσθαι, ἧσσον δ᾽ ἂν ἐκπλίσσοιτο·[35] ἧσσον δ᾽ ἂν αἱμορραγοίη· ἧσσον δ᾽ ἂν σπασμὸς ἐπιγένοιτο. ὅταν δὲ δὴ καθαρὰ γένηται τὰ ἕλκεα, ἐς ὠτειλάς τε ἴῃ, τότε δὴ καὶ παντάπασι χρὴ αἰεὶ τὸν βραχίονα πρὸς τῇσι πλευρῇσι δεδέσθαι, καὶ νύκτα καὶ ἡμέρην. ἀτὰρ καὶ ὅταν ὑγιέα γένηται τὰ ἕλκεα, ὁμοίως ἐπὶ πολὺν χρόνον χρὴ προσδεῖν τὸν βραχίονα πρὸς τὰς πλευράς· οὕτω γὰρ ἂν μάλιστα ἐπουλωθείη καὶ ἀποληφθείη ἡ εὐρυχωρίη, καθ᾽ ἣν μάλιστα ὀλισθάνει ὁ βραχίων.

12. Οἷσι δ᾽ ἂν ὦμος κατηπορηθῇ ἐμβληθῆναι, ἢν μὲν ἔτι ἐν αὐξήσει ἔωσιν, οὐκ ἐθέλει συναύξεσθαι τὸ

Outside the axilla there are only two locations where you might place eschars effectively against the malady: one at the front between the head of the arm and the tendon at the axilla, where you should burn right through the skin, but no deeper, since there is a wide blood vessel in the neighborhood besides the cords, none of which should be heated. Again, another eschar can be placed at the back, well above the tendon at the axilla, but a little below the head of the arm. You should burn all the way through the skin, while still not making the cauterization very deep either, since fire is very hostile to cords. During the whole treatment, the lesions must be attended to without ever forcefully lifting the arm, but only such a moderate distance as the care of the lesions requires. In this way, the lesions will be less exposed to cold (it is helpful to cover all burns, if they are to be properly treated) and less drawn apart; they will hemorrhage less, and convulsions will be less likely to supervene. When the lesions become clean and begin to cicatrize, it is absolutely necessary to keep the arm continuously bound against the ribs both night and day; indeed, even when the lesions have healed, you should bind the arm to the ribs like this for a long time; for in this way the space into which the arm was for the most part displaced will be best cicatrized and cut off.

12. In cases where a shoulder has failed to be reduced, if the patients are still growing, the bone of the affected

---

35 ἐκπλίσσ. Littré: ἐκπλήσσ. BV

ὀστέον τοῦ βραχίονος ὁμοίως τῷ ὑγιεῖ, ἀλλὰ αὔξεται
114 μέν,[36] βραχύτερον δὲ τοῦ ἑτέρου | γίνεται· καὶ οἱ καλεόμενοι δὲ ἐκ γενεῆς γαλιάγκωνες, διὰ δισσὰς συμφορὰς γινονται· ἤν τε[37] τοιοῦτον ἐξάρθρημα αὐτοὺς καταλάβῃ ἐν τῇ γαστρὶ ἐόντας, διά τε ἑτέρην[38] ξυμφορήν, περὶ ἧς ὕστερόν ποτε γεγράψεται. ἀτὰρ καὶ οἷσιν ἔτι νηπίοισιν ἐοῦσι κατὰ τὴν κεφαλὴν τοῦ βραχίονος βαθεῖαι καὶ ὑποβρύχιοι ἐκπυήσιες γίνονται, καὶ οὗτοι πάντες γαλιάγκωνες γίνονται· καὶ ἤν τε τμηθῶσιν, ἤν τε καυθῶσιν, ἤν τε αὐτόματόν σφιν ἐκραγῇ, εὖ εἰδέναι χρὴ ὅτι ταῦτα οὕτως ἔχει. χρῆσθαι μέντοι τῇ χειρὶ δυνατώτεροί[39] εἰσιν οἱ ἐκ γενεῆς γαλιάγκωνες, οὐ μὴν οὐδὲ ἐκεῖνοί γε ἀνατεῖναι παρὰ τὸ οὗς τὸν βραχίονα ἐκτανύσαντες τὸν ἀγκῶνα δύνανται, ἀλλὰ πολὺ ἐνδεεστέρως ἢ τὴν ὑγιᾶ χεῖρα. οἷσι δ᾽ ἂν ἤδη ἀνδράσιν ἐοῦσιν ἐκπέσῃ ὁ ὦμος καὶ μὴ ἐμβληθῇ, ἡ ἐπωμὶς ἀσαρκοτέρη γίνεται, καὶ ἡ ἕξις λεπτὴ κατὰ τοῦτο τὸ μέρος· ὅταν μέντοι ὀδυνώμενοι παύσωνται, ὁπόσα μὲν δεῖ ἐργάζεσθαι ἐπάραντας τὸν ἀγκῶνα ἀπὸ τῶν πλευρέων ἐς τὸ πλάγιον, ταῦτα μὲν οὐ δύνανται ἅπαντα[40] ὁμοίως ἐργάζεσθαι· ὁπόσα δὲ δεῖ ἐργάζεσθαι, παραφέροντας τὸν βραχίονα παρὰ τὰς πλευράς, ἐς τοὐπίσω ἢ ἐς τοὔμπροσθεν, ταῦτα δὲ δύνανται ἐργάζεσθαι· καὶ γὰρ ἂν ἀρίδα ἑλκύσειαν
116 καὶ | πρίονα, καὶ πελεκήσειαν ἄν, καὶ σκάψειαν ἄν, μὴ κάρτα ἄνω αἴροντες τὸν ἀγκῶνα καὶ τἄλλα ὅσα ἐκ τῶν τοιούτων σχημάτων ἐργάζονται.

13. Οἷσι δ᾽ ἂν τὸ ἀκρώμιον ἀποσπασθῇ, τούτοισι

arm (humerus) will refuse to grow like the healthy one: although it does grow a little, it remains shorter than the other one. And those who are called weasel-armed from birth, may owe this infirmity to two different causes: either a dislocation of this kind befell them while they were still in the uterus, or another kind of accident which will be described a little later. But those who, while they are children suffer from a deep-seated suppuration around the head of an arm, all become weasel-armed, too. And whether these are treated by incision or cautery, or the pus breaks out spontaneously, you can be sure that this will happen. Persons who are weasel-armed from birth are quite able to use this arm—although they, too, cannot hold the arm up beside the ear with the elbow extended—but to a much more limited extent than their healthy one. In adults, when the shoulder has been dislocated and not reduced, its point becomes less fleshy than usual and this limb assumes a lean habit. Still, when they cease to suffer pain, any kind of work that requires raising the elbow outward from the ribs on an angle they are unable to do as before, whereas any work that involves moving the arm either backward or forward along the ribs they can execute. For they could work a bow drill or saw, can use a pick or spade without much raising of the elbow, and can carry out all other activities that are performed in positions like this.

13. In cases of avulsion of the acromion, the bone torn

---

[36] Add. ἐπί τι V [37] Add. τι V
[38] ἑτέρην B: ἄλλην V [39] -τώτεροί B: -τώτατοι MV
[40] ἅπαντα V: om. B

φαίνεται ἐξέχον τὸ ὀστέον τὸ ἀπεσπασμένον· ἔστι δὲ τοῦτο ὁ σύνδεσμος τῆς κληῖδος καὶ τῆς ὠμοπλάτης· ἑτεροίη γὰρ ἡ φύσις ἀνθρώπου ταύτῃ ἢ τῶν ἄλλων ζώων. οἱ οὖν ἰητροὶ ἐξαπατῶνται μάλιστα ἐν τούτῳ τῷ τρώματι· ἅτε γὰρ ἀνασχόντος τοῦ[41] ὀστέου τοῦ ἀποσπασθέντος, ἡ ἐπωμὶς φαίνεται χαμαιζήλη καὶ κοίλη ὥσπερ[42] τῶν ὤμων τῶν ἐκπεπτωκότων. πολλοὺς οὖν οἶδα ἰητροὺς τἄλλα οὐ φλαύρους ἐόντας, οἳ πολλὰ ἤδη ἐλυμήναντο, ἐμβάλλειν πειρώμενοι τοὺς τοιούτους ὤμους, οὕτως οἰόμενοι ἐκπεπτωκέναι, καὶ οὐ πρόσθεν παύονται πρὶν ἢ ἀπογνῶναι ἢ ἀπορῆσαι, δοκέοντες αὐτοὶ σφᾶς αὐτοὺς ἐμβάλλειν τὸν ὦμον. τούτοισιν ἰητρείη μέν, ἥ περ καὶ τοῖσιν ἄλλοισι τοῖσι τοιούτοισι, κηρωτὴ καὶ σπλῆνες καὶ ὀθόνια, καὶ ἡ ἐπίδεσις τοιαύτη. καταναγκάζειν μέντοι τὸ ὑπερέχον χρὴ καὶ τοὺς σπλῆνας κατὰ τοῦτο τιθέναι πλείστους, καὶ πιέζειν ταύτῃ μάλιστα· καὶ τὸν βραχίονα πρὸς τῇσι πλευρῇσι προσηρτημένον ἐς τὸ ἄνω μέρος ἔχειν· οὕτω γὰρ ἂν μάλιστα πλησιάζοι τὸ ἀπεσπασμένον. τάδε μὴν εὖ χρὴ εἰδέναι, καὶ προλέγειν ὡς ἀσφαλέα, εἰ ἄλλως ἐθέλοις, ὅτι βλάβη μὲν οὐδεμία οὔτε σμικρὴ οὔτε μεγάλη τῷ ὤμῳ γίνεται ἀπὸ τούτου τοῦ τρώματος, αἴσχιον δὲ τὸ χωρίον· οὐδὲ γὰρ τοῦτο τὸ ὀστέον
118 ἐς τὴν ἀρχαίην ἕδρην ὁμοίως ἂν | ἱδρυθείη, ὡς ἐπεφύκει, ἀλλὰ ἀνάγκη πλέον ἢ ἔλασσον ὀγκηρότερον εἶναι ἐς τὸ ἄνω. οὐδὲ γὰρ ἄλλο ὀστέον οὐδὲν ἐς τὸ αὐτὸ καθίσταται, ὅ τι ἂν κοινωνέον ᾖ ἑτέρῳ ὀστέῳ καὶ προσπεφυκὸς ἀποσπασθῇ ἀπὸ τῆς ἀρχαίης φύσιος.

off makes an obvious projection. This (sc. joint) is the bond between the clavicle and the shoulder blade, for human anatomy is here different from that of other animals. Now practitioners are especially deceived by this injury, since, due to the detached bone being raised up, the point of the shoulder looks depressed and hollow, the way it is in dislocated shoulders. I know many otherwise competent practitioners who have done much damage in attempting to reduce shoulders of this kind, which they thought were dislocated; in fact, they did not cease their efforts until they finally recognized either their error or their impotence, if they went on supposing they were reducing the shoulder joint. The treatment in these, as in other like cases, consists of cerate, compresses, bandages, and the same kind of dressings. The projecting part must, then, be forced down, a great bulk of compresses placed over it, and the strongest pressure made there. Also, the arm should be fixed to the ribs and kept elevated, for in this way it will best be brought close to the part torn off. You must be well aware and predict with assurance, if you think it proper, that no harm, small or great, befalls the shoulder from this injury, although the part will be deformed. This bone cannot be connected as it was in its old natural position, but there will necessarily be more or less of a tuberosity on the upper part. Nor can any other bone be brought back to its original position, if, after uniting as the annex or outgrowth of another bone, it is torn away

[41] The text in M recommences with *ὀστέον*.

[42] *ὥσπερ* B: *ὥστε* (*καὶ*) *προμηθεῖσθαι* MV

ἀνώδυνον δὲ τὸ ἀκρώμιον ἐν ὀλίγῃσιν ἡμέρῃσι γίνεται, ἢν χρηστῶς ἐπιδέηται.

14. Κληῒς δὲ καταγεῖσα, ἢν μὲν ἀτρεκέως ἀποκαυλισθῇ, εὐΐητοτέρη ἐστίν· ἢν δὲ παραμηκέως, δυσιητοτέρη. τἀναντία δὲ τούτοις ἐστὶν ἢ ὡς ἄν τις οἴοιτο. τὴν μὲν γὰρ ἀτρεκέως ἀποκαυλισθεῖσαν προσαναγκάζοι ἄν τις μᾶλλον ἐς τὴν φύσιν ἐλθεῖν· καὶ γὰρ εἰ πάνυ προθυμηθείη, τὸ ἀνωτέρω κατωτέρω ἂν ποιήσειε σχήμασί τε ἐπιτηδείοισι καὶ ἐπιδέσει ἁρμοζούσῃ. εἰ δὲ μὴ τελέως ἱδρυθείη, ἀλλ᾽ οὖν τὸ ὑπερέχον γε τοῦ ὀστέου οὐ κάρτα ὀξὺ γίνεται. ὧν δ᾽ ἂν παραμηκὲς τὸ ὀστέον κατηγῇ, ἱκέλη ἡ ξυμφορὴ γίνεται τοῖσιν ὀστέοισι τοῖσιν ἀπεσπασμένοισι, περὶ ὧν πρόσθεν γέγραπται· οὔτε γὰρ ἱδρυθῆναι αὐτὸ πρὸς ἑωυτὸ κάρτα ἐθέλει, ἥ τε ὑπερέχουσα ὀκρὶς τοῦ ὀστέου ὀξείη γίνεται κάρτα. τὸ μὲν οὖν σύμπαν, εἰδέναι χρὴ ὅτι βλάβη οὐδεμία τῷ ὤμῳ οὐδὲ τῷ ἄλλῳ σώματι γίνεται διὰ τὴν κάτηξιν τῆς κληῗδος, ἢν μὴ ἐπισφακελίσῃ· ὀλιγάκις δὲ τοῦτο γίνεται. αἶσχός γε μὴν προσγίνεται περὶ τὴν κάτηξιν τῆς κληῗδος, καὶ τούτοισι τὸ πρῶτον
120 αἴσχιστον, ἔπειτα μὴν | ἐπὶ ἧσσον γίνεται. ξυμφύεται δὲ ταχέως κληῒς καὶ τἆλλα πάντα ὅσα χαῦνα ὀστέα· ταχείην γὰρ τὴν ἐπιπώρωσιν ποιέεαι ταῦτα. ὅταν μὲν οὖν νεωστὶ καταγῇ, οἱ τετρωμένοι σπουδάζουσιν οἰόμενοι μέζον τὸ κακὸν εἶναι ἢ ὅσον ἐστίν· οἵ τε ἰητροὶ προθυμέονται[43] δῆθεν ὀρθῶς ἰῆσθαι. προϊόντος δὲ τοῦ χρόνου οἱ τετρωμένοι, ἅτε οὐκ ὀδυνώμενοι οὐδὲ κωλυόμενοι οὔτε ὁδοιπορίης οὔτε ἐδωδῆς, καταμελέουσι·

from its original natural position. The acromion becomes painless in a few days, if it is properly bandaged.

14. A fractured collarbone, if broken straight across, is more easily treated, but if fractured obliquely, with more difficulty, which is the opposite to what might be imagined. For you can more readily force a collarbone fractured straight across into its natural position; and if carefully treated, the upper piece can be brought down by means of suitable postures and appropriate bandaging. And even if the reduction is not complete, at least the bone that projects will not be very pointed. But patients in whom the bone is fractured obliquely suffer an accident like the avulsions of bones described above; for the bone refuses to come together, and the projecting bony ridge becomes very sharp. Now in general you must bear in mind that no harm to the shoulder or the rest of the body results from the fracture of a collarbone, unless necrosis supervenes, and this rarely happens. Deformity, it is true, accompanies a fracture of the clavicle, and in these cases is at first very marked, although afterward it decreases. The collarbone unites quickly, as do all spongy bones, for with these the formation of callus is rapid. Thus, when the fracture is recent, patients take it seriously, thinking the damage is worse than it is, and practitioners on their side are careful in applying proper treatment. But as time passes the patients, since they feel no pain and are not hindered either in walking or eating, neglect the matter, and the physi-

---

43 -*μέονται* MV: -*μέους* B

οἵ τε αὖ ἰητροί, ἅτε οὐ δυνάμενοι καλὰ τὰ χωρία ἀποδεικνύναι, ἀποδιδράσκουσι,[44] καὶ οὐκ ἄχθονται τῇ ἀμελείῃ τῶν τετρωμένων· ἐν τούτῳ δὲ ἡ ἐπιπώρωσις συνταχύνεται.

Ἐπιδέσιος μὲν οὖν τρόπος καθέστηκε παραπλήσιος τοῖσι πλείστοισι κηρωτῇ καὶ σπλήνεσι καὶ ὀθονίοισι μακροῖσιν[45] ἰητρεύειν. καὶ τάδε δεῖ προσιητρεύειν, καὶ τάδε δεῖ[46] προσξυνιέναι, καὶ μάλιστα ἐν τούτῳ τῷ χειρίσματι, ὅτι τούς τε σπλῆνας πλείστους κατὰ τὸ ἐξέχον χρὴ τιθέναι, καὶ τοῖσιν ἐπιδέσμοισι πλείστοισι καὶ μάλιστα κατὰ τοῦτο[47] πιέζειν. εἰσὶ δὲ δή τινες, οἳ ἐπεσοφίσαντο ἤδη μολύβδιον βαρὺ προσεπικαταδεῖν, ὡς καταναγκάζειν τὸ ὑπερέχον· συνιᾶσι μὲν οὖν ἴσως οὐδὲ οἱ ἁπλῶς ἐπιδέοντες· ἀτὰρ δὴ οὐδὲ οὗτος ὁ τρόπος κληῖδος κατήξιός ἐστιν· οὐ γὰρ δυνατὸν τὸ ὑπερέχον καταναγκάζεσθαι οὐδέν, ὅ τι ἄξιον λόγου. ἄλλοι δ᾽ αὖ τινες εἰσίν, οἵτινες, καταμαθόντες τοῦτο, ὅτι αὗται αἱ ἐπιδέσιες παράφοροί
122 εἰσι καὶ οὐ | κατὰ φύσιν καταναγκάζουσι τὰ ὑπερέχοντα, ἐπιδέουσι μὲν αὐτοὺς σπλήνεσι καὶ ὀθονίοισι χρώμενοι, ὥσπερ καὶ οἱ ἄλλοι, ζώσαντες δὲ τὸν ἄνθρωπον ταινίῃ τινί, ᾗ εὐζωστότατος αὐτὸς ἑωυτοῦ ἐστιν. ὅταν ἐπιθέωσι τοὺς σπλῆνας ἐπὶ τὰ ὑπερέχοντα τοῦ κατήγματος, ἐξογκώσαντες ἐπὶ τὰ ἐξέχοντα, τὴν ἀρχὴν τοῦ ὀθονίου προσέδησαν πρὸς τὸ ζῶμα ἐκ τοῦ ἔμπροσθεν, καὶ οὕτως ἐπιδέουσιν, ἐπὶ τὴν ἴξιν τῆς κληῖδος ἐπιτανύοντες, ἐς τοὔπισθεν ἄγοντες. καὶ ἔπειτα περιβάλλοντες περὶ τὸ ζῶμα, ἐς τοὔμπροσθεν

cians, since they cannot make the parts look well, gradually withdraw, and are not displeased by the patients' indifference; meanwhile the callus formation quickly proceeds.

Now, the established mode of dressing is like that used for most fractures: treat with cerate, compresses, and soft bandages. Also, the following extra treatment is required, and it should be kept in mind, especially in the manual procedure, that you must place the bulk of the compresses on the projecting part and exert pressure with very many bandages, especially at the point. There are some practitioners, indeed, who have contrived something further and bind on a heavy piece of lead as well, in order to press the projection down. Perhaps those who use a simple bandage are no wiser, yet after all, this is not a suitable plan for a fractured collarbone, for the projecting part cannot be pressed down to any extent worth mentioning. Again, there are other surgeons, who, recognizing that the bandages tend to slip and that they are incapable of pressing the projecting parts down in a natural way, use compresses and bandages like the rest, but encircle the patient with a belt at the most suitable part of his body. When they have put compresses on the part of the fracture that projects, piling them on to the projection, they fix the end of the bandage to the belt in front, and make the application by stretching it in line with the collarbone and bringing it to the back. Then, passing it through the belt, they bring it

---

44 ἀποδιδ. B: ὑπαποδιδ. MV

45 μακροῖσιν B: μαλ(θ)ακοῖσιν MV

46 προσιητρεύειν . . . δεῖ V: om. BM

47 χρὴ add. MV

ἄγουσι, καὶ αὖτις ἐς τοὔπισθεν. οἱ δέ τινες οὐ περὶ τὸ ζῶμα περιβάλλουσι τὸ ὀθόνιον, ἀλλὰ περὶ τὸν περίνεόν τε καὶ παρ᾽ αὐτὴν τὴν ἕδρην καὶ παρὰ τὴν ἄκανθαν κυκλεύοντες τὸ ὀθόνιον, οὕτω πιέζουσι τὸ κάτηγμα. ταῦτ᾽ οὖν ἀπείρῳ μὲν ἀκοῦσαι φαίνεται ἐγγύς τι τοῦ κατὰ φύσιν εἶναι, χρωμένῳ δὲ ἄχρηστα· οὔτε γὰρ μόνιμα οὐδένα χρόνον, οὐδ᾽ εἰ κατακέοιτό τις—καίτοι ἐγγυτάτω ἂν οὕτως. ἀλλ᾽ ὅμως, εἰ καὶ κατακείμενος ἢ τὸ σκέλος συγκάμψειεν ἢ αὐτὸς καμφθείη, πάντα ἂν τὰ ἐπιδέσματα κινέοιτο. ἄλλως τε ἀσηρὴ ἡ ἐπίδεσις· ἥ τε γὰρ ἕδρη ἀπολαμβάνεται, ἀθρόα τε τὰ ὀθόνια ἐν ταύτῃ τῇ στενοχωρίῃ γίνεται. τά τε αὖ περὶ τὴν ζώνην περιβαλλόμενα οὐχ οὕτως |
124 ἰσχυρῶς ἔζωσται, ὡς οὐκ ἀναγκάσει ἐς τὸ ἄνω τὴν ζώνην ἐπανιέναι, καὶ οὕτως ἀνάγκη ἂν εἴη πάντα χαλᾶν τὰ ἐπιδέσματα. ἄγχιστα δ᾽ ἄν τις δοκέοι ποιέειν, καίπερ οὐ μέγα ποιέων, εἰ τοῖσι μέν τισι τῶν ὀθονίων περὶ τὴν ζώνην περιβάλλοι, τοῖσι δὲ πλείστοισι τῶν ὀθονίων τὴν ἀρχαίην ἐπίδεσιν ἐπιδέοι· οὕτω γὰρ ἂν μάλιστα τὰ ἐπιδέσματα μόνιμά τε εἴη καὶ ἀλλήλοισι τιμωρέοι.

Τὰ μὲν οὖν πλεῖστα εἴρηται, ἅσσα καταλαμβάνει τοὺς τὴν κληῖδα καταγνυμένους. προσξυνιέναι δὲ χρή, ὅτι κληῒς ὡς ἐπὶ τὸ πολὺ κατάγνυται, ὥστε τὸ μὲν ἀπὸ τοῦ στήθεος πεφυκὸς ὀστέον ἐς τὸ ἄνω μέρος ὑπερέχειν, τὸ δὲ ἀπὸ τῆς ἀκρωμίης πεφυκὸς ὀστέον ἐν τῷ κάτω μέρει εἶναι. αἴτια δὲ τούτων τάδε, ὅτι τὸ μὲν στῆθος οὔτε κατωτέρω πολὺ ἂν οὔτε ἀνωτέρω χωto the front and again to the back. There are others who pass the bandage, not through a belt, but around the perineum near the anus, and, completing the circle along the spine, apply pressure on the fracture in this way. To an inexperienced person these methods seem to come near the natural, but if someone uses them they prove to be useless; for they provide no permanent stability, not even if the patient remains in bed, although this would be best. Yet even if while he is recumbent, he bends his leg or curves his body, all the bandages will shift. Besides, the dressing is troublesome since it includes the anus, and all the bandages accumulate in this narrow space. As for the bandages passed through the belt, it is impossible to tie the belt so tightly that it will not yield to the force pulling upward, and so the bandages will necessarily become lax. The bandage will appear to be most effective, although without effecting much, if you make some of its turns through the belt while applying most of them the old-fashioned way, since with this arrangement they stay in place and support one another best.

Almost everything, then, has been said on the subject of patients with broken collarbones. But it must also be borne in mind that the clavicle, as a rule, is fractured in such a way that the bone coming from the sternum is on the top and the bone coming from the shoulder point (acromion) is below. This is because the sternum does not move much either downward or upward, so that the range

ρήσειε· σμικρὸς γὰρ ὁ κιγκλισμὸς τοῦ ἄρθρου τοῦ ἐν
τῷ στήθει· αὐτό τε γὰρ ἑωυτῷ συνεχές ἐστι τὸ στῆθος
καὶ τῇ ῥάχει· ἄγχιστα μὴν ἡ κληῒς τὸ πρὸς τοῦ
ὤμου[48] ἄρθρον πλοώδης ἐστίν· ἠνάγκασται γὰρ πυκι-
126 νοκίνητος εἶναι | διὰ τὴν τῆς ἀκρωμίης σύζευξιν. ἄλ-
λως τε ὅταν τρωθῇ, φεύγει ἐς τὸ ἄνω μέρος τὸ πρὸς
τῷ στήθει προσεχόμενον, καὶ οὐ μάλα ἐς τὸ κάτω
μέρος ἀναγκάζεσθαι ἐθέλει· καὶ γὰρ πέφυκε κοῦφον,[49]
καὶ ἡ εὐρυχωρίη αὐτῷ ἄνω πλείων ἢ κάτω. ὁ δὲ ὦμος
καὶ ὁ βραχίων καὶ τὰ προσηρτημένα τούτοισιν εὐ-
απόλυτά ἐστιν ἀπὸ τῶν πλευρέων καὶ τοῦ στήθεος,
καὶ διὰ τοῦτο δύναται καὶ ἀνωτέρω πολὺ ἀνάγεσθαι
καὶ κατωτέρω. ὅταν οὖν καταγῇ ἡ κληΐς, τὸ πρὸς τῷ
ὤμῳ ὀστέον ἐς τὸ κατωτέρω ἐπιρρέπει· ἐς τοῦτο γὰρ
ἐπιτροχώτερον αὐτὸ ἅμα τῷ ὤμῳ καὶ τῷ βραχίονι
κάτω ῥέψαι μᾶλλον ἢ ἐς τὸ ἄνω. ὁπότε οὖν ταῦτα
τοιαῦτά ἐστιν, ἀσυνετέουσιν ὅσοι τὸ ὑπερέχον τοῦ
ὀστέου ἐς τὸ κάτω καταναγκάσαι οἴονται οἷόν τε εἶ-
ναι.[50] ἀλλὰ δῆλον ὅτι τὸ κάτω πρὸς τὸ ἄνω προσ-
ακτέον ἐστί· τοῦτο γὰρ ἔχει κίνησιν, τοῦτο γάρ ἐστι
καὶ τὸ ἀποστὰν ἀπὸ τῆς φύσιος. δῆλον οὖν ὅτι ἄλλως
μὲν οὐδαμῶς ἔστιν ἀναγκάσαι τοῦτο—αἱ γὰρ ἐπιδέ-
σιες οὐδέν τι μᾶλλον προσαναγκάζουσιν ἢ ἀπαναγ-
κάζουσιν—εἰ δέ τις τὸν βραχίονα πρὸς τῇσι πλευ-
ρῇσι ἐόντα ἀναγκάζοι ὡς μάλιστα ἄνω, ὡς ὅτι
ὀξύτατος ὁ ὦμος φαίνηται, δῆλον ὅτι οὕτως ἂν ἁρμο-
σθείη πρὸς τὸ ὀστέον τὸ ἀπὸ τοῦ στήθεος πεφυκός,
ὅθεν ἀπεσπάσθη. εἰ οὖν τις τῇ μὲν ἐπιδέσει χρῷτο τῇ

of the clavicle joint at the sternum is narrow and there is a continuous connection between the sternum and the spine, whereas the clavicle on the side of its connection with the shoulder joint is especially mobile, since it has great freedom of movement owing to the structure of the acromial junction. Besides, when it is injured, the part connected to the sternum flies upward, and can hardly be pressed down, for it is naturally light and there is a larger open space for it above than below. But the shoulder, upper arm, and parts attached to them are easily separated from the ribs and the sternum, and therefore can be considerably elevated and depressed. Thus, when the collarbone is broken, the part toward the shoulder moves downward, for with the shoulder and arm attached to it, it is more readily disposed to move down than up. Whenever things like this occur, it is unreasonable to think that it is possible to press the projecting part of the bone down, since it is obvious that the lower part, being what is moveable and the part out of its natural position, must be brought up. It is obvious then that other methods are useless in reducing this fracture—for bandaging is no more likely to bring the parts together than to separate them—whereas if one presses the arm upward as much as possible, keeping it against the ribs, so that the shoulder appears very pointed, it is clear that the fragment will thus be brought into connection with the bone arising from the sternum from which it was torn. If, then, someone uses the ordinary mode of dressing for the sake of getting a

48 τὸ π. τ. ὤμου B: πρὸς τοῦ στήθους τὸ MV

49 κοῦφον (κυφὸν) MV: λορδὸν B

50 οἴονται . . . εἶναι: οἴονται εἶναι B: οἷόν τε εἶναι MV

νομίμῃ τοῦ ταχέως συναλθεσθῆναι εἵνεκα, ἡγήσαιτο δὲ τἆλλα πάντα μάταια εἶναι παρὰ τὸ σχῆμα τὸ εἰρημένον, ὀρθῶς τε ἂν συνίοι, ἰητρεύοι τε ἂν τάχιστα καὶ κάλλιστα. κατακεῖσθαι μέντοι τὸν ἄνθρωπον μέγα τὸ
128 *διάφορόν | ἐστι· καὶ ἡμέραι ἱκαναὶ τεσσαρεσκαίδεκα,*
εἰ ἀτρεμέοι, εἴκοσι δὲ πάμπολλαι.

15. Εἰ *μέντοι τινὶ ἐπὶ τἀναντία ἡ κληῒς καταγείη, ὃ οὐ μάλα γίνεται, ὥστε τὸ μὲν ἀπὸ τοῦ στήθεος ὀστέον ὑποδεδυκέναι, τὸ δὲ ἀπὸ τῆς ἀκρωμίης ὀστέον ὑπερέχειν καὶ ἐποχέεσθαι ἐπὶ τοῦ ἑτέρου, οὐδεμιῆς μεγάλης ἰητρείης ταῦτά γ᾽ ἂν δέοιτο· αὐτὸς γὰρ ὁ ὦμος ἀφιέμενος καὶ ὁ βραχίων ἱδρύοι ἂν τὰ ὀστέα πρὸς ἄλληλα, καὶ φαύλη ἄν τις ἐπίδεσις ἀρκέοι, καὶ ὀλίγαι ἡμέραι τῆς πωρώσιος γενοίατ᾽ ἄν.*

16. Εἰ *δὲ μὴ καταγείη μὲν οὕτω, παρολισθάνοι δὲ ἐς τὸ πλάγιον ἢ τῇ ἢ τῇ, ἐς τὴν φύσιν μὲν ἀγαγεῖν*[51] *ἂν δέοι, ἀναγαγόντα τὸν ὦμον σὺν τῷ βραχίονι, ὡς καὶ πρόσθεν εἴρηται· ὅταν δὲ ἵζηται ἐς τὴν ἀρχαίην*[52] *φύσιν, ταχείη ἂν ἡ ἄλλη ἰητρείη εἴη. τὰ μὲν οὖν πλεῖστα τῶν παραλλαγμάτων κατορθοῖ αὐτὸς ὁ βραχίων, καταναγκαζόμενος*[53] *πρὸς τὰ ἄνω· ὅσα δὲ ἐκ τοῦ ἄνωθεν παρολισθάνοντα ἐς τὸ πλάγιον ἦλθεν, ἢ ἐς τὸ κατωτέρω, συμπορσύνοι ἂν τὴν κατόρθωσιν, εἰ ὁ μὲν ἄνθρωπος ὕπτιος κέοιτο, κατὰ δὲ τὸ μεσηγὺ τῶν ὠμοπλατέων ὑψηλότερόν τι ὀλίγῳ ὑποκέοιτο, ὡς περιρρηδὲς ᾖ τὸ στῆθος ὡς μάλιστα· καὶ τὸν βραχίονα*[54] *ἀνάγοι τις παρὰ τὰς πλευρὰς παρατεταμένον, ὁ δὲ*
130 *ἰητρὸς τῇ μὲν ἑτέρῃ χειρὶ ἐς τὴν | κεφαλὴν τοῦ βρα-*

quick cure, and considers everything else of no importance compared with the positioning as described, his opinion will be right and his treatment most speedy and correct. Still, it makes a great difference if the patient lies down; fourteen days suffice if he stays at rest, while twenty are very generous.

15. If, however, a person has their collarbone broken in the opposite way, which rarely happens—so that the piece of bone from the sternum is underneath and the piece from the acromion projects and overrides it—this will not require any great treatment, since the shoulder and the arm, left to themselves, will bring the fragments together. Any ordinary dressing will suffice, and callus will form in a few days.

16. If the fracture is not of this kind, but the displacement is on an angle to one side or the other, it will need to be reduced to its natural position by elevating the shoulder toward the arm as described before. When this has been set into its old natural place, the rest of the cure will be rapid. Now, most displacements on an angle are corrected by the arm itself when it is pressed upward; but in cases where the upper (sternal) fragment is displaced laterally or downward, adjustment will be facilitated if the patient lies flat on his back with some slightly elevated support placed between the shoulder blades, so that his chest falls away as much as possible at the sides. Have an assistant raise the arm while keeping it extended along the side, while the practitioner with one hand on the head of

---

[51] ἀγαγεῖν B: ἀπαγ. MV
[52] ἀρχαίην MV: om. B
[53] καταναγ. B: ἀναγ. MV
[54] Add. εἰ MV

χίονος ἐμβάλλων τὸ θέναρ τῆς χειρὸς ἀπωθέοι, τῇ δὲ ἑτέρῃ τὰ ὀστέα τὰ κατεηγότα εὐθετίζοι· οὕτως ἂν μάλιστα ἐς τὴν φύσιν ἄγοι. ἀτάρ, ὥσπερ ἤδη εἴρηται, εὖ μάλα τὸ ἄνωθεν ὀστέον ἐς τὸ κάτω φιλέει ὑποδύνειν. τοῖσι μὲν οὖν πλείστοισιν, ὅταν ἐπιδεθῶσι, τὸ σχῆμα ἀρήγει, παρ᾽ αὐτὰς τὰς πλευρὰς τὸν ἀγκῶνα ἔχοντα, οὕτως ἐς τὸ ἄνω τὸν ὦμον ἀναγκάζεσθαι. ἔστι δὲ οἷσι μὲν τὸν ὦμον ἀναγκάζειν δεῖ ἐς τὸ ἄνω, ὡς εἴρηται, τὸν δὲ ἀγκῶνα πρὸς τὸ στῆθος παράγειν, ἄκρην δὲ τὴν χεῖρα παρὰ τὸ ἀκρώμιον τοῦ ὑγιέος ὤμου ἴσχειν. ἢν μὲν οὖν κατακεῖσθαι τολμᾷ, ἀντιστήριγμά τι προστιθέναι χρή, ὡς ἂν ὁ ὦμος ἀνωτάτω ᾖ· ἢν δὲ περιίῃ, σφενδόνην χρὴ ἐκ ταινίης περὶ τὸ ὀξὺ τοῦ ἀγκῶνος ποιήσαντα ἀναλαμβάνειν περὶ τὸν αὐχένα.

17. Ἀγκῶνος δὲ ἄρθρον παραλλάξαν μὲν ἢ παραρθρῆσαν πρὸς πλευρὴν ἢ ἔξω, μένοντος τοῦ ὀξέος αὐτοῦ[55] ἐν τῷ κοίλῳ τοῦ βραχίονος, ἐς εὐθὺ κατατείναντα, τὰ ἐξέχοντα ἀπωθεῖν ὀπίσω καὶ ἐς τὸ πλάγιον.

18. Τὰ δὲ τελείως ἐκβάντα ἢ ἔνθα ἢ ἔνθα, κατάτα-
132 σις[56] μέν, ἐν | ᾗ ὁ βραχίων κατεηγεὶς ἐπιδέεται· οὕτω γὰρ ἂν τὸ καμπύλον τοῦ ἀγκῶνος οὐ κωλύσει. ἐκπίπτει δὲ μάλιστα ἐς τὸ πρὸς πλευρὴν[57] μέρος. τὰς δὲ κατορθώσιας, ἀπάγοντα ὅτι πλεῖστον, ὡς μὴ ψαύῃ τῆς κορώνης ἡ κεφαλή, μετέωρον περιάγειν καὶ περι-

[55] αὐτοῦ Potter: τοῦ codd. [56] κατάτασις MV: κατάστασις B [57] πλευρὴν B: πλευρὸν MV

the arm presses it away with his palm, and with the other hand adjusts the broken bones; in this way they will best be brought to the natural position. But as was said before, the upper bone tends very much to slip in underneath. In most cases, the position after bandaging with the elbow against the ribs suffices to force the shoulder up; but sometimes it is necessary to press the shoulder up as described, bring the elbow toward the chest, and fix the hand at the point of the sound shoulder. If, then, the patient takes to his bed, you should supply a prop to keep the shoulder as high as possible, but if he continues to go about you should suspend the part by employing a sling bandage around his neck which includes the point of the elbow.

17.[5] When there is a displacement or dislocation of the elbow joint toward the side or outward, with its point (olecranon) remaining in the cavity (olecranon fossa) of the arm, hold the joint extended in a straight line and force the projecting part obliquely backward.

18. Complete dislocations of the elbow in either direction require extension in the position in which a fractured arm is bandaged; this will prevent the curved part of the elbow from getting in the way; generally, the dislocation is toward the side. For adjustment, separate the bones as far as possible so that the head (of the arm) will not come into contact with the coronoid process, keep the arm raised, employ the movements of circumduction and flex-

[5] The text of chapters 17–19 appears to be taken from *Instruments of Reduction* 7–19.

κάμψαι, καὶ μὴ ἐς εὐθὺ βιάζεσθαι, ἅμα δὲ ὠθεῖν τἀναντία ἐφ᾽ ἑκάτερα, καὶ παρωθεῖν ἐς χώρην· συνωφελοίη δ᾽ ἂν καὶ ἐπίστρεψις ἀγκῶνος ἐν τούτοις, ἐν τῷ μὲν ἐς τὸ ὕπτιον, ἐν τῷ δὲ ἐς τὸ πρηνές. ἴησις δέ, σχήματος μὲν ὀλίγῳ ἀνωτέρω ἄκρην τὴν χεῖρα τοῦ ἀγκῶνος ἔχειν, βραχίονα δὲ κατὰ πλευράς, οὕτω δὲ καὶ ἀνάληψις καὶ θέσις. καὶ εὔφορον καὶ φύσις, καὶ χρῆσις ἐν κοινῷ· ἢν ἄρα μὴ κακῶς πωρωθῇ, πωροῦται δὲ ταχέως. ἴησις δὲ ὀθονίοισι κατὰ τὸν νόμον τὸν ἀρθριτικόν, καὶ τὸ ὀξὺ προσεπιδεῖν.

19. Παλιγκοτώτατον δὲ ἀγκὼν πυρετοῖσιν, ὀδύνησιν,
ἀσώδει, ἀκρητοχόλῳ· ἀγκῶνος δὲ μάλιστα τὸ
ὀπίσω διὰ τὸ ναρκῶδες, δεύτερον δὲ τὸ ἔμπροσθεν.
ἴησις δὲ ἡ αὐτή· ἐμβολαὶ δέ, τοῦ μὲν ὀπίσω, ἐκτεί-
ναντα κατατείνειν. σημεῖον δέ· οὐ γὰρ δύνανται ἐκτεί-
νειν· τοῦ δὲ ἔμπροσθεν, οὐ δύνανται συγκάμπτειν.
134 τούτῳ δὲ ἐνθέντα | τι συνειλιγμένον σκληρόν, περὶ
τοῦτο συγκάμψαι ἐξ ἐκτάσιος ἐξαίφνης.

20. Διαστάσιος δὲ ὀστέων σημεῖον κατὰ τὴν φλέβα τὴν κατὰ βραχίονα σχιζομένην διαψαύοντι.

21. Ταῦτα δὲ ταχέως διαπωροῦται· ἐκ γενεῆς δὲ βραχύτερα τὰ κάτω τοῦ σίνεος ὀστέα, πλεῖστῳ τὰ ἐγγύτατα τοῦ πήχεος· δεύτερον δὲ χειρός· τρίτον δὲ δακτύλων. βραχίων δὲ καὶ ὦμος ἐγκρατέστερα διὰ

ion without forcing the arm back in a straight line, and at the same time press the two bones in opposite directions to bring them around into place. In cases like this it is helpful to turn the elbow alternately between supination and pronation. For treatment, keep the hand positioned somewhat higher than the elbow, and the arm against the ribs, during both suspension and fixation. The position is completely comfortable, natural and practicable; if there is no unfavorable ankylosis, the callus will form quickly. Treat by applying bandages according to the method generally employed for joints, and include the point of the elbow inside them.

19. The elbow is very liable to complications with fevers, pains, nausea, and (sc. the vomiting of) unmixed bile, especially a dislocation backward, owing to numbness, but also a dislocation forward. Treatment is the same. Modes of reduction for backward dislocation: extension and counterextension; a sign for this dislocation is that patients cannot extend the arm; for forward dislocation, that they cannot flex it. In this case treatment is to insert some firm rolled-up material into the bend of the elbow, and, starting from the extended position, suddenly to flex the arm over it.

20. Separation of the bones (sc. of the forearm) is recognized by palpation of the blood vessel of the arm where it bifurcates.

21. In these cases (i.e., unreduced dislocations of the elbow) the callus is rapidly formed; if the dislocation is congenital, the bones below the injury will be shortened, those of the forearm nearest the injury most, secondly those of the hand, third those of the fingers. The upper arm and the shoulder are stronger because they receive

τὴν τροφήν· ἡ δὲ ἑτέρη χεὶρ διὰ τὰ ἔργα ὅτι πλείω ἐγκρατεστέρη. μινύθησις δὲ σαρκῶν, εἰ μὲν ἔξω ἐξέπεσεν, εἴσωθεν· εἰ δὲ μή, ἐς τοὐναντίον ᾗ ἐξέπεσεν.

22. Ἀγκὼν δὲ ἢν εἴσω ἢ ἔξω ἐκβαίῃ, κατάτασις μὲν ἐν σχήματι κοινῷ τῷ πήχει πρὸς βραχίονα. τὴν μὲν μασχάλην ἀναλαβόντα ταινίῃ ἀνακρεμάσαι, ἀγκῶνι δὲ ἄκρῳ ὑποτιθέντα τι[58] παρὰ τὸ ἄρθρον βάρος, ἐκκρεμάσαι, ἢ χερσὶ καταναγκάζειν. ὑπεραιωρηθέντος δὲ τοῦ ἄρθρου, αἱ παραγωγαὶ τοῖσι θέναρσιν, ὡς τὰ ἐν χερσίν· ἐπίδεσις ἐν τούτῳ τῷ σχήματι, καὶ ἀνάληψις καὶ θέσις.

23. Τὰ μὲν ὄπισθεν, ἐξαίφνης ἐκτείναντα διορθοῦν τοῖσι θέναρσιν· ἅμα δὲ δὴ[59] ἐν τῇ διορθώσει, καὶ τοῖσι
136 ἑτέροισιν. ἢν δὲ | ἔμπροσθεν, ἀμφὶ ὀθόνιον συνειλιγμένον, εὔογκον, συγκάμπτοντα ἅμα διορθοῦν.

24. Ἢν δὲ ἑτεροκλινὲς ᾖ, ἐν τῇ διορθώσει ἀμφότερα ἅμα χρὴ ποιέειν. τῆς δὲ μελέτης τῆς θεραπείης κοινόν, καὶ τὸ σχῆμα καὶ ἡ ἐπίδεσις. δύναται δὲ καὶ ἐκ τῆς διατάσιος κοινῇ συμπίπτειν ἅπαντα.

25. Τῶν δὲ ἐμβολέων, αἱ μὲν ἐξ ὑπεραιωρήσιος ἐμβάλλονται, αἱ δὲ ἐκ κατατάσιος, αἱ δὲ ἐκ περισφάλσιος· αὗται δὲ ἐκ τῶν ὑπερβολέων τῶν σχημάτων ἢ τῇ ἢ τῇ σὺν τῷ τάχει.

26. Χειρὸς δὲ ἄρθρον ὀλισθάνει ἢ ἔσω ἢ ἔξω, ἔσω δὲ τὰ πλεῖστα. σημεῖα δὲ εὔσημα· καὶ ἢν μὲν εἴσω συγκάμπτειν[60] τοὺς δακτύλους οὐ δύνανται· ἢν δὲ

[58] τι MV: om. B [59] δὴ B: δεῖ MV
[60] συγκάμπ. MV: κάμπ. B

more nutriment. The opposite arm is all the stronger because of its increased work. Atrophy of the muscles occurs on the inner side, if the dislocation is outward, and otherwise on the side opposite to the dislocation.

22. If the elbow is dislocated inward or outward, extension should be made with the forearm held at the normal angle to the upper arm. Elevate the axilla and suspend it with a band, hang a weight under the point of the elbow near the joint, or press it down with your hands. With the articular end raised, make the adjustments with the palms of your hands, the same way as for dislocations of the hand. Bandaging is applied in this position, together with suspension and fixation.

23. Dislocations backward: while holding the joint extended, suddenly adjust it with the palms of your hands; the two actions must be executed simultaneously, as in the other cases. If the dislocation is forward, make a combined flexion and adjustment around a large rolled bandage.

24. If there is deviation to one side, in the reduction both movements should be combined (extension/flexion and adjustment). Positioning and bandaging follow the common rule of managing treatment. It is also possible to reduce all these cases by the common method of double extension.

25. Some reductions are performed by a lifting over, others by extension, and others by circumduction; all these actions operate by overshooting the natural position in one direction or another with a rapid movement.

26. The joint of the hand is dislocated either inward or outward, but usually inward. The signs are easily recognized: if inward patients cannot flex their fingers, if out-

ἔξω, ἐκτείνειν. ἐμβολὴ δέ· ὑπὲρ τραπέξης τοὺς δακτύλους ἔχων, τοὺς μὲν τείνειν, τοὺς δὲ ἀντιτείνειν· τὸ δὲ ἐξέχον ἢ τῷ θέναρι ἢ τῇ πτέρνῃ ἅμα ἀπωθέειν καὶ
138 ὠθέειν πρόσω καὶ κάτω, κάτωθεν, | δὲ κατὰ τὸ ἕτερον ὀστέον, ὄγκον μαλθακὸν ὑποθείς, ἢν μὲν ἄνω, καταστρέψας τὴν χεῖρα, ἢν δὲ κάτω, ὑπτίην. ἴησις ὀθονίοισιν.

27. Ὅλη δὲ ἡ χεὶρ ὀλισθάνει ἢ ἔσω ἢ ἔξω, ἢ ἔνθα ἢ ἔνθα, μάλιστα δὲ ἔσω· ἔστι δὲ ὅτε ἡ ἐπίφυσις ἐκινήθη· ἔστι δ᾽ ὅτε τὸ ἕτερον τῶν ὀστέων διέστη. τούτοισι κατάτασις ἰσχυρὴ ποιητέη, καὶ τὸ μὲν ἐξέχον ἀπωθέειν, τὸ δὲ ἕτερον ἀντωθέειν, δύο εἴδεα ἅμα ἐς τοὐπίσω καὶ ἐς τὸ πλάγιον, ἢ χερσὶν ἐπὶ τραπέζης ἢ πτέρνῃ. παλιγκοτώτατα δὲ καὶ ἀσχήμονα· τῷ χρόνῳ δὲ κρατύνεται ἐς χρῆσιν. ἴησις, ὀθονίοισι σὺν τῇ χειρὶ καὶ τῷ πήχει· καὶ νάρθηκας μέχρι δακτύλων τιθέναι· ἐν νάρθηξι δὲ τεθέντα ταῦτα πυκινότεραν λύειν ἢ τὰ κατήγματα καὶ καταχύσει πλέονι χρῆσθαι.

28. Ἐκ γενεῆς δὲ βραχυτέρη ἡ χεὶρ γίνεται καὶ μινύθησις σαρκῶν μάλιστα τἀναντία ἢ ᾗ τὸ ἔκπτωμα· ηὐξημένῳ δὲ τὰ ὀστέα μένει.

29. Δακτύλου δὲ ἄρθρον ὀλισθὸν μὲν εὔσημον. ἐμβολὴ δή· κατατείναντα ἐς ἰθύ, τὸ μὲν ἐξέχον ἀπω-
140 θεῖν, τὸ δὲ ἐναντίον ἀντωθεῖν· | ἴησις δὲ[61] ὀθονίοισιν.

[61] Add. ταινίῃσι MV

ward they cannot extend them. Reduction: placing the patient's fingers on a table, have one assistant extend the joint and another make counterextension, while the operator with his palm or heel pushes the projecting part away and presses both forward and downward, having placed some thick soft material under the other bone. The hand should be held prone if the dislocation is upward and supine if it is downward. Treatment is with bandages.

27. The entire hand can dislocate either inward or outward, or laterally toward one side or the other, but mainly inward. And sometimes the (sc. inferior radioulnar) epiphysis is displaced, and sometimes one of the two bones separates. In these cases, forceful extension must be made, and the projecting bone pressed back, while the other bone is forced in the opposite direction—the two movements, backward and on an angle, are made simultaneously on a table, with your hands or a heel. These injuries give rise to serious consequences and cause deformity, but with time the joint becomes strong and regains its function. Treatment is with bandages around the hand and forearm, and also the application of splints reaching to the fingers. While the splints are attached to the limb, loosen them more frequently than you would for fractures, and employ more copious affusions.

28. When the dislocation is congenital, the hand becomes shortened, and there is atrophy of the muscles, especially on the side opposite to where the displacement is. But in an adult the bones are unaltered.

29. Dislocation of a finger joint is easy to recognize. Reduction: while holding the joint extended in a straight line, press the projecting part back, and make counterpressure on the opposite side; treatment is with bandages.

μὴ ἐμπεσὸν δέ, ἐπιπωροῦται ἔξωθεν. ἐκ γενεῆς δὲ ἢ ἐν αὐξήσει ἐξαρθρήσαντα, τὰ ὀστέα βραχύνεται τὰ[62] κάτω τοῦ ὀλισθήματος, καὶ σάρκες μινύθουσι τἀναντία μάλιστα ἢ ὡς τὸ ἔκπτωμα· ηὐξημένῳ δὲ τὰ ὀστέα μένει.

30. Γνάθος δὲ ὀλίγοισιν ἤδη τελέως ἐξήρθρησεν· ὀστέον τε γὰρ τὸ ἀπὸ τῆς ἄνω γνάθου πεφυκὸς ὑπεζύγωται πρὸς τῷ ὑπὸ τὸ οὖς ὀστέῳ προσπεφυκότι, ὅπερ ἀποκλείει τὰς κεφαλὰς τῆς κάτω γνάθου, τῆς μὲν ἀνωτέρω ἐόν, τῆς δὲ κατωτέρω τῶν κεφαλέων. τὰ δὲ ἄκρεα τῆς κάτω γνάθου, τὸ μὲν διὰ τὸ μῆκος οὐκ εὐπαρείσδυτον, τὸ δὲ αὖ κορωνόν τε καὶ ὑπερέχον ὑπὲρ τοῦ ζυγώματος· ἅμα τε ἀπ'[63] ἀμφοτέρων τῶν ἄκρων τούτων νευρώδεες τένοντες πεφύκασιν, ἐξ ὧν ἐξήρτηνται οἱ μύες οἱ κροταφῖται καὶ μασητῆρες καλεόμενοι. διὰ τοῦτο δὲ καλέονται καὶ διὰ τοῦτο κινέον-
142 ται, ὅτι ἐντεῦθεν | ἐξήρτηνται· ἐν γὰρ τῇ ἐδωδῇ καὶ ἐν τῇ διαλέκτῳ καὶ ἐν τῇ ἄλλῃ χρήσει τοῦ στόματος, ἡ μὲν ἄνω γνάθος ἀτρεμέει· συνήρτηται γὰρ τῇ κεφαλῇ καὶ οὐ διήρθρωται· ἡ δὲ κάτω γνάθος κινέεται· ἀπήρθρωται γὰρ ἀπὸ τῆς ἄνω γνάθου καὶ ἀπὸ τῆς κεφαλῆς. διότι μὲν οὖν ἐν σπασμοῖσι καὶ τετάνοισι πρῶτον τοῦτο τὸ ἄρθρον ἐπισημαίνει συντεταμένον, καὶ διότι πληγαὶ καίριοι καὶ καροῦσαι αἱ κροταφίτιδες γίνονται, ἐν ἄλλῳ λόγῳ εἰρήσεται. περὶ δὲ τοῦ μὴ κάρτα ἐξαρθρέειν, τάδε τὰ αἴτια· αἴτιον δὲ καὶ τόδε, ὅτι οὐ μάλα καταλαμβάνουσι τοιαῦται ἀνάγκαι βρωμάτων, ὥστε τὸν ἄνθρωπον χανεῖν μέζον ἢ ὅσον δύ-

If the joint is not reduced, callus will be laid down on the outside. When the dislocation is congenital, or while the person is still growing, the bones below the dislocation become shortened, and the muscles atrophy, especially on the side opposite to the dislocation. But in an adult the bones are unaltered.

30. Complete dislocation of the lower jaw rarely occurs, since the bone that arises from the upper jaw forms a yoke with the bone attached below the ear, and thus shuts off the heads of the lower jaw, being above the one and below the other. As to these extremities of the lower jaw, one is not easily dislocated because of its length, while the other is the coronoid, which projects above the zygoma. And besides, cord-like tendons arise from both these prominences, to which the muscles called temporals and masseters are connected. They derive their names and functions from these attachments, since in eating, speaking, and other functions of the mouth, the upper jaw is at rest, being connected to the head directly without any joint. But the lower jaw moves, for it is articulated with the upper jaw and the head. Now, why this joint is the first to display rigidity in spasms and tetanus, and why wounds of the temporal muscles are dangerous and apt to cause coma, will be explained in another treatise. Why this joint is not very liable to dislocation, these are the reasons. A further reason is that the requirements of eating rarely compel a person to open his mouth wider than he can, and

---

62 *βραχ. τὰ* B: *συνήρθρωται* MV
63 *ἀπ'* B: om. MV

ναται· ἐκπέσοι δ' ἂν ἀπ' οὐδενὸς ἄλλου σχήματος ἢ ἀπὸ τοῦ μέγα χανόντα παραγαγεῖν τὴν γένυν ἐπὶ θάτερα. προσσυμβάλλεται μέντοι καὶ τόδε πρὸς τὸ ἐκπίπτειν· ὅσα γὰρ νεῦρα καὶ ὅσοι μύες παρὰ ἄρθρα εἰσίν, ἢ ἀπὸ ἄρθρων ἀφ' ὧν ξυνδέδενται, τούτων ὅσα ἐν τῇ χρήσει πλειστάκις διακινέεται, ταῦτα καὶ ἐς τὰς κατατάσιας[64] δυνατώτατα ἐπιδιδόναι, ὥσπερ καὶ τὰ δέρματα τὰ εὐδεψητότατα πλείστην ἐπίδοσιν ἔχει. περὶ οὗ οὖν ὁ λόγος, ἐκπίπτει μὲν γνάθος ὀλιγάκις,
144 σχᾶται μέντοι πολλάκις ἐν | χάσμησιν, ὡς καὶ ἄλλαι πολλαὶ[65] μυῶν παραλλαγαὶ καὶ νεύρων τοῦτο ποιέουσι. δῆλον μὲν οὖν ἐκ τῶνδε[66] μάλιστά ἐστιν, ὅταν ἐκπεπτώκῃ· προΐσχει[67] γὰρ ἡ κάτω γνάθος ἐς τοὔμπροσθεν καὶ παρῆκται τἀναντία τοῦ ὀλισθήματος· καὶ τοῦ ὀστέου τὸ κορωνὸν ὀγκηρότερον φαίνεται[68] παρὰ τὴν ἄνω γνάθον καὶ χαλεπῶς συμβάλλουσι τὰς[69] γνάθους.

Τούτοισι ἐμβολὴ πρόδηλος, ἥτις γίνοιτ' ἂν ἁρμόζουσα· χρὴ γὰρ τὸν μέν τινα κατέχειν τὴν κεφαλὴν τοῦ τετρωμένου, τὸν δὲ περιλαβόντα τὴν κάτω γνάθον καὶ ἔσωθεν καὶ ἔξωθεν τοῖσι δακτύλοισι κατὰ τὸ γένειον—χάσκοντος τοῦ ἀνθρώπου ὅσον μετρίως δύναται—πρῶτον μὲν διακινεῖν τὴν γνάθον χρόνον τινά, τῇ καὶ τῇ παράγοντα τῇ χειρί, καὶ αὐτὸν τὸν ἄνθρωπον κελεύειν χαλαρὴν τὴν γνάθον ἔχειν, καὶ συμπαράγειν καὶ συνδιδόναι ὡς μάλιστα· ἔπειτα ἐξαπίνης σχᾶσαι, τρισὶ σχήμασι ὁμοῦ προσέχοντα τὸν νόον· δεῖ[70] μὲν γὰρ παράγεσθαι ἐκ τῆς διαστροφῆς ἐς the dislocation would not occur from any other position than lateral displacement of the chin while widely gaping. The following circumstance too favors dislocation: among the cords and muscles that surround these joints or are connected with them and hold them together, the ones whose functions involve the most frequent movements are most capable of yielding to extension, just as the best tanned skins have the greatest elasticity. To come then to our subject, the jaw is rarely dislocated, but often makes a side-slip in yawning, just as many other displacements of the muscles and tendons do. It becomes clear from the following that a dislocation has occurred. The lower jaw is thrown forward and deviates to the side opposite to the dislocation, the coronoid process of the upper jaw appears more prominent, and patients have difficulty bringing their jaws together.

The appropriate mode of reduction in these cases is obvious. Someone should hold the patient's head, while the operator, taking hold of the jaw with his fingers both inside the mouth and outside around the chin—the patient keeping it as wide open as he reasonably can—should first move the jaw back and forth with his hand for a moment, and direct the patient to keep it relaxed and assist in the movement by yielding to it as much as he can. Then suddenly make an adjustment, having in mind the three steps in the maneuver: first the deviation must be returned

---

64 -*στάσιας* MV: -*τάσιας* B 65 *πολλαὶ* B: om. MV
66 *μὲν οὖν ἐκ τῶνδε* B: *δὲ τοισίδε* MV 67 *προΐσχει* B: *προίσχεται γὰρ* MV 68 *φαίνεται* MV: *γίνεται* B
69 *κάτω* add. MV 70 *δεῖ* B: *χρὴ* MV

τὴν φύσιν, δεῖ δὲ ἐς τοὐπίσω ἀπωσθῆναι τὴν γνάθον τὴν κάτω, δεῖ δὲ ἑπόμενον τούτοισι συμβάλλειν τὰς γνάθους, καὶ μὴ χάσκειν. ἐμβολὴ μὲν οὖν αὕτη, καὶ οὐκ ἂν γένοιτο ἀπ' ἄλλων σχημάτων. ἰητρείη δὲ βραχείη ἀρκέει· σπλῆνα προστιθέντα κεκηρωμένον χαλαρῷ ἐπιδέσμῳ ἐπιδεῖν. ἀσφαλέστερον δὲ χειρίζειν ἐστὶν ὕπτιον κατακλίναντα τὸν ἄνθρωπον, ἐρείσαντα τὴν κεφαλὴν αὐτοῦ ἐπὶ σκυτίνου ὑποκεφαλαίου ὡς πληρεστάτου, ἵνα ὡς ἥκιστα ὑπείκῃ· προσκατέχειν δὲ χρή τινα τὴν κεφαλὴν τοῦ τετρωμένου.

146 31. Ἢν δὲ ἀμφότεραι αἱ γνάθοι ἐξαρθρήσωσιν, ἡ μὲν ἴησις ἡ αὐτή. συμβάλλειν δ' ἔτι[71] ἧσσον οὗτοι τὸ στόμα δύνανται· καὶ γὰρ προπετέστεραι αἱ γένυες τούτοισιν, ἀστραβέες δέ. τὸ δὲ ἀστραβὲς μάλιστα ἂν γνοίης τοῖσιν ὁρίοισι τῶν ὀδόντων τῶν τε ἄνω καὶ τῶν κάτω κατ' ἴξιν. τούτοισι συμφέρει ὡς τάχιστα ἐμβάλλειν· ἐμβολῆς δὲ τρόπος πρόσθεν εἴρηται. ἢν δὲ μὴ ἐμπέσῃ, κίνδυνος περὶ τῆς ψυχῆς ὑπὸ πυρετῶν ξυνεχέων καὶ νωθρῆς καρώσιος· καρώδεες γὰρ οἱ μύες οὗτοι, καὶ ἀλλοιούμενοι καὶ ἐντεινόμενοι παρὰ φύσιν· φιλέει δὲ καὶ ἡ γαστὴρ ὑποχωρέειν τούτοισι χολώδεα ἄκρητα ὀλίγα· καὶ ἢν ἐμέωσιν, ἄκρητα ἐμέουσιν· οὗτοι τοίνυν[72] καὶ θνῄσκουσι δεκαταῖοι μάλιστα.

32. Ἢν δὲ καταγῇ ἡ κάτω γνάθος, ἢν μὲν μὴ ἀποκαυλισθῇ παντάπασιν, ἀλλὰ συνέχηται τὸ ὀστέον, ἐγκεκλιμένον[73] δὲ ᾖ, κατορθῶσαι μὲν χρὴ τὸ ὀστέον, παρά γε τὴν γλῶσσαν πλαγίην ὑπείραντα τοὺς δα-

to the natural position, second the lower jaw must be pressed backward, and following this the patient must bring his jaws together and not gape. This, then, is the reduction, and it will not succeed with other maneuvers. A short treatment will suffice. Apply a compress with cerate and a loose bandage over it. The safest way of operating is with the patient lying on his back, and his head raised on a well-stuffed leather pillow, so that it will yield as little as possible; someone must also keep the patient's head still.

31. If both jaws are dislocated (i.e., both sides of the lower jaw), the treatment is the same. These patients are even less able to close their mouth, since the chin is more projecting, although without deviation. You will best recognize the absence of deviation by observing the vertical correspondence of the upper and lower rows of teeth. It is well to reduce these cases as quickly as possible; and the mode of reduction has been described above. If the jaw does not fall into place, there is a risk of death from acute fever and deep coma—for these muscles when either displaced or abnormally stretched cause coma. In these patients there are scanty evacuations of pure bile; if there is vomiting, it is also unmixed and scanty. These patients usually die about the tenth day.

32. If the lower jaw is fractured, but not broken all the way across, and the bone preserves its continuity although it is distorted, the bone should be adjusted by introducing your fingers along the side the tongue, and exerting suit-

---

71 δ' ἔτι Ermerins: δέ τι codd. 72 τοίνυν B: οὖν MV
73 ἐνκεκ. B: ἐκκεκ. MV

κτύλους, τὸ δὲ ἔξωθεν ἀντερείδοντα, ὡς ἂν ξυμφέρῃ· καὶ ἢν διεστραμμένοι ἔωσιν οἱ ὀδόντες οἱ κατὰ τὸ τρῶμα καὶ κεκινημένοι, ὅταν τὸ ὀστέον κατορθωθῇ, ζεῦξαι τοὺς ὀδόντας χρὴ πρὸς ἀλλήλους, μὴ μοῦνον τοὺς δύο, ἀλλὰ καὶ ἐπὶ πλέονας, μάλιστα μὲν χρυσίῳ, ἔστ' ἂν κρατυνθῇ τὸ ὀστέον, εἰ δὲ μή, λίνῳ· ἔπειτα |
148 ἐπιδεῖν κηρωτῇ καὶ σπλήνεσιν καὶ ὀθονίοισιν ὀλίγοισι, μὴ ἄγαν ἐρείδοντα, ἀλλὰ χαλαροῖσιν. εὖ γὰρ εἰδέναι χρή, ὅτι ἐπίδεσις ὀθονίων γνάθῳ καταγείσῃ μικρὰ μὲν ἂν ὠφελέοι ἄν, εἰ χρηστῶς ἐπιδέοιτο, μεγάλα δ' ἂν βλάπτοι, εἰ κακῶς ἐπιδέοιτο. πυκινὰ δὲ παρὰ τὴν γλῶσσαν ἐσματεῖσθαι χρή, καὶ πολὺν χρόνον ἀντέχειν τοῖσι δακτύλοισι κατορθοῦντα τοῦ ὀστέου τὸ ἐγκλιθέν·[74] ἄριστον δὲ εἰ αἰεὶ δύναιτο, ἀλλ' οὐχ οἷόν τε.

33. Ἢν δὲ ἀποκαυλισθῇ παντάπασιν τὸ ὀστέον—ὀλιγάκις δὲ τοῦτο γίνεται—κατορθοῦν μὲν χρὴ τὸ ὀστέον οὕτως, ὡς εἴρηται. ὅταν δὲ κατορθώσῃς, τοὺς ὀδόντας χρὴ ζευγνύναι, ὡς πρόσθεν εἴρηται· μέγα γὰρ ἂν συλλαμβάνοι ἐς τὸ ἀτρεμεῖν· προσέτι καὶ ἤν τις ὀρθῶς ζεύξῃ ὥσπερ χρή, τὰς ἀρχὰς ῥάψας. ἀλλὰ γὰρ οὐ ῥηΐδιον χειρουργίην πᾶσαν ἐν γραφῇ[75] διηγεῖ-
150 σθαι, ἀλλὰ καὶ αὐτὸν ὑποτοπέεσθαι[76] χρὴ ἐκ | τῶν γεγραμμένων. ἔπειτα χρὴ δέρματος Καρχηδονίου· ἢν μὲν νεώτερος[77] ᾖ ὁ τρωθείς, ἀρκέει τῷ λοπῷ[78] χρῆσθαι, ἢν δὲ τελειότερος ᾖ, αὐτῷ τῷ δέρματι· ταμόντα δὲ χρὴ εὖρος ὡς τριδάκτυλον, ἢ ὅπως ἂν ἁρμόζῃ, ὑπαλείψαντα κόμμι τὴν γνάθον—εὐμενέστερον γάρ

able counterpressure from without. If the teeth at the point of injury are displaced or loosened, when the bone is adjusted fasten them to one another, not merely the two, but several, preferably with gold wire, but failing that, with linen thread, until consolidation takes place. Afterward dress with cerate and a few compresses, and bandages, also few, which should not be too tight, but preferably lax. For you must bear in mind that dressing a fractured jaw with bandages will do little good if it is done well, but will do great harm if it is done badly. You should make frequent palpation on the tongue side, and hold the distorted part of the bone adjusted with your fingers for a long time. In fact, it would be best if you could do so continuously, but that is impossible.

33. If the jaw is broken right across, which rarely happens, you should adjust it in the manner described. After adjustment you should fasten the teeth together as was described above, for this will contribute greatly to immobility, especially if you join them up properly and fasten off the ends of the threads as needed. For the rest, it is not easy to give exact and complete details of the operation in writing; but the reader should gain an idea of it from the description. Next, you should take a piece of Carthaginian leather; if the patient is more of a child, the outer layer alone is sufficient, but if he is more adult, use the skin itself. Cut a three-inch breadth, or as much as may be suitable, and, anointing the jaw with gum—for it is more

---

[74] *ἐγκλιθέν* M: *ἐνκλιθέν* B: *ἐκκλιθέν* V

[75] Add. *ἀτρεκέως* MV

[76] *ὑποτοτ*. Littré: *ὑποτυπ*. codd.

[77] *νεώ-* B: *νηπιώ-* MV

[78] λοπῷ V: λοιπῷ BM

κόλλης—προσκολλῆσαι τὸ δέρμα ἄκρον πρὸς τῆς γνάθου τὸ ἀποκεκαυλισμένον, ἀπολείποντα ὡς δάκτυλον ἀπὸ τοῦ τρώματος ἢ ὀλίγῳ πλέον. τοῦτο μὲν ἐς τὸ κάτω μέρος· ἐχέτω δὲ ἐντομὴν κατὰ τὴν ἴξιν τοῦ γενείου ὁ ἱμάς, ὡς ἀμφιβεβήκῃ ἀμφὶ τὸ ὀξὺ τοῦ γενείου. ἕτερον δὲ ἱμάντα τοιοῦτον ἢ ὀλίγῳ πλατύτερον προσκολλῆσαι χρὴ πρὸς τὸ ἄνω μέρος τῆς γνάθου, ἀπολείποντα καὶ τοῦτον ἀπὸ τοῦ τρώματος, ὅσον περ ὁ ἕτερος ἀπέλιπεν· ἐσχίσθω δὲ καὶ οὗτος ὁ ἱμὰς τὴν ἀμφὶ τὸ οὖς περίβασιν. ἀποξέες δὲ ἔστωσαν οἱ ἱμάντες ἀμφὶ τὴν συναφήν· ἐν δὲ τῇ κολλήσει ἡ σὰρξ τοῦ σκύτεος πρὸς τοῦ χρωτὸς ἔστω, ἐχεκολλότερον γὰρ οὕτως. ἔπειτα κατατείναντα καὶ τοῦτον χρὴ τὸν
152 ἱμάντα, μᾶλλον δέ | τι τὸν παρὰ[79] τὸ γένειον, ὡς ὅτι μάλιστα μὴ ἀπομυλλαίνῃ ἡ γνάθος, ξυνάψαι τοὺς ἱμάντας κατὰ τὴν κορυφήν· καὶ ἔπειτα κατὰ[80] τὸ μέτωπον ὀθονίῳ καταδῆσαι, καὶ κατάβλημα χρὴ εἶναι, ὥσπερ νομίζεται, ὡς ἂν ἀτρεμέῃ τὰ δεσμά. τὴν δὲ κατάκλισιν ποιεέσθω ἐπὶ τὴν ὑγιέα γνάθον, μὴ τῇ γνάθῳ ἐρηρεισμένος, ἀλλὰ τῇ κεφαλῇ. ἰσχναίνειν δὲ χρὴ τὸ σῶμα ἄχρι ἡμέρων δέκα, ἔπειτα ἀνατρέφειν |
154 μὴ βραδέως· ἢν γὰρ ἐν τῇσι πρώτῃσιν ἡμέρῃσι μὴ φλεγμήνῃ, ἐν εἴκοσιν ἡμέρῃσιν ἡ γνάθος κρατύνεται· ταχέως γὰρ ἐπιπωροῦται, ὡς καὶ τὰ ἄλλα τὰ ἀραιὰ ὀστέα, ἢν μὴ ἐπισφακελίσῃ. ἀλλὰ γὰρ περὶ σφακελισμῶν τῶν συμπάντων ὀστέων ἄλλος μακρὸς λόγος λείπεται. αὕτη ἡ διάτασις ἡ ἀπὸ τῶν κολλημάτων εὐμενὴς καὶ εὐταμίευτος, καὶ ἐς πολλὰ καὶ πολλαχοῦ

agreeable than glue—fasten the end of the leather to the broken-off part of the jaw at a finger's breadth or rather more from the fracture. This is for the lower part; and let the strap have a slit in the line of the chin, so as to include the point of the chin. Another strap, similar or a little broader, should be gummed to the upper part of the jaw at the same interval from the fracture as the former one; and let it also be split to go around the ear. Let the straps taper off at their junction, where the ends meet and are tied together. In the gumming, let the fleshy side of the leather be turned toward the skin, since that way it will adhere more firmly. You should then apply traction on this strap, but rather more on the one that goes around the chin, in order to avoid as far as possible any distortion of the jaw. Fasten the straps together at the top of the head, and afterward pass a bandage round the forehead; and there should be the usual outer covering to keep the bands steady. The patient should lie on the side of the healthy jaw, supporting the weight (sc. of his head) not on the jaw, but on the head itself. Keep him on a low diet for ten days, and afterward feed him up without delay; for if there is no inflammation in the first period, the jaw consolidates in twenty days, since callus forms quickly as in other porous bones, unless necrosis supervenes. But a lengthy discussion of necrosis in all the bones will be left for another occasion. This mode of extension by straps that are gummed on is convenient, easy to manage, and very useful for a

---

[79] παρὰ B: περὶ MV
[80] κατὰ B: περὶ MV

διορθώματα εὔχρηστος. τῶν δὲ ἰητρῶν οἱ μὴ σὺν νόῳ εὔχειρες καὶ ἐν ἄλλοισι τρώμασι τοιοῦτοί εἰσι καὶ ἐν γνάθων κατήξεσιν· ἐπιδέουσι γὰρ γνάθον καταγεῖσαν ποικίλως καὶ καλῶς καὶ κακῶς· πᾶσα γὰρ ἐπίδεσις γνάθου οὕτως καταγείσης ἐγκλίνει[81] τὰ ὀστέα τὰ ἐς τὸ κάτηγμα ῥέποντα μᾶλλον ἢ ἐς τὴν φύσιν ἄγει.

34. Ἢν δὲ ἡ κάτω γνάθος κατὰ τὴν ξύμφυσιν τὴν κατὰ τὸ γένειον διασπασθῇ—μόνη δὲ αὕτη σύμφυσις ἐν τῇ κάτω γνάθῳ ἐστίν, ἐν δὲ τῇ ἄνω πολλαί· ἀλλ' οὐ βούλομαι ἀποπλανᾶν τοῦ λόγου, ἐν ἄλλοισι γὰρ εἴδεσι νοσημάτων περὶ τούτων εἴρηται[82]—ἢν οὖν διαστῇ ἡ κατὰ τὸ γένειον σύμφυσις, κατορθῶσαι μὲν παντὸς ἀνδρός ἐστιν. τὸ μὲν γὰρ ἐξεστεὸς ἐσωθέειν χρὴ ἐς τὸ ἔσω μέρος, προσβάλλοντα τοὺς δακτύλους, τὸ δ' ἔσω ῥέπον ἀντειρείδοντ' ὡς ἂν συμφέρῃ[83] ἀνάγειν ἐς τὸ ἔξω μέρος, ἐνερείσαντα τοὺς δακτύλους. ἐς διάστασιν μέντοι διατεινάμενον ταῦτα χρὴ ποιέειν· ῥᾷον γὰρ ἐς τὴν φύσιν ἥξει ἢ εἴ τις[84] ἐγχρίμπτοντα
156 ἐς ἄλληλα | τὰ ὀστέα παραναγκάζειν πειρᾶται· τοῦτο παρὰ πάντα τὰ τοιαῦτα χαρίεν εἰδέναι. ὁκόταν δὲ κατορθώσῃς, ζεῦξαι μὲν χρὴ τοὺς ὀδόντας τοὺς ἔνθεν καὶ ἔνθεν πρὸς ἀλλήλους, ὥσπερ καὶ πρόσθεν ἤδη[85] εἴρηται. ἰᾶσθαι δὲ χρὴ κηρωτῇ καὶ σπλήνεσιν ὀλίγοισι καὶ ὀθονίοισιν. ἐπίδεσιν δὲ βραχείην ἢ ποικίλην μάλιστα τοῦτο τὸ χωρίον ἐπιδέχεται, ἐγγὺς γάρ τι τοῦ ἰσορρόπου ἐστίν, ὡς δὴ μὴ ἰσόρροπον ἐόν. τοῦ δὲ ὀθονίου τὴν περιβολὴν ποιέεσθαι χρή,[86] ἢν μὲν ἡ δεξιὴ γνάθος ἐξεστήκῃ, ἐπὶ δεξιά· ἐπὶ δεξιὰ δὲ νο-

variety of adjustments. Practitioners who have manual skill but lack intelligence expose their inadequacy in fractures of the jaw above all other injuries. They bandage a fractured jaw in a variety of ways, sometimes well, sometimes badly; but any bandaging of a jaw fractured in this way tends to turn the bony fragments at the lesion inward, rather than bring them to their natural position.

34. If the lower jaw is torn apart at the symphysis in the chin—there is only one symphysis in the lower jaw, while in the upper jaw there are many; but I do not want to digress in my narrative, since these details are discussed in the accounts of other maladies. When, therefore, the symphysis at the chin is separated, anyone can make the adjustment. You must press the projecting part inward with your fingers, and force the part inclining inward out by applying counterpressure with your other fingers. This, however, must be done while the parts are held separated by tension: for in this way the bones will return to their natural positions more easily than if you try to force them into position while they are overriding one another. This is a thing it is well to bear in mind in all such cases. After adjustment, you should attach the teeth on either side as described above. Treat with cerate and a few compresses and bandages. A simple dressing rather than a complicated one is especially suited to this part, for it is nearly cylindrical without actually being so. The bandage should be carried around to the right if the right part of the jaw

81 ἐγκλ. B: ἐκκλ. MV 82 εἴρηται B: ῥητέον MV
83 ἀντερειίδοντ᾽ . . . συμφέρῃ B: om. B² MV
84 εἴ τις B: ἐὰν MV 85 ἤδη B: om. MV
86 χρή B: om. MV

μίζεται,[87] ἢν ἡ δεξιὴ χεὶρ ἡγῆται τῆς ἐπιδέσιος· ἢν δὲ ἡ ἑτέρη γνάθος ἐκστῇ, ὡς ἑτέρως χρὴ τὴν ἐπίδεσιν ἄγειν. καὶ ἢν μὲν ὀρθῶς τις κατορθώσηται καὶ ἐπα-
158 τρεμήσῃ ὡς χρή, ταχείη | μὲν ἡ ἄλθεξις, οἱ δὲ ὀδόντες ἀσινέες γίνονται· ἢν δὲ μή, χρονιωτέρη μὲν ἡ ἄλθεξις, διαστροφὴν δὲ ἴσχουσιν οἱ ὀδόντες, καὶ σιναροὶ καὶ ἀχρεῖοι γίνονται.

35. Ἢν δὲ ἡ ῥὶς καταγῇ, τρόπος μὲν οὐχ εἷς ἐστὶ κατήξιος· ἀτὰρ πολλὰ μὲν καὶ ἄλλα λωβέονται οἱ χαίροντες τῇσι καλῇσιν ἐπιδέσεσιν ἄνευ νόου, ἐν δὲ τοῖσι περὶ τὴν ῥῖνα μάλιστα· ἐπιδεσίων τε γὰρ αὕτη ποικιλωτάτη καὶ πλείστους μὲν σκεπάρνους ἔχουσα, διαρρωγὰς δὲ καὶ διαλάμψιας[88] ποικιλωτάτας τοῦ χρωτὸς ῥομβοειδέας. ὡς οὖν εἴρηται, οἱ τὴν ἀνόητον εὐχειρίην ἐπιτηδεύοντες ἄσμενοι ῥινὸς κατηγυίης ἐπιτυγχάνουσνι, ὡς ἐπιδήσωσι. μίαν μὲν οὖν ἡμέρην ἢ δύο ἀγάλλεται ὁ ἰητρός, χαίρει δὲ ὁ ἐπιδεδεμένος· ἔπειτα ταχέως μὲν ὁ ἐπιδεδεμένος κορίσκεται, ἀσηρὸν γὰρ τὸ φόρημα· ἀρκέει δὲ τῷ ἰητρῷ, ἐπειδὴ ἐπέδειξεν
160 | ὅτι ἐπίσταται ποικίλως ῥῖνα ἐπιδεῖν. ποιέει δὲ ἡ ἐπίδεσις ἡ τοιαύτη πάντα τἀναντία τοῦ δέοντος· τοῦτο μὲν γάρ, ὅσοι σιμοῦνται διὰ τὴν κάτηξιν, δῆλον ὅτι εἰ ἄνωθέν τις πιέζοι, σιμώτεροι ἂν ἔτι εἶεν· τοῦτο δέ, ὅσοισι παραστρέφεται ἢ ἔνθα ἢ ἔνθα ἡ ῥίς, ἢ κατὰ τὸν χόνδρον ἢ[89] ἀνωτέρω, δῆλον ὅτι οὐδὲν αὐτοὺς ἡ ἄνωθεν ἐπίδεσις ὠφελήσει, ἀλλὰ καὶ βλάψει μᾶλλον·

is projecting (by convention it is said to be "to the right" if the right hand goes first in bandaging): while if the other side of the jaw projects, do the bandaging the other way. If the bandaging is well done and the patient stays at rest, as he should, recovery is rapid, and the teeth are not damaged; if not, recovery is slow, and the teeth remain distorted and become damaged and useless.

35. If the nose is broken, which happens in more than one way, those who delight in fine bandaging, but lack judgment, will do even more damage than they usually do. For this is the most ornate of bandages, having the most adze-like turns and diverse rhomboid intervals, and leaving the most uncovered spaces of skin. Now, as I said, those who devote themselves to a foolish parade of manual skill are especially delighted to find a fractured nose to bandage. The result is that the practitioner rejoices, and the patient is pleased for one or two days; afterward the patient soon has had enough of it, for the burden is tiresome; and as for the practitioner, he contents himself with having shown that he knows how to apply complicated nasal bandages. But such bandaging acts in every way contrary to what is beneficial; for first, in cases where the nose is rendered concave by the fracture, if more pressure is applied from above, it will obviously become still more concave, and again in cases where the nose is distorted to either side, whether in the cartilaginous part or higher up, bandaging will obviously be useless in either case, and will rather do harm; for the compresses will not be arranged

---

87 εἶναι add. MV

88 διαλάμψιας B: διαλείψ. MV

89 καὶ add. MV

οὐχ οὕτω γὰρ εὖ[90] ἐξαρμόσει σπλήνεσι τὸ ἐπὶ θάτερον τῆς ῥινός· καίτοι οὐδὲ τοῦτο ποιέουσιν οἱ ἐπιδέοντες.

36. Ἄγχιστα δὲ ἡ ἐπίδεσίς μοι δοκέει ἄν τι ποιέειν, εἰ κατὰ μέσην τὴν ῥῖνα κατὰ τὸ ὀξὺ ἀμφιφλασθείη ἡ σὰρξ κατὰ τὸ ὀστέον, ἢ καὶ τὸ ὀστέον σμικρόν τι σίνος ἔχοι καὶ μὴ μέγα· τοῖσι γὰρ τοιούτοισιν ἐπιπώρωμα ἴσχει ἡ ῥίς, καὶ ὀκριοειδεστέρη τινὶ γίνεται. ἀλλ' ὅμως οὐδὲ τούτοισι δή που πολλοῦ ὄχλου δέεται ἡ ἐπίδεσις, εἰ δή τι καὶ δεῖ ἐπιδεῖν. ἀρκέει δὲ ἐπὶ μὲν τὸ φλάσμα σπληνίον ἐπιτείναντα κεκηρωμένον, ἔπειτα ὡς ἀπὸ δύο ἀρχέων ἐπιδεῖται, οὕτως ὀθονίῳ ἐς ἅπαξ περιβάλλειν. ἀρίστη μέντοι ἰητρείη ἡ[91] τῷ ἀλήτῳ, τῷ σητανίῳ, τῷ πλυτῷ, γλίσχρῳ, πεφυρη-
162 μένῳ, ὀλίγῳ, | καταπλάσσειν τὸ τοιοῦτο· χρὴ δέ, ἢν μὲν ἀγαθῶν ᾖ τῶν πυρῶν τὸ ἄλητον καὶ εὐόλκιμον, αὐτῷ χρῆσθαι ἐς πάντα τὰ τοιαῦτα· ἢν δὲ μὴ πάνυ ὅλκιμον ᾖ, πρὸς[92] ὀλίγην μάννην ὕδατι ὡς λειοτάτην διέντα. τούτῳ φυρᾶν τὸ ἄλητον, ἢ κόμμι πάνυ ὀλίγον ὡσαύτως μίσγειν.

37. Ὁκόσοισι μὲν ῥὶς ἐς τὸ κάτω καὶ ἐς τὸ σιμὸν ῥέπουσα καταγῇ, ἢν μὲν ἐκ τοῦ ἔμπροσθεν μέρεος κατὰ τὸν χόνδρον ἵζηται, οἷόν τέ ἐστι καὶ ἐντιθέναι διόρθωμα ἐς τοὺς μυκτῆρας· ἢν δὲ μή, ἀνορθοῦν μὲν χρὴ πάντα τὰ τοιαῦτα, τοὺς δακτύλους ἐς τοὺς μυκτῆρας ἐντιθέντα, ἢν ἐνδέχηται· ἢν δὲ μή, παχὺ ὑπ-
164 άλειπτρον, | μὴ ἐς τοὔμπροσθεν τῆς ῥινὸς ἀνάγοντα[93] τοῖσι δακτύλοισι, ἀλλ' ᾗ ἵδρυται· ἔξωθεν δὲ τῆς ῥινὸς

well on the other side of the nose, and in fact those who put on such bandages do not even seek to do this.

36. Bandaging seems to me to be most directly effective where the soft parts have been contused against the bone in the middle of the nose at the bridge, or when, without great damage, there is some small injury at the bone; for in such cases the nose gets a superficial callus and a somewhat jagged outline. But not even in these cases is there need of very ornate bandaging, if, in fact, bandaging is required at all. It suffices to stretch a small compress soaked in cerate over the contusion and then take one turn of bandage around it, as from a two-headed roller. After all, the best treatment is to use a little spring-wheat flour, worked and kneaded into a glutinous mass, as a plaster for such lesions. If you have wheat flour of good quality forming a ductile paste, you should use it in all such cases; but if it is not very ductile, soak a little frankincense powdered as finely as possible in water, and knead the flour with this, or mix in a very little gum in the same way.

37. In cases where the nose is fractured with depression and flattening, if the depression is in the front part of the cartilage, it is possible to insert some rectifying support into the nostrils. Failing this, you should restore all such cases, if possible, by inserting your fingers into the nostrils, but if not that, you should insert a thick spatula, directing it with your fingers, not to the front of the nose, but to the depressed part: then taking hold of each side of

---

90 εὐ B: om. MV 91 ἡ M: ᾗ V: om. B
92 πρὸς B: ἐς MV 93 ἀνάγοντα B: ἀπάγ. MV

ἔνθεν καὶ ἔνθεν ἀμφιλαμβάνοντα τοῖσι δακτύλοισι, συναναγκάζειν τε ἅμα καὶ ἀναφέρειν[94] *ἐς τὸ ἄνω. καὶ εἰ μὲν πάνυ ἐν τῷ ἔμπροσθεν τὸ κάτηγμα εἴη, οἷόν τέ τι καὶ ἔσω τῶν μυκτήρων ἐντιθέναι, ὥσπερ ἤδη εἴρηται, ἢ ἄχνην τὴν ἀφ᾽ ἡμιτυβίου ἢ ἄλλο τι τοιοῦτον, ἐν ὀθονίῳ εἱλίσσοντα, μᾶλλον δὲ ἐν Καρχηδονίῳ δέρματι ἐρράψαντα· σχηματίσαντα τὸ ἁρμόσσον σχῆμα τῷ χωρίῳ, ἵνα ἐγκείσεται. ἢν μέντοι προσωτέρω ᾖ τὸ κάτηγμα, οὐδὲν οἷόν τε ἔσω ἐντιθέναι· καὶ γὰρ εἰ ἐν τῷ ἔμπροσθεν ἀσηρὸν τὸ φόρημα, πῶς γε δὴ οὐκ ἐν τῷ ἐσωτέρω; τὸ μὲν οὖν πρῶτον καὶ ἔξωθεν ἀναπλάσασθαι καὶ ἔσωθεν ἀφειδήσαντα χρὴ ἀναγαγεῖν ἐς τὴν ἀρχαίην φύσιν καὶ διορθώσασθαι. κάρτα γὰρ οἵη τε ῥὶς κατηγεῖσα ἀναπλάσασθαι, μάλιστα μὲν αὐθήμερος, εἰ δὲ μή, ὀλίγῳ ὕστερον· ἀλλὰ καταβλακεύουσιν οἱ ἰητροί, καὶ ἁπαλωτέρως τὸ πρῶτον ἅπτονται ἢ ὡς χρή· παραβάλλοντα γὰρ τοὺς δακτύλους χρὴ ἔνθεν καὶ ἔνθεν τοὺς κατὰ τὴν φύσιν τῆς ῥινὸς ὡς κατωτάτω, κάτωθεν συναναγκάζειν, καὶ οὕτω μάλι-*
166 *στα ἀνορθοῦντα*[95] | *σὺν τῇ ἔσωθεν διορθώσει. ἔπειτα δὲ ἐς ταῦτα ἰητρὸς οὐδεὶς ἄλλος ἐστὶ τοιοῦτος, εἰ ἐθέλοι καὶ μελετᾶν καὶ τολμᾶν, ὡς οἱ δάκτυλοι αὐτοῦ οἱ λιχανοί· οὗτοι γὰρ κατὰ φύσιν μάλιστά εἰσιν. παραβάλλοντα γὰρ χρὴ τῶν δακτύλων ἑκάτερον παρὰ πᾶσαν τὴν ῥῖνα ἐρείδοντα, ἡσύχως οὕτως ἔχειν· μάλιστα μέν, εἰ οἷόν τε εἴη αἰεί ἔστ᾽ ἂν κρατυνθῇ· εἰ δὲ μή, ὡς πλεῖστον χρόνον, αὐτόν, ὡς εἴρηται· εἰ δὲ μή, ἢ παῖδα ἢ γυναῖκά τινα· μαλθακὰς γὰρ τὰς*

the nose from the outside with your fingers, combine the two movements of compression and lifting. If the fracture is quite to the front, it is possible, as was said, to insert something into the nostrils, either lint from linen or something of the kind, rolling it up in a rag, or better, sewing it up in Carthaginian leather and adapting its shape to fit the part where it will lie. But if the fracture is further in, nothing can be inserted; for if it is irksome to endure anything in the front part; how could it not be more so if it was further in? The first thing, then, is to reshape the nose from the outside, and internally to spare no pains in bringing it to its natural position and making it straight; for it is quite possible for a broken nose to be reshaped, especially on the day of the accident, or, failing that, a little later. But practitioners act irresolutely, and at first treat the fracture more gently than they should. For one ought to insert the fingers on each side as far as the conformation of the nose allows, and then lift it up from below, thus best combining elevation with rectification from within. Furthermore, no practitioner is so suitable for the job as are the index fingers of the patient himself, if he is willing to be careful and courageous, for these fingers are especially conformable to the nose. He should insert the fingers alternately, applying pressure along the whole length of the nose, and keeping it steady; especially let him continue this, if he can, until consolidation occurs; failing that, as long as possible. As was said, he should do this for himself; but if not, some child or woman must do it, for the hands should be

94 *ἀναφέρειν* recc.: *ἀναγκάζειν* BMV

95 -*οῦντα* Weber in Kühlewein: -*οῦνται* BMV

χεῖρας δεῖ εἶναι. οὕτω γὰρ ἂν κάλλιστα ἰητρευθείη ὁτέῳ ἡ ῥὶς μὴ ἐς τὸ σκολιόν, ἀλλ᾽ ἐς τὸ κάτω ἱδρυμένη ἰσόρροπος εἴη. ἐγὼ μὲν οὖν οὐδεμίαν πω ῥῖνα εἶδον ἥτις οὕτω καταγεῖσα οὐχ οἵη τε διορθωθῆναι αὐτίκα πρὶν πωρωθῆναι ξυναναγκαζομένη ἐγένετο, εἴ τις ὀρθῶς ἐθέλοι ἰητρεύειν. ἀλλὰ γὰρ οἱ ἄνθρωποι αἰσχροὶ μὲν εἶναι πολλοῦ ἀποτιμῶσι, μελετᾶν δὲ ἅμα μὲν οὐκ ἐπίστανται, ἅμα δὲ οὐ τολμῶσιν, ἢν μὴ ὀδυνῶνται, ἢ θάνατον δεδοίκωσιν. καίτοι ὀλιγοχρόνιος ἡ πώρωσις τῆς ῥινός· ἐν γὰρ δέκα ἡμέρῃσι κρατύνεται, ἢν μὴ ἐπισφακελίσῃ.

38. Ὁκόσοισι δὲ τὸ ὀστέον ἐς τὸ πλάγιον κατάγνυται, ἡ μὲν ἴησις ἡ αὐτή· τὴν δὲ διόρθωσιν δῆλον ὅτι χρὴ ποιέεσθαι οὐκ ἰσόρροπον ἀμφοτέρωθεν, ἀλλὰ τό τε ἐγκεκλιμένον[96] ὠθέειν ἐς τὴν φύσιν, ἔκτοσθεν ἀναγκάζοντα καὶ ἐσματευόμενον ἐς τοὺς μυκτῆρας, καὶ τὰ ἔσω ῥέψαντα διορθοῦν ἀόκνως, ἔστ᾽ ἂν κατορθώσῃς,
168 | εὖ εἰδότα ὅτι, ἢν μὴ αὐτίκα κατορθώσηται, οὐχ οἷόν τε μὴ οὐχὶ διεστράφθαι τὴν ῥῖνα. ὅταν δὲ ἀγάγῃς ἐς τὴν φύσιν, προσβάλλοντα χρὴ ἐς τὸ χωρίον ἢ τοὺς δακτύλους ἢ τὸν ἕνα δάκτυλον, ᾗ ἐξέσχεν, ἀνακωχέειν ἢ αὐτὸν ἢ ἄλλον τινά, ἔστ᾽ ἂν κρατυνθῇ τὸ τρῶμα. ἀτὰρ καὶ ἐς τὸν μυκτῆρα τὸν σμικρὸν δάκτυλον ἀπωθέοντα ἄλλοτε καὶ ἄλλοτε διορθοῦν χρὴ τὰ ἐγκλιθέντα. ὅ τι δ᾽ ἂν φλεγμονῆς ὑπογίνηται τούτοισι, δεῖ τῷ σταιτὶ χρῆσθαι· τοῖσι μέντοι δακτύλοισι προσέχειν χρὴ ὁμοίως καὶ τοῦ σταιτὸς ἐπικειμένου.

Ἢν δέ κατὰ τὸν χόνδρον πῃ ἐς τὰ πλάγια καταγῇ,

soft. This is the best treatment when the nose is not distorted laterally, but remains evenly balanced though depressed. Now, I have never seen a nose fractured in this way which could not be adjusted by immediate forcible manipulation before consolidation set in, provided the treatment is properly executed. But while people will give much to avoid being ugly, they do not know how to persist in taking care of themselves, unless they are suffering pain, or fear death. Yet the formation of callus in the nose takes little time, being consolidated in ten days, unless necrosis supervenes.

38. In cases where the bone is fractured with a deviation, the treatment is the same. Adjustment should obviously not be made evenly by applying equal force on both sides, but by pressing the displaced part into its natural position by force from the outside, and, introducing something into the nostrils boldly to rectify the internal deviation until you get it straight, keeping in mind that if the nose is not straightened at once, it will inevitably end up being distorted. And when you have brought it to its normal position, one or more fingers should be applied at the place where it is protruding, and either the patient or someone else should support it until the injured bone is consolidated. You should also insert the little finger from time to time into the nostril and adjust the depressed part. If inflammation arises in these cases, use the dough, but keep up the finger application as before, even when the dough is in place.

If a fracture has a deviation somewhere in the cartilage,

---

96 ἐγκεκ. B: ἐκκεκ. MV

ἀνάγκη τὴν ῥῖνα ἄκρην παρεστράφθαι. χρὴ οὖν τοῖσι τοιούτοισιν ἐς τὸν μυκτῆρα ἄκρον διόρθωμά τι τῶν εἰρημένων ἢ ὅ τι τούτοισιν ἔοικεν ἐντιθέναι. πολλὰ δ' ἄν τις εὕροι τὰ ἐπιτήδεια, ὅσα μήτε ὀδμὴν ἔχει μήτε ἄλλως τε προσηνέα ἐστίν. ἔγωγέ ποτε πλεύμονος προβάτου ἀπότμημα ἐνέθηκα· τοῦτο γάρ πως παρέτυχεν. οἱ γὰρ σπόγγοι ἐντιθέμενοι ὑγράσματα δέχονται· ἔπειτα χρὴ Καρχηδονίου λοπὸν πλάτος ὡς τοῦ μεγάλου δακτύλου τετμημένον, ἢ ὅκως ἂν συμφέρῃ, προσκολλῆσαι ἐς τὸ ἔκτοσθεν πρὸς τὸν μυκτῆρα τὸν ἐγκεκλιμένον.[97] ἔπειτα κατατεῖναι τὸν ἱμάντα ὅκως ἂν
170 συμφέρῃ· | μᾶλλον δὲ ὀλίγῳ τείνειν χρή, ὥστε ὀρθὴν καὶ ἀπαρτῆ τὴν ῥῖνα εἶναι. ἔπειτα—μακρὸς μὲν ἔστω ὁ ἱμάς—κάτωθεν τοῦ ὠτὸς ἀγαγόντα αὐτὸν ἀναγαγεῖν περὶ τὴν κεφαλήν· καὶ ἔξεστι μὲν κατὰ τὸ μέτωπον προσκολλῆσαι τὴν τελευτὴν τοῦ ἱμάντος· ἔξεστι δὲ μακρότερον[98] ἐπιπεριελίσσοντα περὶ τὴν κεφαλὴν καταδεῖν. τοῦτο ἅμα μὲν δικαίην τὴν διόρθωσιν ἔχει, ἅμα δὲ εὐταμίευτον, καὶ μᾶλλον, ἢν ἐθέλῃ, καὶ ἧσσον τὴν ἀντιρροπίην ποιῆσαι τῆς ῥινός. ἀτὰρ καὶ ὁκόσοισιν ἐς τὸ πλάγιον ἡ ῥὶς κατάγνυται, τὰ μὲν ἄλλα ἰητρεύειν χρὴ ὡς προείρηται· προσδέεται δὲ τοῖσι πλείστοισι καὶ τοῦ ἱμάντος πρὸς ἄκρην τὴν ῥῖνα προσκολληθῆναι τῆς ἀντιρροπίης εἵνεκα.

39. Ὅσοισι δὲ σὺν κατήξει καὶ ἕλκεα προσγίνεται, οὐδὲν δεῖ ταράσσεσθαι διὰ τοῦτο· ἀλλ' ἐπὶ μὲν τὰ ἕλκεα ἐπιτιθέναι πισσήρην ἢ τῶν ἐναίμων τι· εὐαλθέα γὰρ τῶν τοιούτων τὰ πλεῖστά ἐστιν[99] ὁμοίως, καὶ ἢν

the end of the nose will always be distorted. In such cases, insert one of the internal props described above, or something of the kind, into the end of the nostril. You will be able to find many suitable substances, that are both odorless and otherwise appropriate. I once inserted a slice from a sheep's lung which happened to be handy; for when sponges are put in, they absorb moisture. Then you should take the outer skin from a piece of Carthaginian leather, cut a strip one inch wide or whatever is suitable, and gum it to the outer part of the nostril which is bent aside. Next exert suitable tension on this strap; in fact, you should pull rather more than suffices to make the nose straight and full. Then bring the strap, which should be long, under the ear and up around the head. You may gum the end of the strap on to the forehead, but you may also carry it further, and after making a turn around the head, fasten it off. This provides an adjustment which is at once normal and easily arranged; and you can make the counterdeviation of the nose more or less as you choose. Again, when the nose is fractured with a deviation to one side, besides the other treatment mentioned, it is also necessary in most cases that a piece of leather should be gummed on to the tip of the nose in order to make a counterdeviation.

39. In cases where the fracture is complicated by lesions, there should be no alarm on that account, but you should apply to them ointment containing pitch or some other remedy for fresh wounds, since the majority of such

---

97 ἐγκεκ. B: ἐκκεκ. MV

98 ἄγειν add. MV

99 τῶν τοιούτων . . . ἐστιν B: τὰ τοιαῦτα· MV

ὀστέα μέλλῃ ἀπιέναι. τὴν δὲ διόρθωσιν τὴν πρώτην ἀόκνως χρὴ ποιέεσθαι, μηδὲν ἐπιλείποντα, καὶ τὰς διορθώσιας τοῖσι δακτύλοισιν ἐν τοῖσιν ἔπειτα χρό-
172 νοις χαλαρωτέροισι μὲν | χρεώμενον, χρεώμενον δέ· εὐπλαστότατον γὰρ[100] παντὸς τοῦ σώματος ἡ ῥίς ἐστιν. τῶν δὲ ἱμάντων τῇ κολλήσει καὶ ἀντιρροπίῃ παντάπασιν οὐδὲν κωλύει χρῆσθαι, οὔτ᾽ ἢν ἕλκος ᾖ, οὔτ᾽ ἢν ἐπιφλεγμήνῃ· ἀλυπόταται γάρ εἰσιν.

40. Ἢν δὲ οὖς καταγῇ, ἐπιδέσιες μὲν πᾶσαι πολέμιαι· οὐ γὰρ οὕτω τις χαλαρὸν περιβάλλει· ἢν δὲ μᾶλλον πιέσῃ, πλέον κακὸν ἐργάσεται, ἐπεὶ καὶ ὑγιὲς οὖς ἐπιδέσει πιεχθέν, ὀδυνηρὸν καὶ σφυγμῶδες καὶ πυρετῶδες γίνεται. ἀτὰρ καὶ ἐπιπλάσματα, κάκιστα μὲν τὰ βαρύτατα ἐπίπαν· ἀτὰρ καὶ τὰ πλεῖστα φλαῦρα καὶ ἀποστατικά, καὶ μύξαν τε ὑποποιέει,[101] καὶ ἔπειτα ἐκπυήσιας ἀσηράς· τούτων δὲ ἥκιστα οὖς καταγὲν προσδέεται· ἄγχιστα μήν, εἴ περ χρή, τὸ γλίσχρον ἄλητον· χρὴ δὲ μηδὲ τοῦτο βάρος ἔχειν. ψαύειν δὲ ὡς ἥκιστα συμφέρει· ἀγαθὸν γὰρ φάρμακόν ἐστιν ἐνίοτε καὶ τὸ μηδὲν προσφέρειν, καὶ πρὸς τὸ οὖς καὶ πρὸς ἄλλα πολλά. χρὴ δὲ καὶ τὴν ἐπικοίμησιν φυλάσσεσθαι· ἰσχναίνειν δὲ τὸ σώμα, ᾧ ἂν κίνδυνος ᾖ ἔμπυον τὸ οὖς γενέσθαι. ἄμεινον δὲ καὶ
174 μαλθάξαι τὴν κοιλίην· ἢν δὲ καὶ εὐήμετος | ᾖ, καὶ ἐμέειν ἀπὸ συρμαισμοῦ. ἢν δὲ ἐς ἐκπύησιν[102] ἔλθῃ,

[100] Add. τοι M: τι V
[101] Add. πλείω MV
[102] ἐκπ. BV: ἐμπ. M

cases heal no less readily, even if bony fragments are going to come away. The first adjustment should be made without delay or reserve; the later rectifications with the fingers are to be done more gently, but they are indispensable, since of all parts of the body the nose is the most easily modeled. There is absolutely no objection to gumming on straps and exerting counterdeviation, not even if there is a lesion, or inflammation supervenes, for the manipulations are quite painless.

40. If the ear is fractured, all bandaging is harmful, since no circular bandage can be applied that is sufficiently lax; and if you apply a bandage that exerts more pressure, further damage will be done; indeed, even a healthy ear under the pressure of a bandage becomes painful, throbs, and is heated. Besides, as far as plasters are concerned, the heaviest are generally the worst; but almost all kinds are harmful and produce abscesses, the formation of mucus, and later troublesome discharges of pus. A fractured ear has the least need of all these measures. If necessary, the best application is the glutinous flour plaster; but even this should not be heavy. It is well to touch the part as little as possible, for it is the best remedy sometimes to apply nothing, both to the ear and to many other parts. Care must be taken that the patient does not lie on the ear. Keep him on a low diet, especially if there is danger of an abscess forming in the ear. It is also better to loosen the bowels, and, if the person vomits easily, use the emetic *syrmaism*.[6] If it comes to suppuration, do not be in a hurry

[6] "Syrmaism" is an emetic made from radishes and salt water (Erotian).

ταχέως μὲν οὐ χρὴ στομοῦν· πολλὰ γὰρ καὶ τῶν δοκεόντων ἐκπυέειν ἀναπίνεταί ποτε, κἢν μηδέν τις καταπλάσῃ. ἢν δὲ ἀναγκασθῇ στομῶσαι, τάχιστα μὲν ὑγιὲς γίνεται, ἤν τις πέρην διακαύσῃ· εἰδέναι μέντοι χρὴ σαφῶς ὅτι κυλλὸν ἔσται τὸ οὖς, καὶ μεῖον τοῦ ἑτέρου, ἢν πέρην διακαυθῇ. ἢν δὲ μὴ πέρην καίηται, τάμνειν χρὴ τὸ μετέωρον, μὴ πάνυ σμικρὴν τομήν· διὰ παχυτέρου μὴν καὶ τὸ πύον εὑρίσκεται ἢ ὡς ἄν τις δοκέοι. ὡς δ' ἐν κεφαλαίῳ εἰρῆσθαι, καὶ τἆλλα πάντα τὰ μυξώδεα καὶ μυξοποιά· ἅτε γὰρ γλίσχρα ἐόντα, ὑποθιγγανόμενα διολισθάνει ταχέως ὑπὸ τοὺς δακτύλους καὶ ἔνθα καὶ ἔνθα· διὰ τοῦτο διὰ παχυτέρου εὑρίσκουσι τὰ τοιαῦτα οἱ ἰητροὶ ἢ ὡς οἴονται. ἐπεὶ καὶ τῶν γαγγλιωδέων ἔνια, ὅσα ἂν πλαδαρὰ ᾖ, καὶ μυξώδεα σάρκα ἔχῃ, πολλοὶ στομοῦσιν, οἰόμενοι ῥεῦμα ἀνευρήσειν ἐς τὰ τοιαῦτα. ἡ μὲν οὖν γνώμη τοῦ ἰητροῦ ἐξαπατᾶται· τῷ δὲ πρήγματι τῷ τοιούτῳ οὐδεμία βλάβη στομωθέντι. ὅσα δὲ ὑδατώδεα χωρία ἐστὶν ἢ μύξης πεπληρωμένα, καὶ ἐν οἵοισι χωρίοισιν ἕκαστα θάνατον φέρει στομούμενα ἢ καὶ[103] ἀλλοίας βλάβας, περὶ τούτων ἐν ἄλλῳ λόγῳ γεγράψεται. ὅταν οὖν τάμῃ τις τὸ οὖς, πάντων μὲν καταπλασμάτων,
176 πάσης δὲ μοτώσιος ἀπέχεσθαι χρή· ἰητρεύειν | δὲ ἐναίμῳ ἢ ἄλλῳ τῳ,[104] ὅ τι μήτε βάρος μήτε πόνον παρασχήσει· ἢν γὰρ ὁ χόνδρος ἄρξηται ψιλοῦσθαι, καὶ ἀποστάσιας[105] ἴσχῃ, ὀχλῶδες μοχθηρόν·[106] γίνεται δὲ ταῦτα δι' ἐκείνας τὰς ἰήσιας. πάντων δὲ τῶν παλιγκοτησάντων ἡ πέρην διάκαυσις αὐταρκέστατον.

to open the abscess, for in many cases where there seems to be an incipient suppuration, this is reabsorbed, and without any cataplasm. If you are forced to open an abscess, it will heal most quickly if you cauterize right through; but bear well in mind that the ear, if cauterized through, will be deformed and become smaller than the other one. If it is not cauterized through, you should make an incision in the raised part, and not a very small one, since the pus will be found under a thicker covering than you might expect. And, speaking generally, all other parts of a mucous nature, or which secrete mucus, being viscous, slip around freely under the fingers in one direction or another; for this reason, practitioners find them thicker to penetrate than they expected. Thus, in the case of some ganglionic tumors that are flabby and have mucoid tissue, many practitioners open them, expecting to find there has been a flux of humors to such parts; in fact, they are deceived in this opinion, although in practice no harm is done by such a tumor being opened. Now, concerning watery parts, or those filled with mucus, and where opening the parts brings death or various other sufferings, these matters will be discussed in another treatise. When, then, someone incises the ear, all plasters and all lint plugs should be avoided. Treat with an application suitable for fresh wounds, or something that is neither heavy nor painful. For if the cartilage begins to be denuded and has abscesses, it will become still more troublesome as the result of the treatment. Perforating cautery by itself is the most effective treatment for all supervening complications.

---

103 καὶ B: κατὰ MV 104 τῳ recc.: om. BMV
105 ἀποστάσιας B: ὑποστ. B^2MV 106 ὠχολωδεας (καὶ B^2) μοχθηρόν B: πυρώδεας ἢ χολώδεας ὄχληρόν MV

41. Σπόνδυλοι δὲ οἱ κατὰ ῥάχιν, οἷσι μὲν ὑπὸ νο-
σημάτων ἕλκονται ἐς τὸ κυφόν, τὰ μὲν πλεῖστα ἀδύ-
νατα λύεσθαι, ποτὶ καὶ ὅσα ἀνωτέρω τῶν φρενῶν τῆς
προσφύσιος κυφοῦται. τῶν δὲ κατωτέρων μετεξέτερα
λύουσι κιρσοὶ γενόμενοι ἐν τοῖς σκέλεσι, μᾶλλον δ'
ἔτι ἐγγινόμενοι κιρσοὶ ἐν τῇ κατὰ ἰγνύην φλεβί· οἷς
178 ἂν τὰ | κυφώματα λύηται,[107] ἐγγίνονται δὲ καὶ ἐν τῇ
κατὰ βουβῶνα. ἤδη δέ τισιν ἔλυσε καὶ δυσεντερίη
πολυχρόνιος γενομένη.[108] καὶ ὅσοισι μὲν κυφοῦται
ῥάχις παισὶν ἐοῦσι, πρὶν ἢ τὸ σῶμα τελεωθῆναι ἐς
αὔξησιν· τούτοισι μὲν οὐδὲ ξυναύξεσθαι ἐθέλει κατὰ
τὴν ῥάχιν τὸ σῶμα, ἀλλὰ σκέλεα μὲν καὶ χεῖρες τε-
λεοῦνται· ταῦτα δὲ ἐνδεέστερα γίνεται. καὶ ὅσοισι μὲν
ἂν ᾖ ἀνωτέρω τῶν φρενῶν τὸ κυφόν, τούτοισι μὲν αἵ
τε πλευραὶ οὐκ ἐθέλουσιν ἐς τὸ εὐρὺ αὔξεσθαι, ἀλλὰ
ἐς τοὔμπροσθεν τό τε στῆθος ὀξὺ γίνεται, ἀλλ' οὐ
πλατύ. αὐτοί τε δύσπνοοι γίνονται καὶ κερχνώδεες·
ἧσσον γὰρ εὐρυχωρίην ἔχουσιν αἱ κοιλίαι αἱ τὸ
πνεῦμα δεχόμεναι καὶ προπέμπουσαι. καὶ γὰρ δὴ καὶ
ἀναγκάζονται κατὰ τὸν μέγαν σφόνδυλον λορδὸν καὶ
αὐχένα ἔχειν, ὡς μὴ προπετὴς ᾖ αὐτοῖσιν ἡ κεφαλή·
στενοχωρίην μὲν οὖν πολλὴν τῇ φάρυγγι παρέχει καὶ
τοῦτο συμβάλλεται[109] ἐς τὸ ἔσω ῥέπον· καὶ γὰρ τοῖσιν
180 ὀρθοῖσι φύσει δύσπνοιαν παρέχει | τοῦτο τὸ ὀτσέον,
ἢν ἔσω ῥέψῃ, ἔστ' ἂν ἀναπιεχθῇ. δι' οὖν τὸ τοιοῦτο
σχῆμα ἐξεχέβρογχοι οἱ τοιοῦτοι τῶν ἀνθρώπων μᾶλ-
λον φαίνονται ἢ οἱ ὑγιέες. φυματίαι τε ὡς ἐπὶ τὸ πολὺ
κατὰ τὸν πλεύμονά εἰσιν οἱ τοιοῦτοι σκληρῶν φυ-

41. When the spinal vertebrae are drawn together into a hump by diseases, most cases cannot be cured, especially when the hump is formed above the attachment of the diaphragm. Some of those lower down do resolve when varicosities form in the legs, and still more when these are in the vessel in the ham. In cases where curvatures disappear, varicosities may also arise in the groin. In some cases, prolonged dysentery has also brought about resolution. When humpback occurs in children before the body has completed its growth, the body refuses to grow normally in the spine, whereas the legs and arms attain full size, although the parts are defective. And where the hump is above the diaphragm, the ribs do not expand in width, but rather forward, and the chest becomes pointed instead of broad; such patients also have difficulty in breathing, and wheeze, for the cavities which receive and send out the breath have a reduced capacity. Besides, they are also obliged to hold their neck concave at the great vertebra (C2), to prevent their head from being thrown forward. This, then, causes a great constriction in the pharynx, since it inclines inward; for this bone, if it inclines inward, causes difficult breathing even in persons without any deformity, until it is pushed back. In consequence of this attitude, such people seem to have more projecting necks than those who are healthy. They have also, as a rule, hard

---

107 *οἷς ἂν . . . λύηται* B: *οἷσι δ' ἄν τι κύφωμα ᾖ, λύουσιν* MV

108 *πολυχρ. γενομένη* B: om. MV

109 *συμβάλλεται* B: om. MV

μάτων καὶ ἀπέπτων. καὶ γὰρ ἡ πρόφασις τοῦ κυφώματος καὶ ἡ ξύντασις τοῖσι πλείστοισι διὰ τοιαύτας συστροφὰς γίνεται, ᾗσιν ἂν κοινωνήσωσιν οἱ τόνοι οἱ ξύνεγγυς. ὅσοισι δὲ κατωτέρω τῶν φρενῶν τὸ κύφωμά ἐστι, τούτοισι νοσήματα μὲν ἐνίοισι προσγίνεται νεφριτικὰ καὶ κατὰ κύστιν. ἀτὰρ καὶ ἀποστάσιες ἐμπυηματικαὶ κατὰ κενεῶνας καὶ κατὰ βουβῶνας, χρόνιαι καὶ δυσαλθέες, καὶ τούτων οὐδετέρη λύει τὰ κυφώματα. ἰσχία δὲ τοῖσι τοιούτοισιν ἔτι ἀσαρκότερα γίνεται ἢ τοῖσιν ἄνωθεν κυφοῖσιν· ἡ μέντοι σύμπασα ῥάχις μακροτέρη τούτοισιν ἢ τοῖσιν ἄνωθεν κυφοῖσιν. ἥβη δὲ καὶ γένειον βραδύτερα καὶ ἀτελέστερα, καὶ ἀγονώτεροι οὗτοι τῶν ἄνωθεν κυφῶν. ὅσοισι δ' ἂν ηὐξημένοισιν ἤδη τὸ σῶμα ἡ κύφωσις γένηται, τούτοισι ἀπαντικρὺ μὲν τῆς νούσου τῆς τότε παρεούσης κρίσιν ποιέει ἡ κύφωσις· ἀνὰ χρόνον μέντοι ἐπισημαίνει τι τῶν αὐτῶν, ὥσπερ καὶ τοῖσι νέοισιν[110] ἢ
182 πλέον ἢ ἔλασσον· ἧσσον | δὲ κακοήθως ὡς τὸ ἐπίπαν μὴν τοιαῦτα πάντα[111] ἐστίν. πολλοὶ μέντοι ἤδη καὶ εὐφόρως ἤνεγκαν καὶ ὑγιηρῶς τὴν κύφωσιν ἄχρι γήραος, μάλιστα δὲ οὗτοι, οἷσιν ἂν ἐς τὸ εὔσαρκον καὶ πιμελῶδες προτράπηται τὸ σῶμα. ὀλίγοι μὴν ἤδη καὶ τῶν τοιούτων ὑπὲρ ἑξήκοντα ἔτη ἐβίωσαν· οἱ δὲ πλεῖστοι βραχυβιώτεροί εἰσιν. ἔστι δ' οἷσι καὶ ἐς τὰ πλάγια σκολιοῦνται σφόνδυλοι ἢ τῇ ἢ τῇ· πάντα μὴν ἢ τὰ πλεῖστα τὰ τοιαῦτα γίνεται διὰ συστροφὰς τὰς εἴσωθεν ῥάχιος· προσσυμβάλλεται δὲ ἐνίοισι σὺν τῇ νούσῳ καὶ τὰ σχήματα, ἐφ' ὁκοῖα ἂν ἐθισθέωσι κε-

and unconcocted growths in the lungs; for the origin of the curvature and the contraction is mainly due to such tubercles, with which the neighboring cords are connected. Cases where the curvature is below the diaphragm are sometimes complicated with affections of the kidneys and parts around the bladder. Besides there are purulent abscessions in the lumbar region and about the groins, which are chronic and difficult to cure; and neither of these brings about resolution of the curvatures. The hips become even more attenuated in such cases than when the hump is high up; yet the spine as a whole is more elongated in these than in high curvatures. Also, the growth of hair on the pubes and chin begins later and is less full in cases of low curvature, and such patients are less fertile than those in whom the hump is higher up. When a curvature appears in persons whose bodily growth is completed, the curvature usually results in an apparent crisis of the disease then present; in time, however, some of the same symptoms found in younger patients show themselves to a greater or lesser degree; but generally, they are all less malignant. Many patients, too, have borne curvature well and with good health up to old age, especially those whose bodies tend to be fleshy and plump. Still, few even of these survive past sixty years, and the majority are quite short-lived. There are some, in whom the vertebrae are curved laterally to one side or the other. All such affections, or most of them, are due to tubercles on the inner side of the spine, while in some cases the positions the patients are accustomed to

110 *νέοισιν* B: *ἑτέροισιν* M
111 *πάντα* B: om. MV

κλίσθαι. ἀλλὰ περὶ μὲν τούτων ἐν τοῖσι χρονίοισι τοῖσι κατὰ πλεύμονα νοσήμασιν εἰρήσεται· ἐκεῖ γάρ εἰσιν αὐτῶν χαριέσταται[112] προγνώσιες περὶ τῶν μελλόντων ἔσεσθαι.

42. Ὅισι δὲ ἐκ καταπτώσιος ῥάχις κυφοῦται, ὀλίγα δὴ τούτων ἐκρατήθη ὥστε ἐξιθυνθῆναι. τοῦτο μὲν γάρ, αἱ ἐν τῇ κλίμακι κατάσιες[113] οὐδένα πω ἐξίθυναν, ὧν γε ἐγὼ οἶδα· χρέονται δὲ οἱ ἰητροὶ μάλιστα αὐτῇ οἱ ἐπιθυμοῦντες ἐκχαυνοῦν τὸν πολὺν ὄχλον· τοῖσι γὰρ τοιούτοισι ταῦτα θαυμάσιά ἐστιν, ἢν ἢ κρεμάμενον ἴδωσιν ἢ ῥιπτεόμενον, ἢ ὅσα τοῖσι τοιούτοισιν ἔοικεν· καὶ ταῦτα κληΐζουσιν αἰεί, καὶ οὐκέτι αὐτοῖσι μέλει ὁποῖόν τι ἀπέβη ἀπὸ τοῦ χειρίσματος, εἴτε κακὸν εἴτε ἀγαθόν. οἱ μέντοι ἰητροὶ οἱ τὰ τοιαῦτα ἐπιτηδεύοντες σκαιοί εἰσιν, οὕς γε ἐγὼ ἔγνων· τὸ μὲν γὰρ ἐπινόημα ἀρχαῖον, καὶ ἐπαινέω ἔγωγε σφόδρα τὸν πρῶτον ἐπινοήσαντα καὶ τοῦτο καὶ ἄλλο πᾶν, ὅ
184 τι μηχάνημα κατὰ φύσιν | ἐπενοήθη· οὐδὲν γάρ μοι ἄελπτον, εἴ τις καλῶς σκευάσας καλῶς[114] κατασείσειε, κἂν ἐξιθυνθῆναι ἔνια. αὐτὸς μέντοι κατῃσχύνθην πάντα τὰ τοιουτότροπα ἰητρεύειν οὕτω διὰ τοῦτο ὅτι πρὸς ἀπατεώνων μᾶλλον οἱ τοιοῦτοι τρόποι.

43. Ὁκόσοισι μὲν οὖν ἐγγὺς τοῦ αὐχένος ἡ κύφωσις γίνεται, ἧσσον εἰκὸς ὠφελέειν τὰς κατατάσιας ταύτας τὰς ἐπὶ κεφαλήν· σμικρὸν γὰρ τὸ βάρος ἡ κεφαλὴ καὶ τὰ ἀκρώμια καταρρέποντα· ἀλλὰ τούς γε τοιούτους εἰκὸς ἐπὶ πόδας κατασεισθέντας μᾶλλον ἐξιθυνθῆναι· μέζων γὰρ οὕτως ἡ καταρροπίη ἡ ἐπὶ ταῦτα. ὅσοισι

take in bed contribute to the malady. But these will be discussed among chronic diseases of the lung, and the most satisfactory prognoses as to their issue will be given there.

42. When the humpback has been caused by a fall, attempts at straightening it rarely succeed. For, to begin with, successions on a ladder have never straightened any case of which I know, and the practitioners who use this method are chiefly those who want to make the vulgar herd gape, to whom it seems marvelous to see a man suspended or shaken or treated in such ways; and they always applaud these performances, while never troubling themselves about the result of the operation, whether bad or good. As to the practitioners who devote themselves to this kind of thing, those—at least whom I have known—are inept. Yet the device is an ancient one, and for my part I have great admiration for the man who first invented it, or thought out any other mechanism in accordance with nature; for I think it is not hopeless, if one has the proper apparatus and does the succussion properly, that some cases may be straightened. Personally, however, I have always felt ashamed to treat such cases in this way, believing that these methods belong rather to charlatans.

43. In cases where the curvature is near the neck, succussion of this kind with the head downward is less likely to be effective; for the recumbent weight of the head and shoulders is small; such cases are more likely to be straightened by succussion with the feet pendant, since the impulsion downward is greater in this position. Cases

112 χαριέσταται B: -τεραι MV
113 κατάσιες B: κατασείσιες MV
114 καλῶς B: om. MV

δὲ κατωτέρω τὸ ὕβωμα, τούτοις εἰκὸς μᾶλλον ἐπὶ κεφαλὴν κατασείεσθαι. εἰ οὖν τις ἐθέλοι κατασείειν, ὀρθῶς ἂν ὧδε σκευάζοι· τὴν μὲν κλίμακα χρὴ σκυτίνοισιν ὑποκεφαλαίοις πλαγίοισιν, ἢ εἰρινέοισι, καταστορέσαι[115] εὖ προσδεδεμένοισιν,[116] ὀλίγῳ πλέον καὶ ἐπὶ μῆκος καὶ ἔνθεν καὶ ἔνθεν, ἢ ὅσον ἂν τὸ σῶμα τοῦ ἀνθρώπου κατάσχοι. ἔπειτα τὸν ἄνθρωπον ὕπτιον κατακλῖναι ἐπὶ τὴν κλίμακα χρή· καὶ ἔπειτα προσδῆσαι μὲν τοὺς πόδας παρὰ τὰ σφυρὰ πρὸς τὴν κλίμακα μὴ διαβεβῶτας, δεσμῷ εὐόχῳ[117] μέν, μαλθακῷ δέ. προσδῆσαι δὲ κατωτέρω ἑκάτερον τῶν γουνάτων καὶ ἀνωτέρω· προσδῆσαι δὲ καὶ κατὰ τὰ ἰσχία· κατὰ δὲ τοὺς κενεῶνας καὶ τὸ στῆθος χαλαρῇ ταινίῃ περιβα-
186 λέειν οὕτως, ὅπως μὴ κωλύσει | τὴν κατάσεισιν·[118] τὰς δὲ χεῖρας παρὰ τὰς πλευρὰς παρατείναντα προσκαταλαβεῖν πρὸς αὐτὸ τὸ σῶμα, καὶ μὴ πρὸς τὴν κλίμακα. ὅταν δὲ ταῦτα σκευάσῃς οὕτως, ἀνέλκειν τὴν κλίμακα ἢ πρὸς τύρσιν τινὰ ὑψηλὴν ἢ πρὸς ἀέτωμα οἴκου· τὸ δὲ χωρίον, ἵνα κατασείεις, ἀντίτυπον ἔστω· τοὺς δὲ ἀνατείνοντας εὐπαιδεύτους εἶναι χρή, ὅπως ὁμαλῶς καὶ καλῶς καὶ ἰσορρόπως καὶ ἐξαπιναίως ἀφήσουσι, καὶ μήτε ἡ κλῖμαξ ἑτερόρροπος ἐπὶ τὴν γῆν ἀφίξεται, μήτε αὐτοὶ προπετέες ἔσονται. ἀπὸ μέντοι τύρσιος ἀφιεὶς ἢ ἀπὸ ἱστοῦ καταπεπηγότος καρχήσιον ἔχοντος ἔτι κάλλιον ἄν τις σκευάσαιτο, ὥστε ἀπὸ τροχιλίης τὰ χαλώμενα εἶναι ὅπλα ἢ ἀπὸ ὄνου. ἀηδὲς μὴν καὶ μακρολογεῖν περὶ τούτων·

where the hump is lower may more appropriately undergo succussion head downward. If then you desire to perform succussion, the following is the proper arrangement. You should cover the ladder with transverse leather or woolen pillows, well tied on, to a rather greater length and breadth than the patient's body will occupy. Next, the patient should be laid on his back against the ladder; and then his feet should be tied at the ankles to the ladder, without being separated, with a strong but soft band. Fasten besides a band below and above each of the knees, and also at the hips; at the flanks and chest bandages are to be passed loosely around, so as not to interfere with the succussion. Tie the arms too, extended along the sides, to the body itself, and not to the ladder. When you have arranged things in this way, lift the ladder against some high tower or house gable. The ground where you do the succussion should be solid, and the assistants who do the lifting well-trained, so that they are able to set the ladder down smoothly, neatly, vertically, and promptly; in this way neither will the ladder meet the ground unevenly, nor will the patients themselves be pulled forward. When the ladder is let down by engaging tackle from a tower or from a ship's mast fixed in the ground, it is a still better arrangement to have the lowering tackle run from a pulley or wheel and axle. It is truly disagreeable to be compelled to enlarge on

---

115 *καταστορέσαι* MV: *-στρῶσαι* B
116 Add. *καὶ* MV
117 *εὐόχῳ* B: *εὐβρόχῳ* MV
118 *κατάσεισιν* MV: *κατάτασιν* B

ὅμως δὲ ἐκ τούτων ἂν τῶν παρασκευῶν μάλιστα ἄν τις κατασεισθείη. |

188 44. Εἰ μέντοι κάρτα ἄνω εἴη τὸ ὕβωμα, δέοι δὲ κατασείειν πάντως, ἐπὶ πόδας[119] λυσιτελέει, ὥσπερ ἤδη εἴρηται· πλείων γὰρ οὕτω γίνεται ἡ καταρροπὴ ἐπὶ ταῦτα. ἑρμάσαι δὲ χρὴ κατὰ μὲν τὸ στῆθος πρὸς τὴν κλίμακα προσδήσαντα ἰσχυρῶς, κατὰ δὲ τὸν αὐχένα ὡς χαλαρωτάτῃ ταινίῃ, ὅσον τοῦ κατορθοῦσθαι[120] εἵνεκα· καὶ αὐτὴν τὴν κεφαλὴν κατὰ τὸ μέτωπον προσδῆσαι πρὸς τὴν κλίμακα· τὰς δὲ χεῖρας παρατανύσαντα πρὸς τὸ σῶμα προσδῆσαι, καὶ μὴ πρὸς τὴν κλίμακα. τὸ μέντοι ἄλλο σῶμα ἄδετον εἶναι χρή, πλήν, ὅσον τοῦ κατορθοῦσθαι εἵνεκα, ἄλλῃ καὶ ἄλλῃ ταινίῃ χαλαρῇ περιβεβλῆσθαι. ὅπως δὲ μὴ κωλύωσιν οὗτοι οἱ δεσμοὶ τὴν κατάσεισιν, σκοπεῖν· τὰ δὲ σκέλεα πρὸς μὲν τὴν κλίμακα μὴ προσδεδέσθω, πρὸς ἄλληλα δέ, ὡς κατὰ τὴν ῥάχιν ἰθύρροπα ᾖ. ταῦτα μέντοι τοιουτοτρόπως ποιητέα, εἰ πάντως δέοι ἐν κλίμακι κατασεισθῆναι. αἰσχρὸν μέντοι ἐν πάσῃ τέχνῃ, καὶ οὐχ ἥκιστα ἐν ἰατρικῇ, πολὺν ὄχλον καὶ πολλὴν ὄψιν καὶ πολὺν λόγον παρασχόντα, ἔπειτα μηδὲν ὠφελῆσαι.

190 45. Χρὴ δὲ πρῶτον μὲν γινώσκειν τὴν φύσιν τῆς ῥάχιος, οἵη τίς ἐστιν· ἐς πολλὰ γὰρ νοσήματα προσδέοι ἂν αὐτῆς. τοῦτο μὲν γάρ, τὸ πρὸς τὴν κοιλίην ῥέπον οἱ σφόνδυλοι ἐντὸς ἄρτιοί εἰσιν ἀλλήλοισι, καὶ δέδενται πρὸς ἀλλήλους δεσμῷ μυξώδει καὶ νευρώδει, ἀπὸ χόνδρων ἀποπεφυκότι ἄχρι πρὸς τὸν νωτιαῖον.

these matters; but all the same, succussion would be best done with the aid of this apparatus.

44. If the hump is very high up, but succussion absolutely required, it is advantageous to do it in the direction of the feet, as was said before; for in this direction the downward impulsion is greater. You should support the patient at the chest, by binding him firmly to the ladder, but at the neck employ the slackest possible band sufficient to keep it straight; also bind the head itself to the ladder at the forehead. Extend the arms along the body and fasten them there, rather than to the ladder. The rest of the body should not be fastened, except in so far as is required to keep it vertical, with a loose band around it here and there. See to it that these attachments do not hinder the succussion. Do not fasten the legs to the ladder, but to one another, so that they will hang down straight in line with the back. This is the way such things must be done if succussion on a ladder is absolutely required. Truly, it is disgraceful in any art, but especially in medicine, to make a parade of great trouble, display, and talk, and then to produce no benefit at all.

45. First it is necessary to understand the structure of the spine, since this is required for many diseases. On the side facing the body's cavity, the vertebrae are fitted exactly into one another, and bound together by a mucous and sinewy connection extending from the cartilages right to the spinal cord. There are also certain sinewy cords

---

119 κατασείειν add. MV
120 κατορθοῦσθαι MV: κατορθοῦν B

ἄλλοι δέ τινες τόνοι νευρώδεες διανταῖοι πρόσφυτοι παρατέτανται ἔνθεν καὶ ἔνθεν αὐτῶν. αἱ δὲ φλεβῶν καὶ ἀρτηρίων κοινωνίαι ἐν ἑτέρῳ λόγῳ δεδηλώσονται, ὅσαι τε καὶ οἷαι, καὶ ὅθεν ὡρμημέναι, καὶ ἐν οἷς οἷα δύνανται, αὐτός τε ὁ νωτιαῖος οἷσιν ἐλύτρωται ἐλύτροισι καὶ ὅθεν ὡρμημένοισι, καὶ ὅπῃ κραίνουσι καὶ οἷσι κοινωνέουσι, καὶ οἷα δυναμένοισιν. ἐν δὲ τῷ ἐπέκεινα ἐν ἄρθροισι γεγιγγλύμωνται πρὸς ἀλλήλους οἱ σφόνδυλοι. τόνοι δὲ κοινοὶ παρὰ πάντας καὶ ἐν τοῖσιν ἔξω μέρεσι καὶ ἐν τοῖσιν εἴσω παρατέτανται· ἀπόφυσίς δέ ἐστιν ὀστέου ἐς τὸ ἔξω μέρος ἀπὸ πάντων τῶν σφονδύλων, μία ἀπὸ ἑνὸς ἑκάστου, ἀπό τε τῶν μεζόνων ἀπό τε τῶν ἐλασσόνων. ἐπὶ δὲ τῇσιν
192 ἀποφύσεσι ταύτῃσι χονδρίων ἐπιφύσιες, καὶ | ἀπ' ἐκείνων νεύρων ἀποβλάστησις ἠδελφισμένη τοῖσιν ἐξωτάτω τόνοισι. πλευραὶ δὲ προσπεφύκασιν, ἐς τὸ ἔσω μέρος τὰς κεφαλὰς ῥέπουσαι μᾶλλον ἢ ἐς τὸ ἔξω· καθ' ἕνα δὲ ἕκαστον τῶν σφονδύλων προσήρθρωνται· καμπυλώταται δὲ πλευραὶ ἀνθρώπου εἰσὶ ῥαιβοειδέα[121] τρόπον. τὸ δὲ μεσηγὺ τῶν πλευρέων καὶ τῶν ὀστέων τῶν ἀποπεφυκότων ἀπὸ τῶν σφονδύλων ἀπο-
194 πληρέουσιν ἑκατέρωθεν | οἱ μύες ἀπὸ τοῦ αὐχένος ἀρξάμενοι, ἄχρι τῆς προσφύσιος τῶν φρενῶν.[122] αὐτὴ δὲ ἡ ῥάχις κατὰ μῆκος ἰθυσκόλιός ἐστιν· ἀπὸ μὲν τοῦ ἱεροῦ ὀστέου ἄχρι τοῦ μεγάλου σφονδύλου, παρ' ὃν προσήρτηται τῶν σκελέων ἡ πρόσφυσις, ἄχρι μὲν τούτου κυφή· κύστις τε γὰρ καὶ γοναὶ καὶ ἀρχοῦ τὸ χαλαρὸν ἐν τούτῳ ἔκτισται. ἀπὸ δὲ τούτου ἄχρι

running the whole length (sc. of the spine) on either side of the vertebrae. The communications of the veins and arteries will be described in another treatise as regards their number, structure, origin, and functions; also, the spinal cord itself with regard to its coverings, their origin, their endings, their connections, and their functions. On the other side (i.e., facing to the outside) the vertebrae are connected with one another by hinge-like joints; cords shared by them all stretch along both their inner and outer sides. From every vertebra there is an outgrowth (apophysis) of bone posteriorly, one from each, both the larger and smaller; and on the apophyses are epiphyses of cartilage from which tendons grow out, which are in turn related to the outermost cords. The ribs are articulated severally with each of the vertebrae, their heads being disposed rather inward (forward) than outward (backward). The ribs in humans are the most curved and crooked shaped. As to the part between the ribs and the bony outgrowths (apophyses) of the vertebrae, it is filled on each side by muscles that begin at the neck and extend to the attachment of the diaphragm. The spine itself is curved vertically through its length. From the sacrum to the great vertebra (L5), near which the origin of the legs is inserted, all this is concave forward, making room for the bladder, generative organs, and loose part of the rectum to be lodged. From this point to the attachment of the

---

121 ῥαιβοειδέα BMV: ῥυκο- (= ῥοικο-) add. M as alternative in marg.

122 τῶν φρ. B: trans. MV before τῆς προσφύσιος

φρενῶν προσαρτήσιος ἰθυλόρδη, καὶ παραφύσιας ἔχει μυῶν τοῦτο μοῦνον τὸ χωρίον ἐκ τῶν εἴσωθεν, ἃς δὴ καλέουσι ψόας. ἀπὸ δὲ τούτου ἄχρι τοῦ μεγάλου σφονδύλου τοῦ ὑπὲρ τῶν ἐπωμίδων ἰθυκύφη· ἔτι δὲ μᾶλλον δοκέει ἤ ἐστιν. ἡ γὰρ ἄκανθα κατὰ μέσον ὑψηλοτάτας τὰς ἐκφύσιας ἔχει τῶν ὀστέων, ἔνθεν δὲ καὶ ἔνθεν ἐλάσσους. αὐτὸ δὲ τὸ ἄρθρον τὸ τοῦ αὐχένος λορδόν ἐστιν. |

196 46. Ὁκόσοισι μὲν οὖν κυφώματα γίνεται κατὰ τοὺς σφονδύλους, ἔξωσις μὲν μεγάλη ἀπορραγεῖσα ἀπὸ τῆς συμφύσιος ἢ ἑνὸς σφονδύλου ἢ καὶ πλειόνων, οὐ μάλα πολλοῖσι γίνεται, ἀλλ᾽ ὀλίγοισιν. οὐδὲ γὰρ τὰ τρώματα τὰ τοιαῦτα ῥηΐδιον γίνεσθαι· οὔτε γὰρ ἐς τὸ ἔξω ἐξωσθῆναι ῥηΐδιόν ἐστιν, εἰ μὴ ἐκ τοῦ ἔμπροσθεν ἰσχυρῷ τινὶ τρωθείη διὰ τῆς κοιλίης—οὕτω δ᾽ ἂν ἀπόλοιτο—ἢ εἴ τις ἀφ᾽ ὑψηλοῦ του χωρίου πεσὼν ἐρείσειε τοῖσιν ἰσχίοισιν ἢ τοῖσιν ὤμοισιν. ἀλλὰ καὶ οὕτως ἂν ἀποθάνοι, παραχρῆμα δὲ οὐκ ἂν ἀποθάνοι. ἐκ δὲ τοῦ ὄπισθεν οὐ ῥηΐδιον τοιαύτην ἔξαλσιν γενέσθαι ἐς τὸ ἔσω, εἰ μὴ ὑπερβαρύ τι ἄχθος ἐμπέσοι·[123] τῶν τε γὰρ ὀστέων τῶν ἐκπεφυκότων ἔξω ἓν[124] ἕκαστον τοιοῦτόν ἐστιν, ὥστε πρόσθεν ἂν αὐτὸ καταγῆναι πρὶν ἢ μεγάλην ῥοπὴν εἴσω ποιῆσαι, τούς τε συνδέσμους βιησάμενον[125] καὶ τὰ ἄρθρα τὰ ἐνηλλαγμένα. ὅ τε αὖ νωτιαῖος πονοίη ἄν, εἰ ἐξ ὀλίγου χωρίου τὴν περικαμπὴν ἔχοι, τοιαύτην ἔξαλσιν ἐξαλλομένου σφονδύλου· ὅ τε ἐκπηδήσας σφόνδυλος πιέζοι ἂν τὸν νωτιαῖον, εἰ μὴ καὶ ἀπορρήξειε. πιεχθεὶς δ᾽ ἂν καὶ

diaphragm, it is convex forward, and to this area, alone in the abdomen, there are attachments of the muscles they call *psoai*. From here to the great vertebra (C7) over the shoulder blades the spine is convex forward, and appears to be even more so than it is; for the ridge has the outgrowths of bone highest here, while above and below they are lower. The articulation of the neck itself is concave forward.

46. Now in cases of outward curvature involving the vertebrae, any great protrusion and rupture of the articulation of one or more of them does not very often occur, but is rare. Such injuries, in fact, are not easy to produce, since it would not be easy for a blow in the outward direction to occur, unless a person was wounded by some violent force coming from the front through the bodily cavity—which would be fatal—or a person falling from a height landed on his hips or his shoulders. But in that case too he would die, although not perhaps at once. From behind, too, it would not be easy for such a luxation to take place inward, unless some very heavy weight fell on the spine; for each of the external bony epiphyses is of such a nature that it would first be fractured itself before making any great deviation inward by overpowering the ligaments and interconnecting joints. The spinal cord, too, would suffer, if there was a sharp curve from a small area caused by such a dislocation of a vertebra. The vertebra in springing out would press on the spinal cord, even if it did not break it. The cord, then, being compressed and inter-

123 *ἐκπέσοι* MV: *ἐμπέσοιτο* B 124 *ἔξω ἓν* B: *ἔξωθεν* M
125 *βιησάμενον* B (M add. as alternative in marg.): *ποιησάμενον* MV

ἀπολελαμμένος πολλῶν ἂν καὶ μεγάλων καὶ ἐπικαίρων ἀπονάρκωσιν ποιήσειεν· ὥστε οὐκ ἂν μέλοι τῷ ἰητρῷ ὅπως χρὴ τὸν σφόνδυλον κατορθῶσαι, πολλῶν
198 καὶ βιαίων ἄλλων κακῶν παρεόντων. ὥστε | δὴ οὐδ' ἐμβαλεῖν οἷόν τε πρόδηλον τὸν τοιοῦτον οὔτε κατασείσει οὔτε ἄλλῳ τρόπῳ οὐδενί, εἰ μή τις διαταμὼν τὸν ἄνθρωπον, ἔπειτα ἐσμασάμενος ἐς τὴν κοιλίην, ἐκ τοῦ εἴσωθεν τῇ χειρὶ ἐς τὸ ἔξω ἀντωθέοι—καὶ ταῦτα νεκρῷ μὲν οἷόν τε ποιέειν, ζῶντι δὲ οὐ πάνυ. διὰ τί οὖν ταῦτα γράφω; διότι οἴονταί[126] τινες ἰητρευκέναι ἀνθρώπους οἷσιν ἔσωθεν ἔπεσον σφόνδυλοι, τελέως ὑπερβάντες τὰ ἄρθρα· καίτοι γε ῥηΐστην ἐς τὸ περιγενέσθαι τῶν διαστροφέων ταύτην ἔνιοι νομίζουσι καὶ οὐδὲν δεῖσθαι ἐμβολῆς, ἀλλὰ αὐτόματα ὑγιέα γίνεσθαι τὰ τοιαῦτα. ἀγνοέουσι δὲ πολλοί, καὶ κερδαίνουσιν ὅτι ἀγνοέουσι· πείθουσι γὰρ τοὺς πέλας. ἐξαπατῶνται δὲ διὰ τόδε· οἴονται τὴν ἄκανθαν τὴν ἐξέχουσαν κατὰ τὴν ῥάχιν ταύτην τοὺς σφονδύλους αὐτοὺς εἶναι, ὅτι στρογγύλον αὐτῶν ἕκαστον φαίνεται ψαυόμενον, ἀγνοέοντες ὅτι τὰ ὀστέα ταῦτά ἐστι τὰ ἀπὸ τῶν σφονδύλων πεφυκότα, περὶ ὧν ὁ λόγος ὀλίγῳ πρόσθεν εἴρηται. οἱ δὲ σφόνδυλοι πολὺ προσωτέρω ἄπεισιν· στενοτάτην γὰρ πάντων ζῴων ἄνθρωπος κοιλίην ἔχει, ὡς ἐπὶ τῷ μεγέθει, ἀπὸ τοῦ ὄπισθεν ἐς τοὔμπροσθεν, ποτὶ καὶ κατὰ τὸ στῆθος. ὅταν οὖν τι τούτων τῶν ὀστέων τῶν ὑπερεχόντων ἰσχυρῶς καταγῇ, ἥν τε ἓν ἥν τε πλείω, ταύτῃ ταπεινότερον τὸ χωρίον γίνεται ἢ τὸ ἔνθεν καὶ ἔνθεν, καὶ διὰ τοῦτο

cepted, would produce complete insensibility of many large and important parts (sc. of the body), so that the physician would not have to care about how to adjust the vertebra, in the presence of many other urgent complications. So, then, the impossibility of reducing such a dislocation either by succussion or any other method is obvious, unless after cutting open the patient, someone were to inserted his hand into the body cavity and apply pressure from within outward. This he might do with a corpse, but hardly with a living patient! Why then am I writing this? Because some practitioners believe they have cured people whose vertebrae have fallen inward with complete disarticulation; and there are even some who think this is the easiest distortion to recover from, not even requiring reduction, but that such injuries get well by themselves. There are many ignorant practitioners who profit from this ignorance, and gain credit among their neighbors. Their deception is as follows. They think that the projecting ridge along the spine represents the vertebrae themselves, because each of the processes feels rounded on palpation, not knowing that these bones are natural outgrowths from the vertebrae as I have discussed a little above. But the vertebrae lie much farther to the front, since humans have the narrowest body cavity of all animals relative to their size and measured from behind forward, especially in the chest region. Whenever, therefore, there is a violent fracture of these projecting processes, either one or more, the part is more depressed there than on either side; thus, the

[126] οἴονται MV: οἷόν τε B

ἐξαπατῶνται, οἰόμενοι τοὺς σφονδύλους εἴσω οἴχε-
200 σθαι. προσεξαπατᾷ δὲ ἔτι αὐτοὺς καὶ | τὰ σχήματα
τῶν τετρωμένων· ἢν μὲν γὰρ πειρῶνται καμπύλλε-
σθαι, ὀδυνῶνται, περιτενέος γενομένου ταύτῃ τοῦ δέρ-
ματος ᾗ τέτρωνται, καὶ ἅμα τὰ ὀστέα τὰ κατεηγότα
ἐνθράσσει[127] οὕτω μᾶλλον τὸν χρῶτα. ἢν δὲ λορδαί-
νωσι, ῥᾴους εἰσίν· χαλαρώτερον γὰρ τὸ δέρμα κατὰ
τὸ τρῶμα ταύτῃ γίνεται, καὶ τὰ ὀστέα ἧσσον ἐνθράσ-
σει·[128] ἀτὰρ καὶ ἤν τις ψαύῃ αὐτῶν, κατὰ τοῦτο ὑπεί-
κουσι λορδοῦντες· καὶ τὸ χωρίον κενεὸν καὶ μαλθακὸν
ψαυόμενον ταύτῃ φαίνεται. ταῦτα πάντα τὰ εἰρημένα
προσεξαπατᾷ τοὺς ἰητρούς. ὑγιέες δὲ ταχέως καὶ
ἀσινέες αὐτόματοι οἱ τοιοῦτοι γίνονται· ταχέως γὰρ
πάντα τὰ τοιαῦτα ὀστέα ἐπιπωροῦται, ὅσα χαῦνά
ἐστιν.

47. Σκολιαίνεται μὲν οὖν ῥάχις καὶ ὑγιαίνουσι κατὰ
πολλοὺς τρόπους· καὶ γὰρ ἐν τῇ φύσει καὶ ἐν τῇ χρή-
σει οὕτως ἔχει· ἀτὰρ καὶ ὑπὸ γήραος καὶ ὑπὸ ὀδύνης
ξυνδοτική ἐστιν. αἱ δὲ δὴ κυφώσιες αἱ ἐν τοῖσι
πτώμασιν ὡς ἐπὶ τὸ πολὺ γίνονται, ἢν ἢ τοῖσιν ἰσχί-
οισιν ἐρείσῃ ἢ ἐπὶ τοὺς ὤμους πέσῃ. ἀνάγκη γὰρ ἔξω
φαίνεσθαι ἐν τῷ κυφώματι ἕνα μέν τινα ὑψηλότερον
202 τῶν σφονδύλων, | τοὺς δὲ ἔνθεν καὶ ἔνθεν ἐπὶ ἧσσον·
οὔκουν εἷς ἐπὶ πολὺ ἀποπεπηδηκὼς ἀπὸ τῶν ἄλλων
ἐστίν, ἀλλὰ μικρὸν ἕκαστος ξυνδιδοῖ, ἀθρόον δὲ πολὺ
φαίνεται.[129] διὰ οὖν τοῦτο καὶ ὁ νωτιαῖος μυελὸς εὐ-
φόρως φέρει τὰς τοιαύτας διαστροφάς, ὅτι κυκλώδης
αὐτῷ ἡ διαστροφὴ γίνεται, ἀλλ᾽ οὐ γωνιώδης.

practitioners are deceived, and think the vertebrae have been dislocated inward. And the postures of the patients help to deceive them still more; for if they try to bend forward, they suffer pain due to the skin being stretched at the site of the injury, while at the same time the fractured bones tear their flesh more; but if they hollow their backs, they feel more comfortable, since the skin becomes more relaxed at the wound, and the bones lacerate less. Again, if one touches them, they shrink and bend inward; and the region appears hollow and soft on palpation. All these things I have mentioned contribute to deceive physicians, while such patients recover of themselves quickly and without damage; for callus forms rapidly on all bones of this kind, by reason of their being porous.

47. There are many kinds of curvature of the spine, even in healthy persons, occurring both due to its natural configuration and to how it is employed; and besides, it may also collapse in old age, or as the result of pain. Outward curvatures usually occur when a patient lands on his hip or falls on his shoulder. In this curvature, one of the vertebrae necessarily appears to stand out more prominently, but those on either side of it less so. It is not that one has sprung out to a distance from the rest; but each gives way a little, and the displacement taken altogether appears great. This is why the spinal marrow tolerates such distortions well, because the distortion affecting it is curved and not angular.

---

127 ἐνθράσ(σ)ει MV: ἐκθρ. B
128 ἐνθράσσει M: ἐκθρ. BV
129 φαίνεται B: γίνεται MV

Χρὴ δὲ τὴν κατασκευὴν τοῦ διαναγκασμοῦ τοιήνδε κατασκευάσαι. ἔξεστι μὲν ξύλον ἰσχυρὸν καὶ πλατύ, ἐντομὴν παραμηκέα ἔχον, κατορύξαι· ἔξεστι δὲ ἀντὶ τοῦ ξύλου ἐν τοίχῳ ἐντομὴν παραμηκέα ἐνταμέειν, ἢ πήχει ἀνωτέρω τοῦ ἐδάφεος, ἢ ὅπως ἂν μετρίως ἔχῃ. ἔπειτα οἷον στύλον δρύϊνον τετράγωνον πλάγιον
204 παραβάλλειν, ἀπολείποντα ἀπὸ | τοῦ τοίχου ὅσον παρελθεῖν τινά, ἢν δέῃ· καὶ ἐπὶ μὲν τὸν στύλον ἐπιστορέσαι ἢ χλαίνας ἢ ἄλλο τι, ὃ μαλθακὸν μὲν ἔσται, ὑπείξει δὲ μὴ μέγα. τὸν δὲ ἄνθρωπον πυριῆσαι, ἢν ἐνδέχηται, ἢ[130] πολλῷ καὶ θερμῷ λοῦσαι· ἔπειτα πρηνέα κατακλῖναι κατατεταμένον, καὶ τὰς μὲν χεῖρας αὐτοῦ παρατείναντα κατὰ φύσιν προσδῆσαι πρὸς τὸ σῶμα. ἱμάντι δὲ μαλθακῷ, ἱκανῶς πλατεῖ τε καὶ μακρῷ, ἐκ δύο διανταίων[131] ξυμβεβλημένῳ μέσῳ, κατὰ μέσον τὸ στῆθος δὶς περιβεβλῆσθαι χρή, ὡς ἐγγυτάτω τῶν μασχαλέων· ἔπειτα τὸ περισσεῦον τῶν ἱμάντων κατὰ τὴν μασχάλην ἑκάτερον περὶ τοὺς ὤμους περιβεβλήσθω· ἔπειτα αἱ ἀρχαὶ πρὸς ξύλον ὑπεροειδές τι προσδεδέσθωσαν, ἁρμόζουσαι τὸ μῆκος τῷ ξύλῳ τῷ ὑποτεταμένῳ, πρὸς ὅ τι πρόσβαλλον τὸ ὑπεροειδὲς ἀντιστηρίζοντα κατατείνειν. τοιούτῳ δέ τινι ἑτέρῳ δεσμῷ χρὴ ἄνωθεν τῶν γουνάτων δήσαντα καὶ ἄνωθεν τῶν πτερνέων τὰς ἀρχὰς τῶν ἱμάντων πρὸς τοιοῦτόν τι ξύλον προσδῆσαι. ἄλλῳ δὲ ἱμάντι πλατεῖ καὶ μαλθακῷ καὶ δυνατῷ, ταινιοειδεῖ, πλάτος ἔχοντι καὶ μῆκος ἱκανόν, ἰσχυρῶς περὶ τὰς ἰξύας κύκλῳ περιδεδέσθαι ὡς ἐγγυτάτω τῶν ἰσχίων. ἔπειτα

The apparatus for forcible reduction in this case should be arranged as follows. Fix in the ground a strong broad plank, having in it a transverse groove; or, instead of the plank, a horizontal groove may be cut in a wall, a cubit above the ground, or as may be convenient. Then place some kind of quadrangular oak bench parallel with the wall, and far enough away from it that it will be possible for a person to pass between if necessary; on the board spread cloaks or something else that is soft, but not very yielding. Give the patient a vapor bath if possible, or bathe him in plenty of hot water; then make him lie stretched out in a prone position, and fasten his arms, extending them naturally, against the body. A soft band, sufficiently broad and long, composed of two strands, should be applied at its middle to the middle of the chest, and passed twice around it as near as possible to the axillae; then let what remains of the two bands be passed around the shoulders on each side; next, the ends should be attached to a pestle-shaped pole, adjusting their length to that of the underlying board against which the pestle-shaped pole is placed, and using it as a fulcrum make the extension. A second similar band should be attached above the knees and above the heels, and the ends of the bands fastened to a similar pole. With another soft, strong band, like a headband, of sufficient breadth and length, the patient should be bound firmly around the loins, as near as pos-

130 ἢ B: om. MV
131 διανταίων Littré: -εών BMV

206 τὸ | περισσεῦον τοῦ ταινιοειδέος, ἅμα ἀμφοτέρας τὰς ἀρχὰς τῶν ἱμάντων, προσδῆσαι πρὸς τὸ ξύλον τὸ πρὸς ποδῶν· ἔπειτα κατατείνειν ἐν τούτῳ τῷ σχήματι ἔνθα καὶ ἔνθα, ἅμα μὲν ἰσορρόπως, ἅμα δὲ ἐς ἰθύ. οὐδὲν γὰρ ἂν μέγα κακὸν ἡ τοιαύτη κατάτασις ποιήσειεν, ἢν χρηστῶς σκευασθῇ, εἰ μὴ ἄρα ἐξεπίτηδές τις βούλοιτο σίνεσθαι. τὸν δὲ ἰητρὸν χρὴ ἢ ἄλλον, ὅστις ἰσχυρὸς καὶ μὴ ἀμαθής, ἐπιθέντα τὸ θέναρ τῆς χειρὸς ἐπὶ[132] τὸ ὕβωμα, τὴν δὲ ἑτέρην χεῖρα προσεπιθέντα τὸ θέναρ τῆς ἑτέρης χειρὸς[133] καταναγκάζειν, προσξυνιέντα, ἤν τε ἐς ἰθὺ ἐς τὸ κάτω πεφύκῃ καταναγκάζεσθαι, ἤν τε πρὸς τῆς κεφαλῆς, ἤν τε πρὸς τῶν ἰσχίων. καὶ ἀσινεστάτη μὲν αὕτη ἡ ἀνάγκη· ἀσινὲς δὲ καὶ ἐπικαθέζεσθαί τινα ἐπὶ τὸ κύφωμα, αὐτοῦ ἅμα κατατεινομένου, καὶ ἐνσεῖσαι μετεωρισθέντα. ἀτὰρ καὶ ἐπιβῆναι τῷ ποδὶ καὶ ὀχηθῆναι ἐπὶ τὸ κύφωμα, ἡσύχως τε ἐπενσεῖσαι οὐδὲν κωλύει. τὸ τοιοῦτο δὲ ποιῆσαι μετρίως ἐπιτήδειος ἄν τις εἴη τῶν ἀμφὶ παλαίστρην εἰθισμένων. δυνατωτάτη μέντοι τῶν ἀναγκέων ἐστίν, εἰ ὁ μὲν τοῖχος ἐντετμημένος ἢ τὸ ξύλον τὸ κατορωρυγμένον, ᾗ ἐντέτμηται, κατωτέρω εἴη τῆς ῥάχιος τοῦ ἀνθρώπου, ὅκως ἂν δοκέῃ μετρίως ἔχειν, σανὶς δὲ φιλυρίνη, μὴ λεπτή, ἐνείη, ἢ καὶ ἄλλου τινὸς ξύλου· ἔπειτα ἐπὶ μὲν τὸ ὕβωμα ἐπιτεθείη ἢ τρύχιόν τι πολύπτυχον ἢ μικρόν τι σκύτινον ὑποκεφάλαιον· ὡς ἐλάχιστα μὴν ὑποκεῖσθαι ξυμφέρει,
208 μόνον προμηθεόμενον ὡς | μὴ ἡ σανὶς ὑπὸ σκληρότητος ὀδύνην παρὰ καιρὸν προσπαρέχῃ. κατὰ ἴξιν δὲ

sible to the hips. Then fasten what is left over of this band, as well as the ends of the two other straps, to the pole at the foot end; next, make extension in this position toward both ends at the same time, equally, and in a straight line. Such extension would do no great harm, if well arranged, unless indeed someone wanted deliberately to do harm. The surgeon, or an assistant who is strong and not untrained, should put the palm of one of his hands on the hump, and the palm of the other on the first hand, to reduce it forcibly, taking into consideration whether the reduction should naturally be made straight downward, or toward the head, or toward the hips. This method of reduction too is very safe; indeed, it will do no harm even if someone sits on the hump while the extension is being applied, and makes succussion by lifting himself up. Indeed, nothing speaks against putting your foot on the hump and making gentle succussion by bringing your weight to bear there. A suitable person to perform such an operation properly would be one with experience in the wrestling school. But the most effective method of reduction is to arrange the incision in a wall, or to use a post embedded in the ground, at an appropriate level somewhat below that of the patient's spine, with a sturdy plank of lime or other wood inserted in it. Then place many thicknesses of cloth or a small leather pillow on the hump: this should be as small as possible, only sufficient to prevent the plank from causing needless additional pain by its hardness. Let the hump come as nearly as possible into

---

132 ἐπὶ BV: ὑπὸ M

133 θέναρ . . . χειρὸς B: ἐπὶ τὴν ἑτέρην MV

ἔστω ὡς μάλιστα τῇ ἐντομῇ τῇ ἐς τὸν τοῖχον τὸ ὕβωμα, ὡς ἂν ἡ σανίς, ᾗ μάλιστα ἐξέστηκε, ταύτῃ μάλιστα πιέζῃ ἐπιτεθεῖσα. ὅταν δὲ ἐπιτεθῇ, τὸν μέν τινα καταναγκάζειν χρὴ τὸ ἄκρον τῆς σανίδος, ἤν τε ἕνα δέῃ ἤν τε δύο, τοὺς δὲ κατατανύειν τὸ σῶμα κατὰ μῆκος, ὡς πρόσθεν εἴρηται, τοὺς μὲν τῇ, τοὺς δὲ τῇ. ἔξεστι δὲ καὶ ὀνίσκοισι τὴν κατάτασιν ποιέεσθαι, ἢ παρακατορύξαντα παρὰ τὸ ξύλον, ἢ ἐν αὐτῷ τῷ ξύλῳ τὰς φλιὰς τῶν ὀνίσκων ἐντεκτηνάμενον, ἤν τε ὀρθὰς θέλῃς,[134] μικρὸν ὑπερεχούσας, ἤν τε κατὰ κορυφὴν τοῦ ξύλου ἔνθεν καὶ ἔνθεν. αὗται αἱ ἀνάγκαι εὐταμίευτοί εἰσι καὶ ἐς τὸ ἰσχυρότερον καὶ ἐς τὸ ἧσσον, καὶ ἰσχὺν ἔχουσι τοιαύτην, ὥστε καὶ εἴ τις ἐπὶ λύμῃ βούλοιτο, ἀλλὰ μὴ ἐπὶ ἰητρείῃ, ἐς τοιαύτας ἀνάγκας ἀγαγεῖν, καὶ γὰρ ἂν κατατείνων κατὰ μῆκος μοῦνον ἔνθεν καὶ ἔνθεν οὕτω καὶ ἄλλην ἀνάγκην οὐδεμίαν προστιθείς, ὅμως κατατείνειεν ἄν τις· ἀλλὰ μὴν καὶ εἰ μὴ κατατείνων, αὐτῇ δὲ μοῦνον τῇ σανίδι οὕτως ἰποίη[135] τις, καὶ οὕτως ἂν καταναγκάσειε. καλαὶ οὖν αἱ τοιαῦται ἰσχύες εἰσίν, ᾗσιν ἔξεστι καὶ ἀσθενεστέρῃσι καὶ ἰσχυροτέρῃσι χρῆσθαι αὐτὸν ταμιεύοντα.
210 καὶ μὲν δὴ καὶ κατὰ φύσιν γε ἀναγκάζουσι· | τὰ μὲν γὰρ ἐξεστεῶτα ἐς τὴν χώρην ἀναγκάζει ἡ ἴπωσις ἰέναι, τὰ δὲ ξυνεληλυθότα κατὰ φύσιν κατατείνουσιν αἱ κατὰ φύσιν κατατάσιες. οὔκουν ἔχω τούτων ἀνάγκας καλλίους οὐδὲ δικαιοτέρας· ἡ γὰρ κατ᾽ αὐτὴν τὴν ἄκανθαν ἰθυωρίη τῆς κατατάσιος κάτωθέν τε καὶ κατὰ τὸ ἱερὸν ὀστέον καλεόμενον οὐκ ἔχει ἐπιλα-

line with the groove in the wall, so that the plank, when in place, exerts the greatest pressure on the part projecting most. When it is put in place, an assistant, or two if necessary, should press down the extremity of the plank, while others extend the body lengthwise, some at one end, some at the other, as was described above. But it is possible to make the extension by a wheel and axle, either embedded in the earth beside the board, or with the supports of the axle carpentered on to the board itself, either projecting upward a little, if you like, or on the top of the board at each end. This reduction apparatus is easy to regulate as regards the amount of force, and has so much power that, if someone wanted to use such forcible maneuvers for harm rather than healing, it is able to act effectively for this purpose, too. For even by making traction lengthwise only at both ends and without any other additional force, sufficient extension could be produced. On the other hand, if, without applying traction, you only pressed downward with the plank in this way, this would bring about reduction. Such forces, then, are good where it is possible for the operator to regulate their use with regard to strength, and, what is more, they act in accordance with nature; for the pressure forces the protruding parts into place, and the extensions, being according to nature, draw apart naturally the parts which have come together. For my part, then, I know of no better or more correct modes of reduction than these, since straight-line extension of the spine itself from below at the so-called "sacred bone"

134 ἑκατέρωθεν add. MV
135 ἰποίῃ Littré: εἰ ποιέῃ BM: -ὶ ποιέει V

βὴν[136] *οὐδεμίαν· ἄνωθεν δὲ καὶ κατὰ τὸν αὐχένα καὶ κατὰ τὴν κεφαλὴν ἐπιλαβὴν μὲν ἔχει, ἀλλ᾽ ἐσιδέειν τε ἀπρεπὴς ταύτῃ γινομένη ἡ κατάτασις καὶ ἄλλας βλάβας ἂν προσπαράσχοι πλεονασθεῖσα. ἐπειρήθην δὲ δή*[137] *ποτε ὕπτιον τὸν ἄνθρωπον κατατείνειν, ἀσκὸν ἀφύσητον ὑποθεὶς ὑπὸ τὸ ὕβωμα· καὶ ἔπειτα αὐλῷ ἐκ χαλκείου ἐς τὸν ἀσκὸν τὸν ὑποκείμενον φυσᾶν ἐνιέναι· ἀλλά μοι οὐκ εὐπορεῖτο· ὅτε μὲν γὰρ εὖ κατατείνοιμι τὸν ἄνθρωπον, ἡσσᾶτο ὁ ἀσκός, καὶ οὐκ ἠδύνατο ἡ φῦσα ἐσαναγκάζεσθαι·*[138] *καὶ ἄλλως ἕτοιμον ἐς τὸ περιολισθάνειν ἦν, ἅτε ἐς τὸ αὐτὸ ἀναγκαζόμε-*
212 *νον τό τε τοῦ ἀνθρώπου* | *ὕβωμα καὶ τὸ τοῦ ἀσκοῦ πληρουμένου κύρτωμα. ὅτε δ᾽ αὖ μὴ κάρτα κατατείνοιμι τὸν ἄνθρωπον, ὁ μὲν ἀσκὸς ἐκυρτοῦτο ἂν ὑπὸ τῆς φύσης· ὁ δὲ ἄνθρωπος πάντῃ μᾶλλον ἐλορδαίνετο ἢ ᾗ ξυνέφερεν. ἔγραψα δὲ ἐπίτηδες τοῦτο· καλὰ γὰρ καὶ ταῦτα τὰ μαθήματά ἐστιν, ἃ πειρηθέντα ἀπορηθέντα ἐφάνη, καὶ δι᾽ ἅσσα ἠπορήθη.*

48. *Ὁκόσοισι δὲ ἐς τὸ ἔσω σκολιαίνονται οἱ σφόνδυλοι ὑπὸ πτώματος, ἢ καὶ ἐμπεσόντος τινὸς βαρέος, εἷς μὲν ἢ οὐδεὶς τῶν σφονδύλων μέγα ἐξίσταται κάρτα ὡς ἐπὶ τὸ πολὺ ἐκ τῶν ἄλλων, ἢν δὲ ἐκστῇ μέγα ἢ εἷς ἢ πλείονες, θάνατον φέρουσιν· ὥσπερ δὲ καὶ πρόσθεν εἴρηται, κυκλώδης καὶ αὕτη*[139] *γίνεται ἡ παραλλαγή. οὖρα μὲν οὖν τοῖσι τοιούτοισι καὶ ἀπόπατος μᾶλλον ἵσταται ἢ τοῖσιν ἔξω κυφοῖσι· καὶ πόδες καὶ ὅλα τὰ σκέλεα ψύχεται μᾶλλον, καὶ θανατηφόρα ταῦτα μᾶλλον ἐκείνων, καὶ ἢν περιγένωνται δέ, ῥυώ-*

gets no grip, whereas from above, at the neck and head, it does get a grip, although extension made there looks unseemly, and might well also cause other injuries if carried to excess. I once tried to make extension with the patient on his back, and, after putting an uninflated bladder under the hump, then tried to blow air into the bladder with a bronze tube. But my attempt was not a success, for when I got the patient well stretched, the bladder collapsed, and air could not be forced into it; and besides the patient's hump was apt to slip around when it was brought forcibly together with the convexity of the inflated bladder; while when I made no great extension of the patient, but got the bladder well inflated, the person's back was hollowed as a whole rather than where it should have been. I relate this on purpose, for hypothetical methods are also instructive, which, on being tried, reveal themselves to be failures, and demonstrate why they have failed.

48. In cases where the vertebrae curve inward from a fall or the impact of some heavy weight, no single vertebra is generally much displaced from the others; and if there is great displacement of one or more, it brings death. But, as was said before, such a dislocation is also circular. In these cases, then, retention of urine and stools is more frequent than in outward curvatures; the feet and whole lower limbs more usually lose heat, and these injuries are more generally fatal. Even if they survive, such patients

---

[136] ἐπιλαβὴν B: ἐπὶ βλάβην MV (B add. as alternative in marg.) [137] δή MV: δεῖ B [138] ἐσαναγκ. MV: ἀναγκ. B [139] Add. καὶ οὐ γωνιώδης MV

δεες τὰ οὖρα μᾶλλον οὗτοι, καὶ τῶν σκελέων ἀκρατέστεροι καὶ ναρκωδέστεροι. ἢν δὲ καὶ ἐν τῷ ἄνω μέρει μᾶλλον τὸ λόρδωμα γένηται, παντὸς τοῦ σώματος ἀκρατέες καὶ καταναρκωμένοι γίνονται. μηχανὴν δὲ
214 οὐκ ἔχω οὐδεμίαν ἔγωγε, ὅπως χρὴ τὸν | τοιοῦτον ἐς τὸ αὐτὸ καταστῆσαι, εἰ μή τινα ἡ διὰ[140] τῆς κλίμακος κατάσεισις ὠφελέειν οἵη τε εἴη, ἢ καὶ ἄλλη τις τοιαύτη ἴησις ἢ κατάτασις,[141] οἵη περ ὀλίγῳ πρόσθεν εἴρηται. κατανάγκασιν δὲ σὺν τῇ κατατάσει οὐδεμίαν ἔχω, ἥ τις ἂν γίνοιτο ὡς τῷ κυφώματι τὴν κατανάγκασιν ἡ σανὶς ἐποιέετο. πῶς γὰρ ἄν τις ἐκ τοῦ ἔμπροσθεν τῆς κοιλίης ἀναγκάσαι δύναιτο; οὐ γὰρ οἷόν τε. ἀλλὰ μὴν οὔτε βῆχες οὔτε πταρμοὶ οὐδεμίαν δύναμιν ἔχουσιν, ὥστε τῇ κατατάσει ξυντιμωρέειν· οὐ μὴν οὐδὲ ἔνεσις φύσης ἐνιεμένης ἐς τὴν κοιλίην οὐδὲν ἂν δυνασθείη. καὶ μὴν αἱ μεγάλαι σικύαι προσβαλλόμεναι ἀνασπάσιος εἵνεκα δῆθεν τῶν εἴσω ῥεπόντων σφονδύλων μεγάλη ἁμαρτὰς τῆς γνώμης ἐστίν· ἀπωθέουσι γὰρ μᾶλλον ἢ ἀνασπῶσι· καὶ οὐδὲ αὐτὸ τοῦτο γινώσκουσι οἱ προσβάλλοντες. ὅσῳ γὰρ ἄν τις μέζω προσβάλλῃ, τοσούτῳ μᾶλλον λορδοῦνται οἱ προσβληθέντες, συναναγκαζομένου ἄνω τοῦ δέρματος. τρόπους δὲ ἄλλους κατατασίων, ἢ οἷοι πρόσθεν εἴρηνται, ἔχοιμι ἂν εἰπεῖν ἁρμόζειν ἂν δοκέοντας μᾶλλον τῷ παθήματι· ἀλλ᾽ οὐ κάρτα πιστεύω αὐτοῖσι· διὰ
216 τοῦτο οὐ γράφω. ἀθρόον δὴ ξυνιέναι χρὴ περὶ | τῶν τοιούτων, ὧν ἐν κεφαλαίῳ εἴρηται, ὅτι τὰ μὲν ἐς τὸ λορδὸν ῥέψαντα ὀλέθριά ἐστι καὶ σινάμωρα, τὰ δὲ ἐς

are more liable to have incontinence of urine, as well as greater weakness and torpor of the legs; if the incurvation occurs higher up, there is weakness and complete torpor of the whole body. For my part, I know of no method for reducing such an injury, unless succussion on the ladder may possibly be of use, or some other such treatment or extension, as was described a little above. I have no means of applying pressure combined with extension, which might bring about reduction, as the plank did in the case of humpback. For how could you apply force from the front through the body cavity? It is impossible. Certainly, neither coughs nor sneezings have any power to assist extension; nor indeed would the inflation of air into the body cavity be able to do anything. And indeed, large cupping instruments applied with the idea of drawing out the depressed vertebrae, is a great error of judgment, since in practice they push it in rather than draw it out. And it is just this which the practitioners who apply them fail to see, for the larger the instrument applied, the more patients hollow their backs, as the skin is drawn together and upward. I could mention other modes of extension, besides the ones discussed above, that might seem to be more suitable for the condition, but I have no great faith in any of them, and therefore do not record them. As to cases similar to the ones briefly described, you must bear in mind that generally inward deviations cause death or grievous injury, while deviations in the form of a hump do

140 διὰ B: κατὰ MV
141 κατάτασις B: κατάσεισις MV

τὸ κυφὸν ἀσινέα θανάτου, καὶ οὔρου σχεσίων καὶ ἀποναρκωσίων τὸ ἐπίπαν· οὐ γὰρ ἐντείνει τοὺς ὀχετοὺς τοὺς κατὰ τὴν κοιλίην, οὐδὲ κωλύει εὐρόους εἶναι ἡ ἐς τὸ ἔξω κύφωσις. ἡ δὲ λόρδωσις ταῦτά τε ἀμφότερα ποιέει καὶ ἐς ἄλλα πολλὰ προσγίνεται. ἐπεί τοι πολὺ πλείονες σκελέων τε καὶ χειρῶν ἀκρατέες γίνονται, καὶ καταναρκοῦνται τὸ σῶμα, καὶ οὖρα ἴσχεται αὐτοῖσιν, οἷς ἂν μὴ ἐκστῇ μὲν τὸ ὕβωμα μήτε ἔσω μήτε ἔξω, σεισθέωσι δὲ ἰσχυρῶς ἐς τὴν ἰθυωρίην τῆς ῥάχιος· οἷσι δ᾿ ἂν ἐκστῇ τὸ ὕβωμα, ἧσσον τοιαῦτα πάσχουσι.

49. Πολλὰ δὲ καὶ ἄλλα ἐν ἰητρικῇ ἄν τις τοιαῦτα κατίδοι, ὧν τὰ μὲν ἰσχυρὰ ἀσινέα ἐστὶ καὶ καθ᾿ ἑωυτὰ τὴν κρίσιν ὅλην λαμβάνοντα τοῦ νοσήματος, τὰ δὲ ἀσθενέστερα σινάμωρα, καὶ ἀποτόκους νοσημάτων χρονίους ποιέοντα καὶ κοινωνέοντα τῷ ἄλλῳ σώματι ἐπὶ πλέον. ἐπεὶ καὶ πλευρέων κάτηξις τοιοῦτό τι πέπονθεν· οἷσι μὲν γὰρ ἂν καταγῇ πλευρή, ἢ μία ἢ
218 πλέονες, ὡς τοῖσι πλείστοισι | κατάγνυται, μὴ διασχόντα τὰ ὀστέα ἐς τὸ ἔσω μέρος μηδὲ ψιλωθέντα, ὀλίγοι μὲν ἤδη ἐπυρέτηναν· ἀτὰρ οὐδὲ αἷμα πολλοὶ ἤδη ἔπτυσαν, οὐδὲ ἔμπυοι πολλοὶ γίνονται, οὐδὲ ἔμμοτοι, οὐδὲ ἐπισφακελίσιες τῶν ὀστέων· δίαιτά τε φαύλη ἀρκέει· ἢν γὰρ μὴ πυρετὸς συνεχὴς ἐπιλαμβάνῃ αὐτούς, κενεαγγέειν κάκιον τοῖσι τοιούτοισιν ἢ μὴ κενεαγγέειν, καὶ ἐπωδυνέστερον καὶ πυρετωδέστερον καὶ βηχωδέστερον· τὸ γὰρ πλήρωμα τὸ μέτριον τῆς κοιλίης, διόρθωμα τῶν πλευρέων γίνετα. ἡ δὲ

not usually threaten death, retention of urine, or loss of sensation; for these external curvatures do not stretch the channels passing down through the cavity, nor do they hinder the free flow (sc. of fluids). Inward curvature, however, does both these things, and brings many other complications besides. In fact, many more patients have paralysis of the legs and arms, loss of sensation in the body, and retention of urine when there is no displacement either inward or outward, but only a severe concussion in the line of the backbone; while those who have a displacement backward are less liable to such conditions.

49. In medicine, many similar examples of violent injuries may be observed which are without danger and contain in themselves the whole resolution of the condition, whereas other apparently slighter injuries can be malignant, producing a chronic succession of diseases which spread widely into the rest of the body. Fracture of the ribs is such a (sc. self-limiting) affection; in cases of fractured ribs, whether one or more, as usually occurs, since the bones are not separated and driven inward or laid bare, patients rarely have fever; nor do many of them cough up blood, suppurate internally, or develop wounds requiring lint plugs; nor is there necrosis of the bones. A simple diet suffices; for if the patients are not attacked by chronic fever, fasting is worse in such cases than not fasting, and would involve a greater liability to pain, fever, and coughing; for a moderate fullness of the cavity tends to straighten

κένωσις κρεμασμὸν μὲν τῇσι πλευρῇσι· ποιέει, ὁ δὲ κρεμασμός, ὀδύνην. ἔξωθέν τε αὖ φαύλη ἐπίδεσις ἀρκέει τοῖσι τοιούτοισιν· κηρωτῇ καὶ σπλήνεσι καὶ ὀθονίοισιν ἡσύχως ἐρείδοντα, ὁμαλὴν τὴν ἐπίδεσιν ἐπιδεῖν ἢ και ἐριῶδές τι προσεπιτιθέντα. κρατύνεται δὲ πλευρὴ ἐν εἴκοσιν ἡμέρῃσι· ταχεῖαι γὰρ αἱ ἐπιπωρώσιες τῶν τοιούτων ὀστέων.

50. Ἀμφιφλασθείσης μέντοι τῆς σαρκὸς ἀμφὶ τῇσι πλευρῇσιν ἢ ὑπὸ πληγῆς ἢ ὑπὸ πτώματος ἢ ὑπὸ ἀντερείσιος ἢ ὑπὸ ἄλλου τινὸς τοιουτοτρόπου, πολλοὶ πολὺ ἤδη αἷμα ἔπτυσαν· οἱ γὰρ ὀχετοὶ οἱ παρὰ[142] τὸ λαπαρὸν τῆς πλευρῆς ἑκάστης παρατεταμένοι, καὶ οἱ τόνοι ἀπὸ τῶν ἐν τῷ σώματι ἐπικαιροτάτων τὰς ἀφορμὰς ἔχουσιν. πολλοὶ οὖν ἤδη βηχώδεες καὶ φυματίαι καὶ ἔμπυοι ἐγένοντο καὶ ἔμμοτοι, καὶ ἡ πλευρὴ ἐπεσφακέλισεν αὐτοῖσιν. ἀτὰρ καὶ οἷσι μηδὲν τοιοῦτο προσεγένετο, ἀμφιφλασθείσης τῆς σαρκὸς ἀμφὶ τῇσι πλευρῇσιν, ὅμως δὲ βραδύτερον ὀδυνώμενοι παύονται οὗτοι ἢ οἷσιν ἂν πλευρὴ καταγῇ, καὶ ὑποστροφὰς
220 μᾶλλον | ἴσχει ὀδυνημάτων τὸ χωρίον ἐν τοῖσι τοιούτοισι τρώμασιν ἢ τοῖσιν ἑτέροισιν. μάλα μὲν οὖν μετεξέτεροι καταμελέουσι τῶν τοιούτων σινέων, μᾶλλον ἢ ἢν πλευρὴ κατεαγῇ αὐτοῖσιν· ἀτὰρ καὶ ἰήσιος σκεθροτέρης οἱ τοιοῦτοι δέονται, εἰ σωφρονοῖεν· τῇ τε γὰρ διαίτῃ ξυμφέρει ξυνεστάλθαι, ἀτρεμέειν τε τῷ σώματι ὡς μάλιστα, ἀφροδισίων τε ἀπέχεσθαι, βρωμάτων τε λιπαρῶν καὶ κερχνωδέων, καὶ ἰσχυρῶν πάντων, φλέβα τε κατ' ἀγκῶνα τέμνεσθαι, σιγᾶν τε ὡς

the ribs, while emptiness leaves them hanging, and this suspension causes pain. Externally, again, a simple dressing suffices in such cases, with cerate, compresses, and bandages; apply them smoothly with gentle pressure, or also add a little wool. A rib consolidates in twenty days, for new callus forms rapidly in bones of this kind.

50. When the flesh about the ribs is contused, either by a blow, a fall, a bruise, or something else of the sort, many patients have considerable hemoptysis. For the channels extending along the interstices between the ribs, as well as the cords coming from the most critical parts of the body, have their origin there. Thus, many patients develop coughs, tubercles, and internal suppurations, and require plugging with lint; also, necrosis of the ribs is found in these patients. But even when nothing of this kind occurs after contusion of the flesh around the ribs, such patients still recover from the pain more slowly than in cases where a rib is broken; and the part is more liable to relapses of pain after such injuries than in the other cases. Now it is true that many practitioners tend to neglect such injuries, in comparison to when there is a broken rib; yet in fact they would need to provide more careful treatment, if they wanted to be prudent. It is well to reduce the diet, keep the body at rest as far as possible, and avoid sexual activity, rich foods and any that excite coughing, and all strong nourishment. Incise a vein at the elbow, observe silence as

---

142 παρὰ B: κατὰ MV

μάλιστα, ἐπιδέεσθαί τε τὸ χωρίον τὸ φλασθὲν σπλήνεσι μὴ πολυπτύχοισι, συχνοῖσι δὲ καὶ πολὺ πλατυτέροισι πάντη τοῦ φλάσματος. κηρωτῇ τε ὑποχρίειν,[143] ὀθονίοισί τε πλατέσι σὺν ταινίῃσι πλατείῃσι καὶ μαλθακῇσιν ἐπιδεῖν· ἐρείδειν δὲ μετρίως, ὥστε μὴ κάρτα πεπιέχθαι φάναι τὸν ἐπιδεδεμένον, μηδ᾽ αὖ χαλαρόν· ἄρχεσθαι δὲ τὸν ἐπιδέοντα κατὰ τὸ φλάσμα, καὶ ἐρηρεῖσθαι ταύτῃ μάλιστα· τὴν δὲ ἐπίδεσιν ποιέεσθαι ὡς ἀπὸ δύο ἀρχέων, ἐπιδεῖν τε, ἵνα μὴ περιρρεπὲς τὸ δέρμα τὸ περὶ τὰς πλευρὰς εἴη, ἀλλ᾽ ἰσόρροπον. ἐπιδεῖν δὲ ἢ καθ᾽ ἑκάστην ἡμέρην ἢ παρ᾽ ἑτέρην. ἄμεινον δὲ καὶ τὴν κοιλίην μαλθάξαι κούφῳ τινί, ὅσον κενώσιος εἵνεκεν τοῦ σίτου, καὶ ἐπὶ μὲν δέκα ἡμέρας ἰσχναίνειν, ἔπειτα ἀναθρέψαι τὸ σῶμα καὶ ἁπαλῦναι. τῇ δὲ ἐπιδέσει, ἔστ᾽ ἂν μὲν ἰσχναίνῃς, ἐρηρεισμένῃ μᾶλλον χρῆσθαι· ὁκόταν δὲ ἐς τὸν ἁπαλυσμὸν ἄγῃς, ἐπιχαλαρωτέρῃ. καὶ ἢν μὲν αἷμα ἀπο-
222 πτύσῃ | καταρχάς, τεσσαρακονθήμερον τὴν μελέτην καὶ τὴν ἐπίδεσιν ποιέεσθαι χρή· ἢν δὲ μὴ πτύσῃ τὸ αἷμα, ἀρκέει ἐν εἴκοσιν ἡμέρῃσιν ἡ μελέτη ὡς ἐπὶ τὸ πολύ· τῇ ἰσχύϊ δὲ τοῦ τρώματος τοὺς χρόνους προτεκμαίρεσθαι[144] χρή. ὅσοι δ᾽ ἂν ἀμελήσωσι τῶν τοιούτων ἀμφιφλασμάτων, ἢν καὶ ἄλλο μηδὲν αὐτοῖσι μέζον φλαῦρον γένηται, ὅμως τό γε χωρίον, ᾗ[145] ἀμφεφλάσθη,[146] μυξωδεστέρην τὴν σάρκα ἴσχει ἢ πρόσθεν εἶχεν. ὅκου δέ τι τοιοῦτον ἐγκαταλείπεται, καὶ μὴ εὖ ἐξιποῦται τῇ ἀλθέξει, φλαυότερον μέν, ἢν παρ᾽ αὐτὸ τὸ ὀστέον ἐγκαταλειφθῇ τὸ μυξῶδες· οὔτε

much as possible, and dress the contused part with compresses not much folded, but numerous, and that extend in every direction a good way beyond the contusion. Anoint with cerate, and bandage with broad, soft linen bands, making them suitably firm, so that the patient says there is no great pressure, but also that the dressing is not slack. The dresser should begin at the contusion, and apply the most pressure there; and the bandaging should be done as with a two-headed roller, in such a way that the skin at the ribs does not form folds, but lies smooth. Change the dressing every day or every other day. It is better to relax the bowels with something light, sufficient to clear out the food, and to give a mild diet for ten days; then nourish the body and build it up. Bandaging during the period of attenuation should be rather tighter, but more relaxed when you come to the building up of flesh. If there is hemoptysis to begin with, the treatment and bandaging must be kept up for forty days; if there is no hemoptysis, a twenty-day course of treatment usually suffices. The (sc. critical) times are to be forecast from the gravity of the wound. In cases where such contusions have been neglected, even if nothing worse happens to them, still the tissue in the contused part becomes infiltrated with more mucus than was there before. When anything of this kind is left behind and not well wrung out by the curative process, it is worse if the mucus is left in the region of the

143 ὑπαλείφθειν B: ὑποχρίειν MV
144 προτεκ. V: προστεκ. BM
145 ᾗ B: ἣν MV
146 ἀμφεφλ. Ermerins: ἀμφιφλ. BMV

γὰρ ἔτι ἡ σὰρξ ὁμοίως ἅπτεται τοῦ ὀστέου, τό τε ὀστέον νοσηρότερον γίνεται, σφακελισμοί τε χρόνιοι ὀστέου πολλοῖσιν ἤδη ἀπὸ τοιούτων προφασίων ἐγένοντο. ἀτὰρ καὶ ἢν μὴ παρὰ τὸ ὀστέον, ἀλλ᾿ αὐτὴ ἡ σὰρξ μυξώδης ᾖ, ὅμως[147] ὑποστροφαὶ γίνονται καὶ ὀδύναι ἄλλοτε καὶ ἄλλοτε, ἤν τις τῷ σώματι τύχῃ πονήσας· καὶ διὰ τοῦτο τῇ ἐπιδέσει χρῆσθαι[148] χρή, ἅμα μὲν ἀγαθῇ, ἅμα δὲ ἐπὶ πολὺ προηκούσῃ, ἕως ἂν ξηρανθῇ μὲν καὶ ἀναποθῇ τὸ ἐκχύμωμα τὸ ἐν τῇ φλάσει ἐγγενόμενον, αὐξηθῇ δὲ σαρκὶ ὑγιέϊ τὸ χωρίον, ἅψηται δὲ τοῦ ὀστέου ἡ σάρξ. οἷσι δ᾿ ἂν ἀμεληθεῖσι χρονιωθῇ καὶ ὀδυνῶδες τὸ χωρίον γένηται[149] καὶ ἡ σὰρξ ὑπόμυξος, τούτοισι καῦσις ἴησις ἀρίστη. καὶ ἢν μὲν αὐτὴ ἡ σὰρξ ᾖ μυξώδης, ἄχρι τοῦ ὀστέου καίειν χρή, μὴ μὴν διαθερμανθῆναι τὸ ὀστέον. ἢν δὲ μεσηγὺ τῶν πλευρῶν ᾖ, ἐπιπολῆς μὲν οὐδὲ οὕτω χρὴ
224 καίειν, φυλάσσεσθαι μέντοι μὴ διακαύσῃς πέρην. | ἢν δὲ πρὸς τῷ ὀστέῳ δοκῇ εἶναι τὸ φλάσμα, καὶ ἔτι νεαρὸν ᾖ, καὶ μή πω σφακελίζῃ τὸ ὀστέον, ἢν μὲν κάρτα ὀλίγον ᾖ, οὕτω καίειν χρὴ ὥσπερ εἴρηται· ἢν μέντοι παραμηκὴς ᾖ ὁ μετεωρισμὸς ὁ κατὰ τὸ ὀστέον, πλέονας ἐσχάρας ἐμβάλλειν χρή. περὶ δὲ σφακελισμοῦ πλευρῆς ἅμα τῇ τῶν ἐμμότων ἰητρείῃ εἰρήσεται.

51. Ἢν δὲ μηροῦ ἄρθρον ἐξ ἰσχίου ἐκπέσῃ, ἐκπίπτει δὲ κατὰ τέσσαρας τρόπους· ἐς μὲν τὸ ἔσω πολὺ[150] πλειστάκις· ἐς δὲ τὸ ἔξω τῶν ἄλλων πλειστάκις· ἐς δὲ τὸ ὄπισθεν καὶ τὸ ἔμπροσθεν ἐκπίπτει μέν, ὀλιγάκις δέ. ὁκόσοισι μὲν οὖν ἂν ἐς τὸ ἔσω ἐκβῇ, μακρότερον

bone itself; for then the flesh no longer adheres so closely to the bone, and the bone becomes more subject to disease. Chronic necroses of bone are found to arise in many cases from causes like these. Besides, even if the mucoid part is not along the bone, but involves the flesh itself, relapses still occur, as well as periodical pains, if a patient happens to strain his body. For this reason, you should use bandaging, both careful and prolonged, for some time, until the exudation that formed in the bruise is dried up and consumed, the part is reconstituted with healthy flesh, and the flesh is firmly attached to the bone. In neglected cases which have become chronic, when the part is painful and the flesh rather mucoid, the best treatment is to cauterize. If the flesh itself is mucous, you must cauterize down to the bone, but avoid overheating the bone itself. If the cautery is between the ribs, even then it should not be superficial, although you should take care not to burn right through. But if the contusion appears to have reached the bone, if it is still fresh, and the bone is not yet necrosed, and if the contusion is quite circumscribed, you should cauterize as directed; if there is an elongated swelling over the bone, you should make several eschars. Necrosis of a rib will be discussed along with the treatment of patients that require lint tents.

51. When the joint of the thigh is dislocated from the hip, it is dislocated in one of four ways: far most frequently inward; and of the others the most frequent is outward. Dislocation backward and forward does occur, but is rare. In cases where the displacement is inward, the leg appears

---

147 γοῦν add. MV 148 χρῆσθαι B: δέεσθαι MV
149 γένηατι B: om. MV 150 πολὺ B: om. MV

τὸ σκέλος φαίνεται, παραβαλλόμενον πρὸς τὸ ἕτερον, διὰ δισσὰς προφάσιας εἰκότως· ἐπί τε γὰρ τὸ ἀπὸ τοῦ ἰσχίου πεφυκὸς ὀστέον, τὸ ἄνω φερόμενον πρὸς τὸν κτένα, ἐπὶ τοῦτο ἡ ἐπίβασις τῆς κεφαλῆς τοῦ μηροῦ γίνεται, καὶ ὁ αὐχὴν τοῦ ἄρθρου ἐπὶ τῆς κοτύλης ὀχέεται· ἔξωθέν τε αὖ ὁ γλουτὸς κοῖλος φαίνεται, ἅτε ἔσω ῥεψάσης τῆς κεφαλῆς τοῦ μηροῦ, τό τε αὖ κατὰ τὸ γόνυ τοῦ μηροῦ ἄκρον ἀναγκάζεται ἔξω ῥέπειν, καὶ ἡ κνήμη καὶ ὁ ποὺς ὡσαύτως. ἅτε οὖν ἔξω ῥέποντος τοῦ ποδός, οἱ ἰητροὶ δι' ἀπειρίην τὸν ὑγιέα πόδα πρὸς τοῦτον προσίσχουσιν, ἀλλ' οὐ τοῦτον πρὸς τὸν ὑγιέα. διὰ τοῦτο πολὺ μακρότερον φαίνεται τὸ σιναρὸν τοῦ
226 ὑγιέος. πολλαχῇ δὲ | καὶ ἄλλῃ τὰ τοιαῦτα παραξύνεσιν ἔχει. οὐ μὴν οὐδὲ ξυγκάμπτειν δύνανται κατὰ τὸν βουβῶνα ὁμοίως τῷ ὑγιέϊ· ἀτὰρ καὶ ψαυομένη ἡ κεφαλὴ τοῦ μηροῦ παρὰ[151] τὸν περίνεον ὑπερογκέουσα εὔδηλός ἐστι. τὰ μὲν οὖν σημεῖα ταῦτά ἐστιν, οἷς ἂν ἔσω ἐκπεπτώκῃ ὁ μηρός.

52. Οἷσι μὲν ἂν οὖν ἐκπεσὼν μὴ ἐμπέσῃ, ἀλλὰ καταπορηθῇ καὶ ἀμεληθῇ, ἥ τε ὁδοιπορίη περιφοράδην τοῦ σκέλεος ὥσπερ τοῖσι βουσὶ γίνεται, καὶ ἡ ὄχησις πλείστη αὐτοῖσιν ἐπὶ τοῦ ὑγιέος σκέλεός ἐστι. καὶ ἀναγκάζονται κατὰ τὸν κενεῶνα καὶ κατὰ τὸ ἄρθρον τὸ ἐκπεπτωκὸς κοῖλοι καὶ σκολιοὶ εἶναι· κατὰ δὲ τὸ ὑγιὲς ἐς τὸ ἔξω ὁ γλουτὸς ἀναγκάζεται περιφερὴς εἶναι· εἰ γάρ τις ἔξω τῷ ποδὶ τοῦ ὑγιέος σκέλεος βαίνοι, ἀπωθέοι ἂν τὸ σῶμα τὸ ἄλλο ἐπὶ τοῦ σιναροῦ σκέλεος[152] τὴν ὄχησιν ποιέεσθαι· τὸ δὲ σιναρὸν οὐκ

longer when compared with the other one, for two obvious reasons: first, it is on to the bone arising from the ischium and passing up to the pubes that the head of the thigh comes to rest, with its neck supported against the cotyloid foramen; second, the buttock looks hollow on the outer side because the head of the thigh is turned inward; again, the end of the thigh at the knee is compelled to turn outward, and the leg and the foot likewise. Thus, as the foot inclines outward, practitioners through inexperience bring the foot of the sound limb to it, instead of bringing it to the sound one. This makes the damaged limb appear much longer than the sound one, and the same situation causes misapprehension in a variety of other cases. The patients, moreover, cannot bend their body at the groin as well as with a sound limb; and for the rest, on palpating the head of the thigh, it is easily discovered as an abnormal prominence in the perineum. These then are the signs in cases of internal dislocation of the thigh.

52. In cases where the dislocation is not reduced, but is given up or neglected, walking is performed, as in oxen, by swinging the leg around; and patients place most of their weight on the healthy leg. The body is also of necessity curved in and twisted in the region of the flank and the dislocated joint, while on the sound side the buttock is necessarily rounded outward. For if someone was to walk with the foot of the sound leg turned outward, he would push the rest of his body over, and put its weight on the injured leg, and the injured limb would not be able to

151 παρὰ B: κατὰ MV
152 ἐπὶ τοῦ . . . σκέλεος B: ἐς τὸ σιναρὸν σκέλος MV

ἂν δύναιτο ὀχέειν· πῶς γάρ; ἀναγκάζεται οὖν οὕτω κατὰ τοῦ ὑγιέος σκέλεος τῷ ποδὶ εἴσω βαίνειν, ἀλλὰ μὴ ἔξω· οὕτω γὰρ ὀχέει μάλιστα τὸ σκέλος τὸ ὑγιὲς καὶ τὸ ἑωυτοῦ μέρος τοῦ σώματος καὶ τὸ τοῦ σιναροῦ σκέλεος μέρος. κοιλαινόμενοι δὲ κατὰ τὸν κενεῶνα καὶ κατὰ τὰ ἄρθρα, σμικροὶ φαίνονται καὶ ξύλῳ τῷ ἀντερείδεσθαι ἀναγκάζονται πλάγιοι κατὰ τὸ ὑγιὲς σκέ-
228 λος· δέονται γὰρ ἀντικοντώσιος ταύτῃ. | ἐπὶ τοῦτο γὰρ οἱ γλουτοὶ ῥέπουσι, καὶ τὸ ἄχθος τοῦ σώματος ἔγκειται[153] ἐπὶ τοῦτο. ἀναγκάζονται δὲ καὶ ἐπικύπτειν· τὴν γὰρ χεῖρα τὴν κατὰ τὸ σκέλος τὸ σιναρὸν ἀναγκάζονται κατὰ πλάγιον τὸν μηρὸν ἐρείδειν· οὐ γὰρ δύναται τὸ σιναρὸν σκέλος ὀχέειν τὸ σῶμα ἐν τῇ μεταλλαγῇ τῶν σκελέων, ἢν μὴ κατέχηται πρὸς τὴν γῆν πιεζόμενον. ἐν τούτοισιν οὖν τοῖσι σχήμασιν ἀναγκάζονται ἐσχηματίσθαι, οἷσιν ἂν εἴσω ἐκβὰν τὸ ἄρθρον μὴ ἐμπέσῃ, οὐ προβουλεύσαντος τοῦ ἀνθρώπου, ὅκως ἂν ῥήϊστα ἐσχηματισμένος ᾖ, ἀλλ᾽ αὐτὴ ἡ ξυμφορὴ διδάσκει ἐκ τῶν παρεόντων τὰ ῥήϊστα αἱρέεσθαι. ἐπεὶ καὶ ὅσοι ἕλκος ἔχοντες ἐν ποδὶ ἢ κνήμῃ κάρτα μὴ δύνανται ἐπιβαίνειν τῷ σκέλει, πάντες, καὶ οἱ νήπιοι, οὕτως ὁδοιπορέουσιν· ἔξω γὰρ βαίνουσι τῷ σιναρῷ·[154] καὶ δισσὰ κερδαίνουσι, δισσῶν γὰρ δέονται· τό τε γὰρ σῶμα οὐκ ὀχέεται ὁμοίως ἐπὶ τοῦ ἔξω ἀποβαινομένου, ὥσπερ ἐπὶ τοῦ εἴσω· οὐδὲ γὰρ κατ᾽ ἰθυωρίην αὐτῷ γίνεται τὸ ἄχθος, ἀλλὰ πολλῷ μᾶλλον ἐπὶ τοῦ ὑποβαινομένου·[155] κατ᾽ ἰθυωρίην γὰρ αὐτῷ γίνεται τὸ ἄχθος, ἔν τε αὐτῇ τῇ ὁδοιπορίῃ καὶ τῇ με-

carry it. How could it? He is thus obliged to walk with the foot of the sound leg turned in and not out; for in this way the sound limb is best able to support both its own share of the body and that of the injured leg. But, owing to the inward curvatures at the loin and the joints, patients appear small in stature, and are compelled to support themselves laterally with a wooden crutch on the side of the sound leg. They need support there, because the buttocks incline that way, and the weight of the body lies in that direction. They are also obliged to stoop; for they have to press their hand on the side of the injured leg laterally against their thigh, since the injured limb cannot support the body during the change of legs, unless it is forced down on to the ground by this pressure. Such then are the postures that patients are obliged to assume in cases of unreduced internal dislocation of the hip—not as a result of previous deliberation by the patient as to what will be the easiest posture to assume; but the impediment itself teaches him to choose the easiest posture available. So too those who, when they have a lesion on the foot or leg, can hardly use the limb—all of them, even young children, walk in this way. They turn the injured side outward in walking, and thereby gain two benefits to suit their two needs; for the body is not carried as equally on the limb turned outward as it is on one turned inward, since the weight is not perpendicular to it, but presses much more on the limb that is held under the body, making the weight perpendicular to the latter both in actual walking, and

[153] ἔγκειται B: ὀχέεται MV [154] Add. σκέλει MV
[155] ὑποβαιν. B: ὑπερβαιν. MV

ταλλαγῇ τῶν σκελέων. ἐν τούτῳ τῷ σχήματι τάχιστα ἂν δύναιτο ὑποτιθέναι τὸ ὑγιὲς σκέλος, εἰ τῷ μὲν σιναρῷ ἐξωτέρω βαίνοι, τῷ δὲ ὑγιέϊ ἐσωτέρω. περὶ οὗ οὖν ὁ λόγος, ἀγαθὸν εὑρίσκεσθαι αὐτὸ ἑωυτῷ τὸ
230 *σῶμα ἐς τὰ ῥήϊστα τῶν | σχημάτων. ὅσοισι μὲν οὖν μήπω τετελεωμένοισιν ἐς αὔξησιν ἐκπεσὼν μὴ ἐμπέσοι, γυιοῦται καὶ ὁ μηρὸς καὶ ἡ κνήμη καὶ ὁ πούς· οὔτε γὰρ τὰ ὀστέα ἐς τὸ μῆκος ὁμοίως αὔξεται, ἀλλὰ βραχύτερα γίνεται, μάλιστα δὲ τὸ τοῦ μηροῦ, ἄσαρκόν τε πᾶν τὸ σκέλος καὶ ἄμυον καὶ ἐκτεθηλυσμένον καὶ λεπτότερον γίνεται, ἅμα μὲν διὰ τὴν στέρησιν τῆς χώρης τοῦ ἄρθρου, ἅμα δὲ ὅτι ἀδύνατον χρῆσθαί ἐστιν, ὅτι οὐ κατὰ φύσιν κέεται· χρῆσις γὰρ μετεξετέρη ῥύεται τῆς ἄγαν ἐκθηλύνσιος. ῥύεται δέ τι καὶ τῆς ἐπὶ μῆκος ἀναυξίης. κακοῦται μὲν οὖν μάλιστα οἷς ἂν ἐν γαστρὶ ἐοῦσιν ἐξαρθρήσῃ τοῦτο τὸ ἄρθρον, δεύτερον δὲ οἷς ἂν ὡς νηπιωτάτοισιν ἐοῦσιν, ἥκιστα δὲ τοῖσι τετελεωμένοισιν. τοῖσι μὲν οὖν τετελεωμένοισιν εἴρηται οἵη τις ἡ ὁδοιπορίη γίνεται. οἷσι δ' ἂν νηπίοισιν ἐοῦσιν ἡ ξυμφορὴ αὕτη γένηται,[156] οἱ μὲν πλεῖστοι καταβλακεύουσι τὴν διόρθωσιν τοῦ σώματος, ἀλλὰ εἰλέονται ἐπὶ τὸ ὑγιὲς σκέλος, τῇ χειρὶ πρὸς τὴν γῆν ἀπερειδόμενοι τῇ κατὰ τὸ ὑγιὲς σκέλος. καταβλακεύουσι δὲ ἔνιοι τὴν ἐς τὸ ὀρθὸν[157] ὁδοιπορίην καὶ οἷς ἂν τετελεωμένοισι ἡ ξυμφορὴ αὕτη γένηται. ὁκόσοι δ' ἂν νήπιοι ἐόντες ταύτῃ τῇ ξυμφορῇ χρησάμενοι ὀρθῶς παιδαγωγηθέωσι, τῷ μὲν ὑγιέϊ σκέλει χρέονται ἐς ὀρθόν, ὑπὸ δὲ τὴν μασχάλην*

during the change of legs. It is in this attitude, with the injured leg rather outward and the sound one rather inward, that the sound limb can most rapidly be put under the body. In this matter, then, it is good that the body discovers for itself the easiest postures. Now, when in persons who have not yet completed their growth the dislocated hip remains unreduced, the thigh is maimed, as well as the leg and foot. The bones do not grow to their normal length, but are shorter, especially the one in the thigh, while the whole leg is deficient in flesh and muscle, and becomes flaccid and thinner. This is due simultaneously both to the loss of the joint's place, and to the patient not being able to use the bone as the result of its not resting in its normal position. A certain amount of exercise does, however, save the leg from excessive flaccidity, and in some degree promotes its growth in length. Thus, the greatest damage is done to those in whom this joint is dislocated before birth; next, to those who are very young; and least to adults. In the case of adults, their mode of walking has been described; but when this accident occurs in those who are very young, most are unable to walk with their body upright, but instead crawl about on the sound leg, supporting themselves with the hand on the sound side held to the ground. Some are unable to walk with their body erect, even of those who were injured in adulthood. But persons afflicted by this accident in early childhood and who receive proper training, use the sound leg to stand up on, but carry a crutch under the axilla on that

156 γένηται B: φαίνηται MV
157 ὀρθὸν MV: ἄρθρον B

232 τὴν κατὰ τὸ ὑγιὲς σκέλος σκίπωνα περιφέρουσι, | μετεξέτεροι δὲ καὶ ὑπ' ἀμφοτέρας τὰς χεῖρας. τὸ δὲ σιναρὸν σκέλος μετέωρον ἔχουσι, καὶ τοσούτῳ ῥηΐους εἰσίν, ὅσῳ ἂν αὐτοῖσιν ἔλασσον τὸ σκέλος τὸ σιναρὸν ᾖ· τὸ δὲ ὑγιὲς ἰσχύει αὐτοῖσιν οὐδὲν ἧσσον ἢ εἰ καὶ ἀμφότερα ὑγιέα ἦν. θηλύνονται δὲ πᾶσι τοῖσι τοιούτοισι αἱ σάρκες τοῦ σκέλεος, μᾶλλον δέ τι[158] θηλύνονται ἐκ τοῦ ἔξω μέρεος ἢ ἐκ τοῦ εἴσω ὡς ἐπὶ πολὺ πᾶσιν.

53. Μυθολογοῦσι δέ τινες, ὡς αἱ Ἀμαζόνες τὸ ἄρσεν γένος τὸ ἑωυτῶν αὐτίκα νήπιον ἐὸν ἐξαρθρέουσιν—αἱ μὲν κατὰ γούνατα, αἱ δὲ κατὰ ἰσχία—ὡς δῆθεν χωλὰ γίνοιτο, καὶ μὴ ἐπιβουλεύοι τὸ ἄρσεν γένος τῷ θήλει. χειρώναξιν ἄρα τούτοισι χρέωνται, ὁκόσα ἢ σκυτείης ἔργα ἢ χαλκείης, ἢ ἄλλο ὅ τι ἑδραῖον ἔργον. εἰ μὲν οὖν ἀληθέα ταῦτά ἐστιν, ἐγὼ μὲν οὐκ οἶδα· ὅτι δὲ γένοιτ' ἂν τοιαῦτα οἶδα, εἴ τις ἐξαρθρέοι αὐτίκα νήπια ἐόντα. κατὰ μὲν οὖν τὰ ἰσχία μέζον τὸ διάφορόν ἐστιν ἐς τὸ εἴσω ἢ ἐς τὸ ἔξω ἐξαρθρῆσαι· κατὰ δὲ τὰ γούνατα διαφέρει μέν τι, ἔλασσον δέ τι διαφέρει. τρόπος δὲ ἑκατέρου τοῦ χωλώματος
234 ἴδιός ἐστιν· κυλλοῦνται | μὲν γὰρ μᾶλλον, οἷς ἂν ἐς τὸ ἔξω ἐξαρθρήσῃ· ὀρθοὶ δὲ ἧσσον ἵστανται οἷσιν ἂν ἐς τὸ εἴσω ἐξαρθρήσῃ. ὡσαύτως δὲ καὶ ἢν παρὰ τὸ σφυρὸν ἐξαρθρήσῃ, ἢν μὲν ἐς τὸ ἔξω μέρος, κυλλοὶ μὲν γίνονται, ἑστάναι δὲ δύνανται. ἢν δὲ ἐς τὸ εἴσω μέρος ἐξαρθρήσῃ, βλαισοὶ γίνονται, ἧσσον δὲ ἑστάναι δύνανται. ἥ γε μὴν ξυναύξησις τῶν ὀστέων τοιήδε

side, and some of them under both arms. As for the injured leg, they keep it raised off the ground, and do so the more easily, the smaller the injured leg is; and their sound leg has as much strength as if both were sound. In all such cases the fleshy parts of the leg are flaccid; and, as a general rule, they are somewhat more flaccid (if the dislocation is) on the outer than on the inner side.

53. Some people tell a tale about how the Amazons dislocate the joints of their male offspring in early infancy—some at the knees and some at the hips—so that they will henceforth be lame, and the males will be unable to conspire against the females. They then make use of them as artisans in all kinds of leather or copper work, or any other sedentary occupation. Now whether this is true I for my part do not know, but I do know that this would be the result if children had their joints dislocated. At the hips there is quite a difference between dislocation inward and outward; but at the knees, though there is some difference, it is less. The character of each lameness is peculiar: those in whom the dislocation at the knee is outward are more bandy-legged, while those in whom it is inward are less able to stand erect. Similarly, when the dislocation is at the ankle, if it is outward, they become clubfooted, but are able to stand, whereas if it is inward, they become splayfooted, and are less able to stand. The growth of the bones is as following: when the bone of the leg at the ankle

158 δέ τι Littré: δ' ἔτι BMV

γίνεται· οἷσι μὲν ἂν τὸ κατὰ τὸ σφυρὸν ὀστέον τὸ τῆς κνήμης ἐκστῇ, τούτοισι μὲν τὰ τοῦ ποδὸς ὀστέα ἥκιστα συναύξεται· ταῦτα γὰρ ἐγγυτάτω τοῦ τρώματός ἐστιν· τὰ δὲ τῆς κνήμης ὀστέα αὔξεται οὐ πολὺ ἐνδεεστέρως, αἱ μέντοι σάρκες μινύθουσιν. οἷσι δ' ἂν κατὰ μὲν τὸ σφυρὸν μένῃ τὸ ἄρθρον κατὰ φύσιν, κατὰ δὲ τὸ γόνυ ἐξεστήκῃ, τούτοισι τὸ τῆς κνήμης ὀστέον οὐκ ἐθέλει ξυναύξεσθαι ὁμοίως, ἀλλὰ βραχύτερον[159] γίνεται· τοῦτο γὰρ ἐγγυτάτω τοῦ τρώματός ἐστιν· τοῦ μέντοι ποδὸς τὰ ὀστέα μινύθει μέν, ἀτὰρ οὐχ ὁμοίως, ὥσπερ ὀλίγον τι πρόσθεν εἴρηται, ὅτι τὸ ἄρθρον τὸ παρὰ τὸν πόδα σῶόν ἐστι. εἰ δέ οἱ χρῆσθαι ἠδύναντο, ὡς καὶ τῷ κυλλῷ, ἔτι ἂν ἧσσον ἐμινύθει τὰ τοῦ ποδὸς ὀστέα τούτοισιν. οἷσι δ' ἂν κατὰ τὸ ἰσχίον ἡ ἐξάρθρησις γένηται, τούτοισι τοῦ μηροῦ τὸ ὀστέον οὐκ ἐθέλει ξυναύξεσθαι ὁμοίως (τοῦτο γὰρ ἐγγυτάτω τοῦ τρώματός ἐστιν) ἀλλὰ βραχύτερον τοῦ ὑγιέος γίνεται. τὰ μέντοι τῆς κνήμης ὀστέα οὐχ ὁμοίως τούτοισιν ἀναυξέα γίνεται, οὐδὲ τὰ τοῦ ποδός,[160] διὰ τοῦτο ὅτι τὸ τοῦ μηροῦ ἄρθρον τὸ παρὰ τὴν κνήμην ἐν τῇ ἑωυτοῦ φύσει μένει, καὶ τὸ τῆς
236 κνήμης τὸ παρὰ τὸν | πόδα. σάρκες μέντοι μινύθουσι παντὸς τοῦ σκέλεος τούτοισιν. εἰ μέντοι χρῆσθαι τῷ σκέλει ἠδύναντο, ἔτι ἂν μᾶλλον τὰ ὀστέα ξυνηυξάνετο, ὡς καὶ πρόσθεν εἴρηται, πλὴν ἢ τὸ τοῦ μηροῦ, κἂν ἧσσον ἄσαρκα εἴη, ἀσαρκότερα δὲ πολλῷ ἢ εἰ ὑγιέα ἦν. σημεῖον δὲ ὅτι ταῦτα τοιαῦτά ἐστιν· ὁκόσοι τοῦ βραχίονος ἐκπεσόντος, γαλιάγκωνες ἐγένοντο ἢ

is dislocated, the bones of the foot show the least growth, since they are nearest the injury, but growth of the leg bones is not very deficient; the muscles however become wasted. In cases where the ankle joint keeps its natural position while there is dislocation at the knee, the bone of the leg will not grow like the sound one, but become shortened; for this is nearest to the injury. The bones of the foot are atrophied, but not to the same extent as was noted just above, because the joint at the foot is intact. If patients are able to use the part, as is the case even in clubfoot, the bones of the foot would be still less atrophied. When the dislocation occurs at the hip, the affected thigh bone will not grow like the other one, since it is closest to the injury, but becomes shorter than the sound one; the bones of the leg, however, do not stop growing in the same way, nor do those of the foot, since the joint of the thigh at the knee remains in its natural place, as also the bone of the leg at the foot. But the muscles of the whole leg are atrophied in these cases. If, however, patients were able to use their leg, the bones would correspond in their growth to a still greater extent, as was said before, except for the thigh; and they would be less deficient in flesh, though much more so than if the limb was sound. Here is a proof that these things are so: those who become weasel-armed owing to

---

159 βραχύτερον B: -χύτατον MV
160 Add. ὀστέα B

ἐκ γενεῆς ἢ ἐν αὐξήσει καὶ πρὶν τελεωθῆναι, οὗτοι τὸ μὲν ὀστέον τοῦ βραχίονος βραχὺ ἴσχουσι, τὸν δὲ πῆχυν καὶ ἄκρην τὴν χεῖρα ὀλίγῳ ἐνδεεστέρην τοῦ ὑγιέος, διὰ ταύτας τὰς προφάσιας τὰς εἰρημένας, ὅτι ὁ μὲν βραχίων ἐγγυτάτω [τοῦ ἄρθρου][161] τοῦ τρώματός ἐστιν, ὥστε διὰ τοῦτο βραχύτερος ἐγένετο. ὁ δὲ αὖ πῆχυς διὰ τοῦτο οὐχ ὁμοίως ἐνακούει τῆς ξυμφορῆς, ὅτι τὸ τοῦ βραχίονος ἄρθρον τὸ πρὸς τοῦ πήχεος ἐν τῇ ἀρχαίῃ φύσει μένει, ἥ τε αὖ χεὶρ ἄκρη ἔτι τηλοτέρω ἄπεστιν ἢ ὁ πῆχυς ἀπὸ τῆς συμφορῆς. διὰ ταύτας μὲν τὰς εἰρημένας προφάσιας, τῶν ὀστέων τά τε μὴ ξυναυξανόμενα οὐ ξυναύξεται, τά τε συναυξανόμενα συναύξεται. ἐς δὲ τὸ εὔσαρκον τῇ χειρὶ καὶ τῷ βραχίονι ἡ ταλαιπωρίη τῆς χειρὸς μέγα προσωφελέει· ὅσα γὰρ χειρῶν ἔργα ἐστί, τὰ πλεῖστα προθυμέονται οἱ γαλιάγκωνες ἐργάζεσθαι[162] τῇ χειρὶ ταύτῃ, ὅσα περ καὶ τῇ ἑτέρῃ δύνανται οὐδὲν ἐνδεεστέρως τῆς ἀσινέος. οὐ γὰρ δεῖ ὀχέεσθαι τὸ σῶμα ἐπὶ τῶν χειρῶν, ὡς ἐπὶ τῶν σκελέων, ἀλλὰ κοῦφα αὐτοῖσι τὰ ἔργα ἐστί. διὰ δὲ τὴν χρῆσιν οὐ μινύθουσιν αἱ σάρκες αἱ κατὰ τὴν χεῖρα καὶ κατὰ τὸν πῆχυν τοῖσι γα-
238 λιάγκωσιν· | ἀλλὰ καὶ ὁ βραχίων τι προσωφελέεται ἐς εὐσαρκίην διὰ ταύτην. ὅταν δὲ ἰσχίον ἐκπαλὲς γένηται ἐς τὸ ἔσω μέρος ἐκ γενεῆς ἢ καὶ ἔτι νηπίῳ ἐόντι, μινύθουσιν αἱ σάρκες διὰ τοῦτο μᾶλλον ἢ τῆς χειρός, ὅτι οὐ δύνανται χρῆσθαι τῷ σκέλει. μαρτύριον δέ τι ἐνέσται καὶ ἐν τοῖσιν ὀλίγον ὕστερον εἰρησομένοισιν, ὅτι ταῦτα τοιαῦτά ἐστιν.

dislocation of the arm either congenitally or during adolescence, and before they become adults, have shortness of the bone of the upper arm, but the forearm and hand are little inferior to those on the sound side, for the reasons that have been given, *viz*. that the upper arm is nearest the injury, and on that account is shorter. The forearm, on the contrary, is not equally affected by the injury, because the joint of the arm that articulates with the forearm retains its old position. And the hand, again, is still further away from the injury than the forearm. These are the reasons why bones that do not grow normally are defective in growth, and those that do grow maintain their growth. Exertion of the hand contributes greatly to muscularity in the hand and arm. In fact, in most sorts of handiwork carried out with the arms, the weasel-armed strive to accomplish with the impaired limb no less than with the other arm, and do the work no less effectively than with their sound limb. For the body's weight does not need to be supported on the arms as it is on the legs, and the work done by the weasel-armed is light. As the result of exercise, the flesh of the hand and forearm is not atrophied in the weasel-armed; and even the upper arm gains some further development from this. But when the hip is dislocated inward, either congenitally or in childhood, there is more atrophy of flesh than in the arm, just because they cannot use the leg. A special piece of evidence that this is the case will also be found in what is said a little below.

---

161 Del. Kühlewein Withington
162 ἐργάζ. MV: συνεργάζ. B

54. Οἷσι δ' ἂν ἐς τὸ ἔξω ἡ τοῦ μηροῦ ἡ κεφαλὴ ἐκβῇ, τούτοισι βραχύτερον μὲν τὸ σκέλος φαίνεται παρατεινόμενον παρὰ τὸ ἕτερον, εἰκότως· οὐ γὰρ ἐπ' ὀστέον ἡ ἐπίβασις τῆς κεφαλῆς τοῦ μηροῦ ἐστιν, ὡς ὅτε εἴσω ἐκπέπτωκεν, ἀλλὰ παρ' ὀστέον παρακεκλιμένην[163] τὴν φύσιν ἔχον, ἐν σαρκὶ δὲ στηρίζεται ὑγρῇ καὶ ὑπεικούσῃ· διὰ τοῦτο μὲν βραχύτερον γίνεται. ἔσωθεν δὲ ὁ μηρὸς παρὰ τὴν πλιχάδα καλεομένην κοιλότερος καὶ ἀσαρκότερος φαίνεται· ἔξωθεν δὲ ὁ γλουτὸς κυρτότερος, ἅτε ἐς τὸ ἔξω τῆς κεφαλῆς τοῦ μηροῦ ὠλισθηκυίης. ἀτὰρ καὶ ἀνωτέρω φαίνεται ὁ γλουτὸς, ἅτε ὑπειξάσης τῆς σαρκὸς τῆς ἐνταῦθα τῇ τοῦ μηροῦ κεφαλῇ· τὸ δὲ παρὰ τὸ γόνυ τοῦ μηροῦ ἄκρον ἔσω ῥέπον φαίνεται καὶ ἡ κνήμη καὶ ὁ πούς. ἀτὰρ οὐδὲ συγκάμπτειν ὥσπερ τὸ ὑγιὲς σκέλος δύνανται. τὰ μὲν οὖν σημεῖα ταῦτα τοῦ ἔξω ἐκπεπτωκότος μηροῦ εἰσίν.

55. Οἷσι μὲν οὖν τετελεωμένοισιν ἤδη ἐκπεσὸν τὸ
240 ἄρθρον | μὴ ἐμπέσῃ, τούτοισι βραχύτερον μὲν φαίνεται τὸ σύμπαν σκέλος· ἐν δὲ τῇ ὁδοιπορίῃ τῇ μὲν πτέρνῃ οὐ δύνανται καθικνέεσθαι[164] τῆς γῆς, τῷ δὲ στήθει τοῦ ποδὸς βαίνουσιν ἐπὶ τὴν γῆν· ὀλίγον δὲ ἐς τὸ ἔσω μέρος ῥέπουσι τοῖσι δακτύλοισιν ἄκροισιν. ὀχέειν δὲ τὸ σῶμα δύναται τὸ σιναρὸν σκέλος τούτοισι πολλῷ μᾶλλον ἢ οἷσιν ἂν ἐς τὸ ἔσω μέρος ἐκπεπτώκῃ, ἅμα μὲν ὅτι ἡ κεφαλὴ τοῦ μηροῦ καὶ ὁ αὐχὴν τοῦ ἄρθρου πλάγιος φύσει πεφυκὼς ὑπὸ συχνῷ μέρει τοῦ ἰσχίου τὴν ὑπόστασιν πεποίηται, ἅμα δὲ

54. In cases where the head of the thigh is dislocated outward, the leg is seen to be shorter, when placed beside the other one. Naturally so, for it is no longer on bone that the head of the thigh has its support, as when it was displaced inward, but it lies along the natural slope of the hip bone, and is held up in soft and yielding flesh, making its shortness obvious. The thigh on the inside at what is called the fork (perineum) appears hollower and more emaciated, while the buttock is rather more rounded on the outside, since the head of the thigh is displaced outward; besides this, the buttock seems to be higher, since the flesh at that part recedes from the head of the thigh. But the end of the thigh at the knee appears to turn inward, and with it the leg and foot; furthermore, patients are unable to bend this leg in the same way as the sound leg. These then are the signs of dislocation of the thigh outward.

55. Now in adults, when the joint is not reduced after (sc. outward) dislocation, the whole leg seems to be shorter; and in walking patients are unable to reach the ground with their heel, but walk with the ball of their foot on the ground, and turn the tips of their toes a little inward. But the injured leg can still bear the weight of the body much better in these cases than where there has been a dislocation inward, partly because the head of the thigh and the neck of the joint, being naturally oblique, have gained a support from a large part of the hip, and

163 *παρακεκ.* B: *παρεγκεκ.* MV
164 *ἐπὶ* add. MV

ὅτι ἄκρος ὁ ποὺς οὐκ ἐς τὸ ἔξω μέρος ἀναγκάζεται ἐκκεκλίσθαι,[165] ἀλλ᾽ ἐγγύς ἐστι τῆς ἰθυωρίης τῆς κατὰ τὸ σῶμα καὶ τείνει καὶ ἐσωτέρω. ὅταν οὖν τρίβον μὲν λάβῃ τὸ ἄρθρον ἐν τῇ σαρκὶ ἐς ἣν ἐξεκλίθη, ἡ δὲ σὰρξ γλισχρανθῇ, ἀνώδυνον τῷ χρόνῳ γίνεται· ὅταν δὲ ἀνώδυνον γένηται, δύνανται μὲν ὁδοιπορέειν ἄνευ ξύλου, ἢν ἄλλως βούλωνται· δύνανται δὲ ὀχέειν τὸ σῶμα ἐπὶ τὸ σιναρὸν σκέλος. διὰ οὖν τὴν χρῆσιν ἧσσον τοῖσι τοιούτοισιν ἐκθηλύνονται αἱ σάρκες ἢ οἷσιν ὀλίγον πρόσθεν εἴρηται· ἐκθηλύνονται δὲ ἢ πλεῖον ἢ ἔλασσον· μᾶλλον δέ τι[166] ἐκθηλύνονται κατὰ τὸ εἴσω μέρος ἢ κατὰ τὸ ἔξω ὡς ἐπὶ τὸ πολύ. τὸ μέντοι ὑπόδημα μετεξέτεροι τούτων ὑποδέεσθαι οὐ δύνανται, διὰ τὴν ἀκαμπίην τοῦ σκέλεος, οἱ δέ τινες καὶ δύνανται. οἷσι δ᾽ ἂν ἐν γαστρὶ ἐοῦσιν ἐξαρθρήσῃ τοῦτο τὸ ἄρθρον, ἢ ἔτι ἐν αὐξήσει ἐοῦσι βίῃ ἐκπεσὸν μὴ ἐμπέσῃ, ἢ καὶ ὑπὸ νόσου ἐξαρθρήσῃ τοῦτο τὸ ἄρθρον καὶ ἐκπαλήσῃ—πολλὰ γὰρ τοιαῦτα γίνεται—καὶ ἐνίων μὲν τῶν τοιούτων ἢν ἐπισφακελίσῃ ὁ μηρός, ἐμπυήματα χρόνια καὶ ἔμμοτα γίνεται, καὶ ὀστέων ψιλώσιες ἐνίοισιν· ὁμοίως δὲ καὶ οἷσιν ἐπισφακελίζει
242 | καὶ οἷσι μὴ ἐπισφακελίζει, τοῦ μηροῦ τὸ ὀστέον πολλῷ βραχύτερον γίνεται, καὶ οὐκ ἐθέλει συναύξεσθαι ὥσπερ τοῦ ὑγιέος. τὰ μέντοι τῆς κνήμης βραχύτερα μὲν γίνεται ἢ τὰ τῆς ἑτέρης, ὀλίγῳ δέ, διὰ τὰς αὐτὰς προφάσιας, αἳ καὶ πρόσθεν εἴρηνται· ὁδοιπορέειν δὲ δύνανται οἱ τοιοῦτοι, οἱ μέν τινες αὐτῶν τοῦτον τὸν τρόπον ὥσπερ οἷσι τετελεωμένοισιν ἐξέπε-

partly because the foot is not obliged to incline outward, but is near the vertical line of the body, and even tends rather inward. As soon, then, as the articular part has worn out a socket for itself in the flesh into which it is dislocated, and the flesh becomes lubricated, in time it becomes painless; and when it becomes painless, patients can walk without a crutch, if they wish to, and can put the weight of the body on the injured leg. Owing to the exercise, the flesh becomes less flaccid in such cases than in those mentioned just above; still a certain amount of flaccidity, more or less, *is* present, which, as a rule, is rather greater on the inner than on the outer side. Some of these patients are unable to put on a shoe, because they cannot bend the leg; but others manage it. In cases where this joint is dislocated before birth, or is forcibly put out and not reduced during adolescence, or when the joint is dislocated and expelled from its socket by disease—such cases occur often—if necrosis of the thigh occurs in some of these cases, chronic abscesses form, requiring tents; and in some of these there is denudation of bones. Likewise, whether necrosis is present or not, the bone of the thigh becomes much shorter, and will not grow the same way the sound one does. The bones of the lower leg, however, although shorter than those of the other side, are only slightly so, for the same reasons as were given above. These patients can walk, some of them in the same fashion as adults who

165 ἐκκεκλ. Littré: ἐγκεκλ. BMV
166 δέ τι M: δ᾽ ὅτι B: δ᾽ ἔτι V

σεν καὶ μὴ ἐνέπεσεν, οἱ δὲ καὶ βαίνουσι μὲν παντὶ τῷ ποδί, διαρρέπουσι δὲ ἐν τῇσιν ὁδοιπορίῃσιν, ἀναγκαζόμενοι διὰ τὴν βραχύτητα τοῦ σκέλεος. ταῦτα δὲ τοιαῦτα γίνεται, ἢν ἐπιμελέως μὲν παιδαγωγηθέωσιν ἐν τοῖσι σχήμασι καὶ ὀρθῶς ἐν οἷσι δεῖ, πρὶν κρατυνθῆναι ἐς τὴν ὁδοιπορίην, ἐπιμελέως δὲ καὶ ὀρθῶς, ἐπὴν κρατυνθῶσιν. πλείστης δὲ ἐπιμελείης δέονται οἷσιν ἂν νηπιωτάτοισιν ἐοῦσιν αὕτη ἡ ξυμφορὴ γένηται· ἢν γὰρ ἀμεληθῶσι νήπιοι ἐόντες, ἀχρεῖον παντάπασι καὶ ἀναυξὲς ὅλον τὸ σκέλος γίνεται. αἱ δὲ σάρκες τοῦ σύμπαντος σκέλεος μινύθουσι μᾶλλον ἢ τοῦ ὑγιέος· πάνυ μὴν πολλῷ ἧσσον τούτοισι μινύθουσιν ἢ οἷσιν ἂν εἴσω ἐκπεπτώκῃ, διὰ τὴν χρῆσιν καὶ τὴν ταλαιπωρίην, οἷον εὐθέως δύνασθαι χρῆσθαι τῷ σκέλει, ὡς καὶ πρόσθεν ὀλίγῳ περὶ[167] τῶν γαλιαγκώνων εἴρηται.

56. Εἰσὶ δέ τινες, ὧν τοῖσι μὲν ἐκ γενεῆς αὐτίκα, τοῖσι δὲ καὶ ὑπὸ νόσου ἀμφοτέρων τῶν σκελέων ἐξέστη τὰ ἄρθρα ἐς τὸ ἔξω μέρος. τούτοισιν οὖν τὰ μὲν
244 ὀστέα ταὐτὰ παθήματα πάσχει· αἱ μέντοι | σάρκες ἥκιστα ἐκθηλύνονται τοῖσι τοιούτοισιν. ἅμα γὰρ εὔσαρκα[168] τὰ σκέλεα γίνεται, πλὴν εἴ τι ἄρα κατὰ τὸ εἴσω μέρος ἐλλείπει ὀλίγον. διὰ τοῦτο δὲ εὔσαρκά ἐστιν, ὅτι ἀμφοτέροισι τοῖσι σκέλεσιν ὁμοίως ἡ χρῆσις γίνεται· ὁμοίως γὰρ σαλεύουσιν ἐν τῇ ὁδοιπορίῃ ἔνθα καὶ ἔνθα· ἐξεχέγλουτοι δὲ ἰσχυρῶς οὗτοι φαίνονται καὶ ῥαιβοὶ οἱ μηροὶ[169] διὰ τὴν ἔκστασιν τῶν ἄρθρων. ἢν δὲ μὴ ἐπισφακελίσῃ αὐτοῖσι τὰ ὀστέα, μηδὲ

have an unreduced dislocation, while others use the whole foot, but sway from side to side in their gait, being compelled to do so by the shortness of the leg. But such results are only attained if patients are carefully instructed in the correct postures before they have acquired strength for walking, and carefully and properly guided when they are already strong. The greatest care is required in cases where this injury has occurred when they were very young; for if they are neglected when infants, the whole leg becomes completely useless and fails to grow. The flesh is attenuated all through the leg, compared with the sound one; but the attenuation is much less in these cases than where the dislocation is inward, owing to use and exercise, since they can use the leg at once, as was said a little before concerning the weasel-armed.

56. There are some cases in which the joints of both legs are dislocated outward, either immediately at birth or from disease. Here the bones are affected in the same way as was described, but there is very little flaccidity of the tissues in such cases; for the legs remain muscular, unless they have a little deficiency on the inner side. The well-developed muscles are due to the fact that both legs are equally well exercised, since the patients have an even gait swaying to both sides. These patients display very prominent buttocks and curved thighs, because of the displacement of the hip joints; but if no necrosis of the bones

---

167 *ὀλίγῳ περὶ* B: *ὀλίγον ἐπὶ* MV

168 *ἅμα γὰρ εὔσαρκα* B: *εὔσαρκα δὲ καὶ* MV

169 *καὶ ῥ. οἱ μηροὶ* B: om. MV

κυφοὶ ἀνωτέρω τῶν ἰσχίων γένωνται—ἐνίους γὰρ καὶ τὰ τοιαῦτα καταλαμβάνει—ἢν οὖν μή τι τοιοῦτον γένηται, ἱκανῶς ὑγιηροὶ τἆλλα διαφέρονται· ἀναυξέστεροι μέντοι τὸ σύμπαν σῶμα οὗτοι γίνονται, πλὴν τῆς κεφαλῆς.

57. Οἷσι δ᾽ ἂν ἐς τοὔπισθεν ἡ κεφαλὴ τοῦ μηροῦ ἐκπέσῃ—ὀλίγοισι δὲ ἐκπίπτει—οὗτοι ἐκτανύειν οὐ δύνανται τὸ σκέλος, οὔτε κατὰ τὸ ἄρθρον τὸ ἐκπεσὸν οὔτε τι κάρτα κατὰ τὴν ἰγνύην· ἀλλ᾽ ἥκιστα τῶν ἐκπαλέων οὗτοι [μᾶλλον][170] ἐκτανύουσι καὶ τὸ κατὰ τὸν βουβῶνα καὶ τὸ κατὰ τὴν ἰγνύην ἄρθρον. προσξυνιέναι μὲν οὖν καὶ τόδε χρή—εὔχρηστον γὰρ καὶ πολλοῦ ἄξιόν ἐστι καὶ τοὺς πλείστους λήθει—ὅτι οὐδ᾽
246 ὑγιαίνοντες δύνανται κατὰ τὴν ἰγνύην ἐκτανύειν | τὸ ἄρθρον, ἢν μὴ συνεκτανύσωσι καὶ τὸ κατὰ τὸν βουβῶνα ἄρθρον, πλὴν ἢν μὴ πάνυ ἄνω ἀείρωσι τὸν πόδα, οὕτω δ᾽ ἂν δύναιντο. οὐ[171] τοίνυν οὐδὲ ξυγκάμπτειν δύνανται τὸ κατὰ τὴν ἰγνύην ἄρθρον ὁμοίως, ἀλλὰ πολὺ χαλεπώτερον, ἢν μὴ συγκάμψωσι καὶ τὸ κατὰ τὸν βουβῶνα ἄρθρον. πολλὰ δὲ καὶ ἄλλα κατὰ τὸ σῶμα τοιαύτας ἀδελφίξιας ἔχει, καὶ κατὰ νεύρων ξυντάσιας καὶ κατὰ μυῶν σχήματα—καὶ πλεῖστά τε καὶ πλείστου ἄξια γινώσκεσθαι ἢ ὥς τις οἴεται—καὶ κατὰ τὴν τοῦ ἐντέρου φύσιν καὶ τὴν τῆς ξυμπάσης κοιλίης, καὶ κατὰ τὰς τῶν ὑστέρων πλάνας καὶ ξυντάσιας. ἀλλὰ περὶ μὲν οὖν τούτων ἑτέρωθι λόγος ἔσται ἠδελφισμένος τοῖσι νῦν λεγομένοισι. περὶ οὗ δὲ ὁ λόγος ἐστίν, οὔτε ἐκτανύειν δύνανται, ὥσπερ ἤδη

supervenes, and they do not become kyphotic above the hips—for this is an affection which attacks some—if nothing of this sort occurs, they are distinguished by very fair health in other respects. Still, these patients have defective growth of the whole body, except the head.

57. In cases where the head of the thigh is dislocated backward—this is a rare dislocation—patients cannot extend the leg at the dislocated joint, nor indeed at the ham; in fact, of all dislocations, those who suffer this one are the least capable of extension, both at the groin and at the ham. You should also bear the following in mind—since it is a useful and valuable matter, of which most people are ignorant—that not even healthy individuals can extend the joint at the ham, unless at the same time they extend the joint at the groin as well, except if they lift the foot very high; then they can do it. Nor can they as readily flex the joint at the ham, but have greater difficulty unless they flex the joint at the groin simultaneously. Many other parts of the body have dependencies of this kind, both as regards contractions of cords and positions of muscles—being very numerous, and more important to recognize than you would think—both as regards the arrangement of the intestine and the whole cavity, and also the displacements and contractions of the uterus. But these matters will be discussed elsewhere in a treatise similar to the present one. To return to our subject, as has already been observed, the patients cannot extend their leg, and also it

---

170 μᾶλλον BMV: om. recc. Littré

171 οὐ Galen Littré: οὗτοι BMV

εἴρηται, βραχύτερον δὲ τὸ σκέλος φαίνεται, διὰ δισσὰς προφάσιας· ὅτι τε οὐκ ἐκτανύεται, ὅτι τε πρὸς τὴν σάρκα ὠλίσθηκε τὴν τοῦ πυγαίου· ἡ γὰρ φύσις τοῦ ἰσχίου τοῦ ὀστέου ταύτῃ, ᾗ καὶ ἡ κεφαλὴ καὶ ὁ αὐχὴν τοῦ μηροῦ γίνεται, ὅταν δὲ ἐξαρθρήσῃ, καταφερής τε πέφυκεν ἐπὶ τοῦ πυγαίου τὸ ἔξω μέρος. ξυγκάμπτειν μέντοι δύνανται, ὅταν μὴ ἡ ὀδύνη κωλύῃ· καὶ ἡ κνήμη τε καὶ ὁ ποὺς ὀρθὰ ἐπιεικέως φαίνεται, καὶ οὔτε τῇ οὔτε τῇ πολὺ ἐκκεκλιμένα· κατὰ δὲ τὸν βουβῶνα δοκέει τι ἡ σὰρξ λαπαρωτέρη εἶναι ποτὶ καὶ ψαυομένη, ἅτε τοῦ ἄρθρου ἐς τὰ ἐπὶ θάτερα ὠλισθη-
248 κότος μέρη· | κατὰ δὲ αὐτὸ τὸ πυγαῖον διαψαυομένη ἡ κεφαλὴ τοῦ μηροῦ δοκέει[172] ἐξογκέειν μᾶλλον. τὰ μὲν σημεῖα ταῦτά ἐστιν, ᾧ ἂν ἐς τοὔπισθεν ἐκπεπτώκῃ ὁ μηρός.

58. Ὅτῳ μὲν οὖν ἂν τετελεωμένῳ ἤδη ἐκπεσὼν μὴ ἐμπέσῃ, ὁδοιπορέειν μὲν δύναται, ὅταν ὁ χρόνος ἐγγένηται καὶ ἡ ὀδύνη παύσηται, καὶ ἐθισθῇ τὸ ἄρθρον ἐν τῇ σαρκὶ ἐνστρωφᾶσθαι. ἀναγκάζεται μέντοι ἰσχυρῶς ξυγκάμπτων κατὰ τοὺς βουβῶνας ὁδοιπορέειν, διὰ δισσὰς προφάσιας, ἅμα μὲν ὅτι πολλῷ βραχύτερον τὸ σκέλος γίνεται διὰ τὰ προειρημένα, καὶ τῇ μὲν πτέρνῃ καὶ πάνυ πολλοῦ δέεται ψαύειν τῆς γῆς· εἰ γὰρ πειρήσαιτο ἐπ᾽ ὀλίγου τοῦ ποδὸς ὀχηθῆναι, μηδενὶ ἄλλῳ ἀντιστηριζόμενος, ἐς τοὐπίσω ἂν πέσοι· ἡ γὰρ ῥοπὴ πολλὴ ἂν εἴη, τῶν ἰσχίων ἐς τοὐπίσω ἐπὶ πολὺ ὑπερεχόντων ὑπὲρ τοῦ ποδὸς τῆς βάσιος καὶ τῆς ῥάχιος ἐς τὰ ἰσχία ῥεπούσης. μόλις δὲ τῷ στήθει

appears shorter, for two reasons; both because it is not extended, and also because it has slipped into the flesh of the buttock; for the form of the bone of the hip at the part where the head and neck of the thigh rest, when dislocated, has a natural inclination toward the outer side of the buttock. Now patients can flex the limb, when pain does not prevent it; and the lower leg and foot appear fairly straight, without much inclination to either side. At the groin the flesh seems rather lax, especially on palpation, since the joint has slipped to the other side; while at the buttock itself the head of the thigh seems, on deep palpation, to project too much. These then are the signs in a case of dislocation of the thigh backward.

58. When the dislocation occurs in an adult, and is not reduced, the patient can walk, indeed, after an interval, when the pain has subsided and the joint of the bone has become accustomed to rotate in the tissue; but he is obliged in walking to flex his body strongly at the groin, for two reasons, both because the leg is much shorter, from the causes explained above, and is held very far from touching the ground with the heel; for if the patient were to try even for a moment to rest his weight on the foot with no other support, he would fall backward; for there would be a great inclination that way, the hips extending far back beyond the sole of the foot, and the spine inclining toward the hips. He hardly reaches the ground with the ball of the

---

[172] τι add. MV

τοῦ ποδὸς καθικνέεται, καὶ οὐδὲ οὕτως, ἢν μὴ κάμψῃ αὐτὸς ἑωυτὸν κατὰ τοὺς βουβῶνας, καὶ τῷ ἑτέρῳ σκέλει κατὰ τὴν ἰγνύην ἐπιξυγκάμψῃ. ἐπὶ δὲ τούτοισιν ἀναγκάζεται ὥστε τῇ χειρὶ τῇ κατὰ τὸ σιναρὸν σκέλος ἐρείδεσθαι ἐς τὸ ἄνω τοῦ μηροῦ ἐφ᾿ ἑκάστῃ ξυμβάσει. ἀναγκάζει οὖν τι καὶ τοῦτο αὐτὸ ὥστε κάμπτεσθαι κατὰ τοὺς βουβῶνας· ἐν γὰρ τῇ μεταλλαγῇ τῶν σκελέων ἐν τῇ ὁδοιπορίῃ οὐ δύναται τὸ σῶμα ὀχέεσθαι ἐπὶ τοῦ σιναροῦ σκέλεος, ἢν μὴ προσκατε-
250 *ρείδηται | τὸ σιναρὸν πρὸς τὴν γῆν ὑπὸ τῆς χειρός, ἅτε οὐχ ὑφεστεῶτος τοῦ ἄρθρου ὑπὸ τῷ σώματι, ἀλλ᾿ ἐς τὸ ὄπισθεν ἐξεστεῶτος κατὰ τὸ ἰσχίον. ἄνευ μὲν οὖν ξύλου δύνανται ὁδοιπορέειν οἱ τοιοῦτοι, ἢν ἄλλως ἐθισθέωσι διὰ τοῦτο, ὅτι ἡ βάσις τοῦ ποδὸς κατὰ τὴν ἀρχαίην ἰθυωρίην ἐστίν, ἀλλ᾿ οὐκ ἐς τὸ ἔξω ἐκκεκλιμένη· διὰ τοῦτο οὖν οὐδὲν δέονται τῆς ἀντικοντώσιος. ὅσοι μέντοι βούλονται ἀντὶ τῆς τοῦ μηροῦ ἐπιλαβῆς ὑπὸ τὴν μασχάλην τὴν κατὰ τὸ σιναρὸν σκέλος ὑποτιθέμενοι σκίπωνα ἀντερείδειν, κεῖνοι, ἢν μὲν μακρότερον τὸν σκίπωνα ὑποτιθέοιντο, ὀρθότεροι μὲν ὁδοιπορήσουσι, τῷ δὲ ποδὶ πρὸς τὴν γῆν οὐκ ἐρείδονται· εἰ δ᾿ αὖ βούλονται ἐρείδεσθαι τῷ ποδί, βραχύτερον μὲν τὸ ξύλον φορητέον,*[173] *κατὰ δὲ τοὺς βουβῶνας ἐπιξυγκάμπτεσθαι ἂν δέοι αὐτούς. τῶν δὲ σαρκῶν αἱ μινυθήσιες κατὰ λόγον γίνονται καὶ τούτοισιν, ὥσπερ πρόσθεν εἴρηται· τοῖσι μὲν γὰρ μετέωρον ἔχουσι τὸ σκέλος καὶ μηδὲν ταλαιπωρέουσι, τούτοισι καὶ μάλιστα μινύθουσιν· οἳ δ᾿ ἂν πλεῖστα χρέωνται τῇ ἐπιβά-*

foot, and cannot do this without a flexure at the groins and a simultaneous flexion of the other leg at the ham. Furthermore, he is forced at every step to exert pressure with his hand on the upper part of the thigh on the side of the injured leg. This of itself would compel him to bend his body somewhat at the groins, since at the change of legs in walking, the body weight cannot be carried by the injured leg unless it is also pressed downward to the ground by the hand, because the joint is not in line under the body, but displaced backward at the hip. Still, such patients can walk without a crutch, at any rate after practice, for this reason, *viz.* that the sole of the foot keeps its old straight line, and is not inclined outward; thus, they have no need for anything to counterbalance them. Those who prefer, instead of taking hold of the thigh, to have the support of a crutch held under the arm on the side of the injured leg, if they have a rather long crutch, will walk more erect, but will not be able to reach the ground with their foot. But if they want to put weight on the foot, a shorter crutch must be carried; and they must also bend themselves at the groin. Proportionate wasting of the flesh takes place in these cases, as was said before. In patients who keep their leg off the ground and give it no exercise, the wasting is greatest, while in those who use it most in walking, it is least. Still, the sound leg gains no benefit, but rather be-

---

173 *φορητέον* B: *ποιητέον* MV

252 σει, τούτοισιν | ἥκιστα μινύθουσι. τὸ μέντοι ὑγιὲς σκέλος οὐκ ὠφελέεται, ἀλλὰ καὶ ἀσχημονέστερον γίνεται, ἢν χρέωνται τῷ σιναρῷ σκέλει ἐπὶ τὴν γῆν· ξυνυπουργέον γὰρ ἐκείνῳ ἐξίσχιόν τε ἀπαναγκάζεται εἶναι, καὶ κατὰ τὴν ἰγνύην ξυγκάμπτειν. ἢν δὲ μὴ προσχρέηται ἐπὶ τὴν γῆν τῷ σιναρῷ σκέλει, ἀλλὰ μετέωρον ἔχων σκίπωνι ἀντερείδηται, οὕτω δὲ καρτερὸν γίνεται τὸ ὑγιὲς σκέλος· ἔν τε γὰρ τῇ φύσει διαιτᾶται, τά τε γυμνάσια προσκρατύνει αὐτό. φαίη μὲν οὖν ἄν τις, ἔξω ἰητρικῆς τὰ τοιαῦτα εἶναι· τί γὰρ δῆθεν δεῖ περὶ τῶν ἤδη ἀνηκέστων γεγονότων ἔτι προσξυνιέναι; πολλοῦ δὲ δεῖ οὕτως ἔχειν· τῆς γὰρ αὐτῆς γνώμης καὶ ταῦτα ξυνιέναι· οὐ γὰρ οἷόν τε ἀπαλλοτριωθῆναι ἀπ' ἀλλήλων. δεῖ μὲν γὰρ ἐς[174] τὰ ἀκεστὰ μηχανᾶσθαι, ὅκως μὴ ἀνήκεστα ἔσται, ξυνιέντα ὅκῃ ἂν μάλιστα κωλυτέα ἐς τὸ ἀνήκεστον ἐλθεῖν· δεῖ δὲ τὰ ἀνήκεστα ξυνιέναι, ὡς μὴ μάτην[175] λυμαίνηται. τὰ δὲ προρρήματα λαμπρὰ καὶ ἀγωνιστικὰ ἀπὸ τοῦ διαγινώσκειν, ὅπῃ ἕκαστον καὶ οἵως καὶ ὁπότε τελευτήσει. ἤν τε ἐς τὸ ἀκεστὸν τράπηται, ἤν τε ἐς τὸ ἀνήκεστον. οἷσι δ' ἂν ἐκ γενεῆς ἢ καὶ ἄλλως ἐν αὐξήσει ἐοῦσιν οὕτως ὀλίσθῃ τὸ ἄρθρον ὀπίσω καὶ μὴ ἐμπέσῃ, ἤν τε βίῃ ὀλίσθῃ, ἤν τε καὶ ὑπὸ νόσου—πολλὰ γὰρ τὰ τοιαῦτα ἐξαρθρήματα γίνεται ἐν νούσοις· οἷαι δέ τινές εἰσιν αἱ νοῦσοι, ἐν ᾗσιν ἐξαρθρέεται τὰ τοιαῦτα, ὕστερον γεγράψεται—ἢν οὖν ἐκστὰν μὴ ἐμπέσῃ, τοῦ μὲν μηροῦ τὸ ὀστέον βραχὺ
254 γίνεται, κακοῦται δὲ καὶ | πᾶν τὸ σκέλος, καὶ ἀναυξέ-

comes even more deformed, if patients use the injured leg on the ground; for in giving assistance to the latter, it is forced outward at the hip, and bends at the ham. But if the patient does not use the injured leg on the ground, but, keeping it up, gets support from a crutch in this way, the sound limb becomes strong; for it is being employed in the natural way, and the exercises strengthen it. One might say that such matters are outside the healing art, for why in the world should you trouble your mind further about cases which have become incurable? But this is far from the right attitude, since the investigation of these matters too belongs to the same science, and it is impossible to separate the two from each other. In the case of curable conditions we must contrive ways to prevent them from becoming incurable, studying the best means for hindering their advance to incurability; while in incurable cases, ways to avoid doing harm by useless efforts must be considered. Brilliant and accurate forecasts are made by ascertaining where, how, and when each case will end, and whether it will take a turn toward recovery or incurability. In cases where such a dislocation backward occurs and is not reduced, whether congenitally or during the period of growth, and whether the displacement is due to violence or disease—many such dislocations occur in diseases, and the diseases which cause such dislocations will be described later—if, then, the displacement is unreduced, the bone of the thigh becomes short, and the whole leg is impaired, failing to grow and losing more of its flesh, be-

---

174 *ἐς* V: om. BV
175 *μάτην* Littré: *μάλιστα* BMV

στερον γίνεται καὶ ἀσαρκότερον πολλῷ διὰ τὸ μηδὲν προσχρῆσθαι[176] αὐτῷ· κακοῦται γὰρ τούτοισι καὶ τὸ κατὰ τὴν ἰγνύην ἄρθρον· τὰ γὰρ νεῦρα ἐντεταμένα γίνεται διὰ τὰ πρόσθεν εἰρημένα. διὸ οὐ δύνανται τὸ κατὰ τὴν ἰγνύην ἄρθρον ἐκτανύειν, οἷς ἂν οὕτως ἰσχίον ἐκπέσῃ. ὡς γὰρ ἐν κεφαλαίῳ εἰρῆσθαι, πάντα τὰ ἐν τῷ σώματι, ὅσα ἐπὶ χρήσει γέγονε, χρεωμένοισι μὲν μέτρια καὶ γυμναζομένοισιν ἐν τῇσι ταλαιπωρίῃσιν, ἐν ᾗσιν ἕκαστα εἴθισται, οὕτω μὲν ὑγιηρὰ καὶ αὔξιμα καὶ εὔγηρα γίνεται· μὴ χρεωμένοισι δέ, ἀλλ᾽ ἐλινύουσι, νοσηρότερα γίνεται καὶ ἀναυξέα καὶ ταχύγηρα. ἐν δὲ τούτοισιν οὐχ ἥκιστα τὰ ἄρθρα τοῦτο πέπονθε καὶ τὰ νεῦρα, ἢν μή τις αὐτοῖσι χρέηται. κακοῦνται μὲν οὖν διὰ ταύτας τὰς προφάσιας μᾶλλόν τι ἐν τούτῳ τῷ τρόπῳ τοῦ ὀλισθήματος ἢ ἐν τοῖσιν ἄλλοισιν· ὅλον γὰρ τὸ σκέλος ἀναυξὲς γίνεται, καὶ τῇ ἀπὸ τῶν ὀστέων φύσει καὶ τῇ ἀπὸ τῶν σαρκῶν. οἱ οὖν τοιοῦτοι ὁκόταν ἀνδρωθῶσι, μετέωρον καὶ ξυγκεκαμμένον[177] τὸ σκέλος ἴσχουσιν, ἐπὶ δὲ τοῦ ἑτέρου ὀχέονται, καὶ τῷ ξύλῳ ἀντιστηριζόμενοι οἱ μὲν ἑνί, οἱ δὲ δυσίν.

59. Οἷσι δ᾽ ἂν ἐς τοὔμπροσθεν ἡ κεφαλη τοῦ μηροῦ ἐκπέσῃ—ὀλιγάκις[178] δὲ τοῦτο γίνεται—οὗτοι ἐκτανύειν μὲν τὸ σκέλος δύνανται τελέως, ξυγκάμπτειν δὲ ἥκιστα οὗτοι δύνανται τὰ κατὰ τὸν βουβῶνα· πονέουσι δὲ καὶ ἢν κατὰ τὴν ἰγνύην ἀναγκάζωνται ξυγκάμπτειν. μῆκος δὲ τοῦ σκέλεος παραπλήσιον φαίνεται
256 καὶ κατὰ | μὲν τὴν πτέρνην καὶ πάνυ· ἄκρος δὲ ὁ ποὺς

cause it gets no exercise. For in these cases, the joint at the ham is also maimed, since the cords become contracted, for the reasons given above. Therefore, patients in whom the hip is dislocated in this way cannot extend the joint at the ham. Speaking generally, all parts of the body that serve some function, if used in moderation and exercised in the labors to which each is accustomed, tend thereby to become healthy and well-developed, and to age slowly; but if they are not used but left idle, they become more liable to disease, defective in growth, and age quickly. This especially is the case with joints and cords, if someone does not use them. For these reasons, patients are more troubled by this sort of dislocation than by the others; for the whole leg fails in the natural growth of both its bones and its muscles. Such patients, then, when they become adults, hold the leg raised and contracted, and walk on the other one, supporting themselves, some with one and some with two crutches.

59. Those in whom the head of the thigh is dislocated forward—a rare occurrence—can extend the leg completely, but are least able to flex it at the groin; and they also suffer pain if they are compelled to bend it at the ham. The length of the leg seems equal, and quite so at the heel;

---

[176] *προσχρίεσθαι* B: *προχρίεσθαι* M: *προχρέεσθαι* V

[177] *-κεκαμμένον* B: *-κεκλιμένον* MV

[178] *ὀλιγάκις* B: *ὀλίγοισι* MV

ἧσσόν τι προκύπτειν ἐθέλει ἐς τοὔπισθεν. ὅλον δὲ τὸ σκέλος ἔχει τὴν ἰθυωρίην τὴν κατὰ φύσιν, καὶ οὔτε τῇ οὔτε τῇ ῥέπει. ὀδυνῶνται δὲ αὐτίκα οὗτοι μάλιστα, καὶ οὖρον ἴσχεται τὸ πρῶτον τούτοισι μᾶλλόν τι ἢ τοῖσιν ἄλλοισιν ἐξαρθρήμασιν· ἐγκέεται γὰρ ἡ κεφαλὴ τοῦ μηροῦ ἐγγυτάτω τούτοισι τῶν τόνων τῶν ἐπικαίρων· καὶ κατὰ μὲν τὸν βουβῶνα ἐξόγκεόν τε[179] καὶ κατατεταμένον τὸ χωρίον φαίνεται, κατὰ δὲ τὸ πυγαῖον στολιδωδέστερον καὶ ἀσαρκότερον.[180] ταῦτα μὲν οὖν ἐστι σημεῖα τὰ εἰρημένα, ὧν ἂν οὕτως ἐκπεπτώκῃ ὁ μηρός.

60. Ὁκόσοισι μὲν οὖν ἂν ἤδη ἠνδρωμένοισι τοῦτο τὸ ἄρθρον ἐκπεσὸν μὴ ἐμπέσῃ, οὗτοι, ὁκόταν αὐτοῖσιν ἡ ὀδύνη παύσηται καὶ τὸ ἄρθρον ἐθισθῇ ἐν τῷ χωρίῳ τούτῳ στρωφᾶσθαι, ἵνα ἐξέπεσεν, οὗτοι δύνανται σχεδὸν ὀρθοὶ ὁδοιπορέειν ἄνευ ξύλου, καὶ πάνυ μέντοι εὐθέες, ἐπὶ δὲ τὸ σιναρόν, ἅτε οὔτε κατὰ τὸν βουβῶνα εὔκαμπτοι ἐόντες, οὔτε κατὰ τὴν ἰγνύην. διὰ γοῦν τοῦ βουβῶνος τὴν ἀκαμπίην εὐθυτέρῳ ὅλῳ τῷ σκέλει ἐν τῇ ὁδοιπορίῃ χρέωνται ἢ ὅτε ὑγίαινον. καὶ σύρουσι δὲ ἐνίοτε πρὸς τὴν γῆν τὸν πόδα, ἅτε οὐ ῥηϊδίως συγκάμπτοντες τὰ ἄνω ἄρθρα, καὶ ἅτε παντὶ βαίνοντες τῷ ποδί· οὐθὲν γὰρ ἧσσον τῇ πτέρνῃ οὗτοι βαίνουσιν ἢ τῷ ἔμπροσθεν· εἰ δέ γε ἠδύναντο μέγα
258 προβαίνειν, κἂν πάνυ πτερνοβάται ἦσαν· καὶ | γὰρ οἱ ὑγιαίνοντες, ὅσῳ ἂν μέζον προβαίνοντες ὁδοιπορέωσι, τοσούτῳ μᾶλλον πτερνοβάται εἰσί, τιθέντες τὸν πόδα, αἴροντες δὲ τὸν ἐναντίον. ὁκόσοισι δὲ δὴ[181] οὕτως ἐκ-

but the tip of the foot is less able to be projected forward. The whole leg preserves its natural straight line, inclining to neither one side nor the other. These patients tend particularly to suffer immediate pain, and the retention of urine sets in from the first more than in other dislocations, since the head of the thigh in these cases lies very close to critical cords. The region of the groin appears prominent and stretched; but at the buttock it is more wrinkled and emaciated. The signs described, then, occur in patients whose thigh is dislocated forward.

60. In cases where this dislocation occurs in persons already adult and is not reduced, these patients, when their pain subsides and the head of the bone has become accustomed to turning in the location to which it was displaced, are able to walk almost erect without a crutch, and even quite straight, so far as the injured part is concerned, seeing that the leg cannot easily bend at either the groin or the ham. Thus, owing to the inflexibility at the groin, they keep the whole leg straighter in walking than when it was sound. And sometimes they drag the foot along the ground, because they cannot easily flex the upper joints, and they walk on the whole foot. In fact, they walk no less on the heel than on the front part of the foot; and if they could take long strides, they would rest solely on the heel in walking. For healthy people, the longer the strides they take in walking, the more they walk on their heels while they are putting down one leg and raising the other one;

179 ἐξογκέον τε Littré: ἐξογκεόνται BMV
180 ἀσαρκότερον V: -κώ(ό)τατον BM
181 δὲ δὴ BM: δ' ἤδη V

*πέπτωκε, καὶ ἔτι μᾶλλον τῇ πτέρνῃ προσεγχρίμπτου-
σιν ἢ τῷ[182] ἔμπροσθεν· τὸ γὰρ ἔμπροσθεν τοῦ ποδός,
ὅταν ἐκτεταμένον ᾖ τὸ ἄλλο σκέλος, οὐχ ὁμοίως δύ-
ναται ἐς τὸ πρόσω καμπύλλεσθαι, ὥσπερ ὅταν ξυγκε-
καμμένον ᾖ τὸ σκέλος· οὔτε αὖ σιμοῦσθαι δύναται ὁ
πούς, ξυγκεκλιμένου τοῦ σκέλεος, ὡς ὅταν ἐκτεταμέ-
νον ἔῃ τὸ σκέλος. ὑγιαίνουσά τε οὖν ἡ φύσις οὕτω
πέφυκεν, ὥσπερ εἴρηται. ὅταν δὲ ἐκπεσὸν μὴ ἐμπέσῃ
τὸ ἄρθρον, οὕτως ὁδοιπορέουσιν ὡς εἴρηται, διὰ τὰς
προφάσιας ταύτας τὰς εἰρημένας· ἀσαρκότερον μέν-
τοι τὸ σκέλος τοῦ ἑτέρου γίνεται, κατά τε τὸ πυγαῖον,
καὶ κατὰ τὴν γαστροκνημίην, καὶ κατὰ τὴν ὄπισθεν[183]
ἴξιν. οἷσι δ᾽ ἂν νηπίοισιν ἔτι ἐοῦσι τὸ ἄρθρον οὕτως
ὀλισθάνον μὴ ἐμπέσῃ, ἢ καὶ ἐκ γενεῆς οὕτω γένηται,
καὶ τούτοισι τὸ τοῦ μηροῦ ὀστέον μᾶλλόν τι μινύθει
ἢ τὰ τῆς κνήμης καὶ τὰ τοῦ ποδός. ἥκιστα μὴν ἐν
τούτῳ τῷ τρόπῳ τοῦ ὀλισθήματος ὁ μηρὸς μειοῦται.
μινύθουσι μέντοι αἱ σάρκες πάντῃ, μάλιστα δὲ κατὰ
τὴν ὄπισθεν ἴξιν, ὥσπερ ἤδη καὶ πρόσθεν εἴρηται.
ὅσοι μὲν οὖν ἂν τιθηνηθῶσιν ὀρθῶς, οὗτοι μὲν δύναν-
ται προσχρῆσθαι τῷ σκέλει αὐξανόμενοι, βραχυτέρῳ
μέν τινι τοῦ ἑτέρου ἐόντι, ὅμως δὲ ἐρειδόμενοι ξύλῳ
ἐπὶ ταῦτα, ᾗ τὸ σιναρὸν σκέλος· οὐ γὰρ κάρτα δύναν-
ται ἄνευ τῆς πτέρνης τῷ στήθει τοῦ ποδὸς χρῆσθαι,*
260 *ἐπικαθιέντες ὥσπερ | ἐν ἑτέροισι χωλεύμασιν ἔνιοι
δύνανται· αἴτιον δὲ τοῦ μὴ δύνασθαι τὸ ὀλίγῳ
πρόσθεν εἰρημένον· διὰ οὖν τοῦτο προσδέονται ξύλου.
ὅσοι δὲ ἂν καταμεληθέωσι καὶ μηδὲν χρέωνται ἐπὶ*

but patients who have this form of dislocation press even more upon their heel than on the front of the foot. For the front of the foot cannot be bent down as well when the leg is extended, as when it is flexed; nor, on the other hand, can the foot be bent upward as well, when the leg is flexed, as when it is extended. This is what happens in the natural healthy state, as was said. But when the joint is dislocated and not reduced, patients walk in the way described, for the reasons given above. The affected leg, however, becomes less muscular than the other one, both at the buttock and calf, and all down its back. In those cases, too, where it is dislocated in childhood and not reduced, or where dislocation occurs congenitally, the bone of the thigh is rather more atrophied than the bones of the calf and the foot; but atrophy of the thigh is least in this form of dislocation. The tissues are atrophied in the whole limb, but especially down the back of it, as was just said. Patients, then, who are properly cared for are able to use the leg when they grow up, although it is a little shorter than the other one; yet they do so by supporting themselves on the side of the injured limb with a crutch, for they cannot use the ball of the foot very well without the heel, by putting it down as some people can in other forms of lameness. The reason for this inability has been mentioned a little above; and this is why they need a crutch. In patients who are neglected, and never set their leg on the ground,

---

182 τῷ V: τὸ BM
183 ὄπισθεν MV: om. B

τὴν γῆν τῷ σκέλει, ἀλλὰ μετέωρον ἔχωσι, τούτοισι μινύθει μὲν τὰ ὀστέα ἐς αὔξησιν μᾶλλον ἢ τοῖσι χρεωμένοισι· μινύθουσι δὲ αἱ σάρκες πολὺ μᾶλλον ἢ τοῖσι χρεωμένοισι. κατὰ δὲ τὰ ἄρθρα ἐς τὸ εὐθὺ πηροῦται τὸ σκέλος τούτοισι μᾶλλόν τι ἢ οἷσιν ἂν ἄλλως ἐκπεπτώκῃ.

61. Ὡς μὲν οὖν ἐν κεφαλαίῳ εἰρέεσθαι, τὰ ἄρθρα τὰ ἐκπίπτοντα καὶ τὰ ὀλισθάνοντα ἀνίσως αὐτὰ ἑωυτοῖσιν ἐκπίπτει καὶ ὀλισθάνει, ἄλλοτε μὲν πολὺ πλέον, ἄλλοτε δὲ ἔλασσον. καὶ οἷσι μὲν ἂν[184] πλέον ὀλίσθῃ ἢ ἐκπέσῃ, χαλεπώτερα ἐμβάλλειν τὸ ἐπίπαν ἐστί, καὶ ἢν μὴ ἐμβιβασθῇ, μέζους καὶ ἐπιδηλοτέρας τὰς πηρώσιας καὶ κακώσιας ἴσχει τὰ τοιαῦτα, καὶ ὀστέων καὶ σαρκῶν καὶ σχημάτων. ὅταν δὲ μεῖον ἐκπέσῃ καὶ ὀλίσθῃ, ῥηΐδιον μὲν ἐμβάλλειν τὰ τοιαῦτα τῶν ἑτέρων γίνεται· ἢν δὲ καταπορηθῇ ἢ καὶ ἀμεληθῇ ἐμπεσεῖν, μείους καὶ ἀσινέστεραι αἱ πηρώσιες γίνονται τούτοισιν ἢ οἷσιν ὀλίγῳ πρόσθεν εἴρηται. τὰ μὲν οὖν ἄλλα ἄρθρα καὶ πάνυ πολὺ διαφέρει ἐς τὸ ὁτὲ μὲν μεῖον, ὁτὲ δὲ μέζον τὸ ὀλίσθημα ποιέεσθαι· μηροῦ δὲ
262 καὶ βραχίονος κεφαλαὶ | παραπλησιώτατα ὀλισθάνουσιν αὐτὴ ἑωυτῇ ἑκατέρη· ἅτε γὰρ στρογγύλαι μὲν αἱ κεφαλαὶ ἐοῦσαι, ἁπλῆν τὴν στρογγύλωσιν καὶ φαλακρὴν ἔχουσι, κυκλοτερέες δὲ αἱ κοιλίαι ἐοῦσαι αἱ δεχόμεναι τὰς κεφαλάς, ἁρμόζουσι δὲ τῇσι κεφαλῇσι· διὰ τοῦτο οὐκ ἔστιν αὐτῇσι τὸ ἥμισυ ἐκστῆναι τοῦ ἄρθρου· ὀλισθάνοι γὰρ ἂν διὰ τὴν περιφερείην, ἢ ἐς τὸ ἔξω ἢ ἐς τὸ ἔσω. περὶ οὗ οὖν ὁ λόγος,[185] ἐκπίπτουσι

but hold it in the air, the bones atrophy more than in those who do use it; and the muscles are much more atrophied than in those who use the leg. At the joints, the leg is maimed in a straight line in these patients, somewhat more than in those who have other forms of dislocation.

61. In summary—dislocations and subluxations of the joints vary among themselves in their seriousness, which is sometimes much greater, and at other times less. In cases where the dislocation or slipping is greater, it is generally harder to reduce; and, if unreduced, the resulting damage and impairment are greater and more manifest in the bones, the muscles, and the postures assumed. But when the displacement or subluxation is less serious, such reductions are easier than in other cases; and if they are not reduced, as a consequence of inability or even neglect, the resulting deformities are smaller and less serious than in the cases mentioned above. Now the other joints also vary very much in the magnitude of their displacements; but the heads of the thigh and arm each slip out in very similar ways, since their heads are rounded and have a smooth and regular spherical surface, and the cavities which receive them are also circular, and adapted to the heads. For this reason, it is impossible for them to be half-dislocated, since due to the circular rim, the head will slip either out or in. As regards our subject, then, they are completely dislocated, since otherwise they are not dislo-

---

[184] πολὺ add. M
[185] Add. ἐστιν MV

τελέως ἤδη, ἐπεὶ ἄλλως γε οὐκ ἐκπίπτουσι· ὅμως δὲ καὶ ταῦτα ὁτὲ μὲν πλεῖον ἀποπηδᾷ ἀπὸ τῆς φύσιος, ὁτὲ δὲ ἔλασσον· μᾶλλον δέ τι μηρὸς τοῦτο βραχίονος πέπονθεν.

62. Ἐπεὶ ἔνια καὶ τῶν ἐκ γενεῆς γενομένων ὀλισθη-
264 μάτων, ἢν | μικρὸν ὀλίσθῃ, οἷά τε ἐς τὴν φύσιν ἄγεσθαι, καὶ μάλιστα τὰ παρὰ τοῦ ποδὸς ἄρθρα. ὁκόσοι ἐκ γενεῆς κυλλοὶ γίνονται, τὰ πλεῖστα τούτων ἰήσιμά ἐστιν, ἢν μὴ πάνυ μεγάλη ἡ ἔκκλισις ᾖ, ἢ καὶ προαυξέων[186] γεγονότων ἤδη τῶν παιδίων ξυμβῇ. ἄριστον μὲν οὖν ὡς τάχιστα ἰητρεύειν τὰ τοιαῦτα, πρὶν πάνυ μεγάλην τὴν ἔνδειαν τῶν ὀστέων τῶν ἐν τῷ ποδὶ γενέσθαι, πρίν τε πάνυ μεγάλην τὴν ἔνδειαν τῶν σαρκῶν τῶν κατὰ τὴν κνήμην εἶναι. τρόπος μὲν οὖν κυλλώσιος οὐχ εἷς, ἀλλὰ πλείονες, τὰ πλεῖστα μὴν οὐκ ἐξηρθρηκότα παντάπασιν, ἀλλὰ διὰ ἔθος σχήματος ἔν τινι ἀπολήψει τοῦ ποδὸς κεκυλλωμένα. προσέχειν δὲ ἐν[187] τῇ ἰητρείῃ τοισίδε χρή· ἀπωθέειν μὲν καὶ κατορθοῦν τὸ τῆς κνήμης τὸ κατὰ τὸ σφυρὸν ὀστέον τὸ ἔξωθεν ἐς τὸ ἔσω μέρος, ἀντωθέειν δὲ ἐς τὸ ἔξω μέρος τὸ τῆς πτέρνης τὸ ἔσωθεν τὸ κατ' αὐτὴν τὴν ἴξιν, ὅπως ἀλλήλοις ἀπαντήσῃ[188] τὰ ὀστέα τὰ ἐξίσχοντα κατὰ μέσον τε καὶ πλάγιον τὸν πόδα· τοὺς δ' αὖ δακτύλους ἀθρόους ξὺν τῷ μεγάλῳ δακτύλῳ ἐς
266 | τὸ εἴσω μέρος ἐγκλίνειν[189] καὶ περιαναγκάζειν οὕτως· ἐπιδέειν δὲ κηρωτῇ ἐρρητινωμένῃ εὖ, καὶ σπλήνεσι, καὶ ὀθονίοισι μαλθακοῖσι μὴ ὀλίγοισι, μηδὲ ἄγαν πιέζοντα. οὕτω δὲ τὰς περιαγωγὰς ποιέεσθαι τῆς ἐπι-

cated at all. Yet even these joints sometimes deviate farther from their natural position, and at other times less; this is more pronounced in the thigh than in the arm.

62. It follows that certain congenital displacements, when they are slight, can be brought back to their natural position, especially in the joints of the foot. Cases of congenital clubfoot are, for the most part, curable, unless the deviation is very great or the children are advanced in their growth. It is therefore best to treat such cases as soon as possible, before there is any very great deficiency in the bones of the foot, and before the same occurs in the muscles of the leg. Now there is not only one kind of clubfoot, but many; and most cases are not the result of complete dislocation, but are deformities due to the constant retention of the foot in a certain position. The things to bear in mind in treatment are the following: push back and adjust the bone of the leg at the ankle from without inward, making counterpressure outward on the inside of the bone of the heel where it comes in line with the leg, so as to bring together the bones which project at the middle and side of the foot; at the same time, bend inward and rotate the toes all together, including the big toe. Dress with cerate well enforced with resin, compresses, and soft bandages, sufficiently numerous, but without too much compression. Bring the turns of the bandaging around in a way corre-

---

186 προαυξ. B: προσαυξ. MV

187 ἐν B: om. MV

188 ὅπως ἀλ. ἀπαντήσῃ Aldina: ὠθέειν δὲ ἐς τὸ εἴσω μέρος BMV

189 ἐγκλ. B: ἐκκλ. MV

δέσιος, ὥσπερ καὶ τῇσι χερσὶν ἡ κατόρθωσις ἦν τοῦ ποδός, ὅκως ὁ ποὺς ὀλίγῳ μᾶλλον ἐς τὸ βλαισὸν ῥέπων φαίνηται. ἴχνος δέ τι χρὴ ποιέεσθαι ἢ δέρματος μὴ ἄγαν σκληροῦ, ἢ μολυβδίου, προσεπιδεῖν δέ, μὴ πρὸς τὸν χρῶτα τιθέντα, ἀλλ' ὅταν ἤδη τοῖσι ὑστάτοισιν ὀθονίοισι μέλλῃς ἐπιδέειν. ὅταν δὲ ἤδη ἐπιδεδεμένος ᾖ, ἑνός τῶν ὀθονίων χρή, οἷσιν ἐπιδέεται, τὴν ἀρχὴν προσράψαι πρὸς τὰ κάτω τοῦ ποδὸς ἐπιδέσματα κατὰ τὴν ἴξιν τοῦ μικροῦ δακτύλου· ἔπειτα ἐς τὸ ἄνω τείνοντα ὅκως ἂν δοκέῃ μετρίως ἔχειν, περιβάλλειν ἄνωθεν τῆς γαστροκνημίης, ὡς μόνιμον ᾖ, κατατεταμένον οὕτως. ἁπλῷ δὲ λόγῳ, ὥσπερ κηροπλαστέοντα, χρὴ ἐς τὴν φύσιν τὴν δικαίην ἄγειν καὶ τὰ ἐκκεκλιμένα καὶ τὰ συντεταμένα παρὰ τὴν φύσιν, καὶ τῇσι χερσὶν οὕτω διορθοῦντα, καὶ τῇ ἐπιδέσει ὡσαύτως, προσάγειν δὲ οὐ βιαίως, ἀλλὰ παρηγορικῶς. προσράπτειν δὲ τὰ ὀθόνια, ὅκως ἂν ξυμφέρῃ τὰς ἀναλήψιας ποιέεσθαι· ἄλλα γὰρ ἄλλης τῶν χωλωμάτων δέεται ἀναλήψιος. ὑποδημάτιον δὲ ποιέειν μο-
268 λύβδινον, ἔξωθεν τῆς ἐπιδέσιος | ἐπιδεδεμένον, οἷον αἱ Χῖαι ῥυθμὸν εἶχον. ἀλλ' οὐδὲν αὐτοῦ δεῖ, ἤν τις ὀρθῶς μὲν τῇσι χερσὶ διορθώσῃ, ὀρθῶς δὲ τοῖσιν ὀθονίοισιν ἐπιδέῃ, ὀρθῶς δὲ τὰς ἀναλήψιας ποιοῖτο. ἡ μὲν οὖν ἴησις αὕτη, καὶ οὔτε τομῆς οὔτε καύσιος οὐθὲν δεῖ, οὔτε ἄλλης ποικιλίης· θᾶσσον γὰρ ἐνακούει τὰ τοιαῦτα τῆς ἰητρείης ἢ ὡς ἂν τις οἴοιτο. προσνικᾶν μέντοι χρὴ τῷ χρόνῳ, ἕως ἂν αὐξηθῇ τὸ σῶμα ἐν τοῖσι δικαίοισι σχήμασιν. ὅταν δὲ ἐς ὑποδήματος λόγον

sponding with the manual adjustment of the foot, so that the latter has some inclination toward the outside. A sole should be made of leather that is not too stiff, or of lead, and should be well attached, not directly on to the skin, but just when you are about to apply the last dressings. When the dressing is completed, the end of one of the bandages used should be sewn on to the underside of the foot dressings, in line with the little toe. Then, making such tension upward as seems suitable, pass the bandage around the calf muscle at the top, so as to keep it firm when it has been arranged. In a word, as in wax modeling, you should bring the parts into their true natural position, both those that are twisted and those that are abnormally contracted, adjusting them in this way both with the hands and by bandaging them in the correct posture; draw them into position not by force, but by gentle means. Sew the bandages on so as to give the appropriate support, for different forms of lameness require different kinds of support. Make a small leaden shoe with the shape of a Chian boot, and fastened it on to the outside of the dressing. But this is quite unnecessary if someone makes the adjustment correctly by hand and dresses correctly with bandages, and the contrivance for providing support is properly executed. This then is the treatment, and there is no need for either incision, cautery, or other complicated methods; for such cases yield to treatment sooner than you would think. Still, time is required for complete success, until the part has acquired growth in its proper positions. When the

εἴη, ἀρβύλαι ἐπιτηδειόταται αἱ πηλοπατίδες καλεόμε-
ναι· τοῦτο γὰρ ὑποδημάτων ἥκιστα κρατεῖται ὑπὸ
τοῦ ποδός, ἀλλὰ κρατέει μᾶλλον. ἐπιτήδειος δὲ καὶ ὁ
Κρητικὸς τρόπος τῶν ὑποδημάτων.

63. Ὁκόσοισι δ᾽ ἂν κνήμης ὀστέα ἐξαρθρήσαντα
270 καὶ ἕλκος | ποιήσαντα τελέως ἐξίσχῃ κατὰ τὰ παρὰ
τὸν πόδα ἄρθρα, εἴτε εἴσω ῥέψαντα, εἴτε μέντοι καὶ
ἔξω, τὰ τοιαῦτα οὐ χρὴ ἐμβάλλειν, ἀλλ᾽ ἐᾶν τὸν βου-
λόμενον τῶν ἰητρῶν ἐμβάλλειν. σαφέως γὰρ εἰδέναι
χρὴ ὅτι ἀποθανεῖται ᾧ ἂν ἐμβληθέντα ἐμμείνῃ, καὶ ἡ
ζωὴ δὲ ὀλιγήμερος τούτοισι γίνεται. ὀλίγοι γὰρ αὐ-
τῶν ἑπτὰ ἡμέρας ἂν ὑπερβάλλοιεν· σπασμὸς γὰρ ὁ
κτείνων ἐστίν. ἀτὰρ καὶ γαγγραινοῦσθαι ἱκνέεται τὴν
κνήμην καὶ τὸν πόδα. ταῦτα βεβαίως εἰδέναι δὲ χρὴ
οὕτως ἐσόμενα· καὶ οὐκ ἄν μοι δοκέει οὐδὲ ἐλλέβορος
ὠφελῆσαι, αὐθημερόν τε δοθεὶς καὶ αὖθις πινόμενος·
ἄγχιστα δέ, εἴπερ τι, τοιοῦτο· οὐ μέντοι γε οὐδὲ τοῦτο
δοκέω. ἢν δὲ μὴ ἐμβληθῇ, μηδὲ ἀπ᾽ ἀρχῆς μηδεὶς
πειρηθῇ ἐμβάλλειν, περιγίνονται οἱ πλεῖστοι αὐτῶν.
χρὴ δὲ ἡρμόσθαι μὲν τὴν κνήμην καὶ τὸν πόδα οὕτως,
ὡς αὐτὸς ἐθέλει, μοῦνον δὲ μὴ ἀπαιωρεύμενα μηδὲ
κινεύμενα ἔστω. ἰητρεύειν δὲ πισσηρῇ καὶ σπλήνεσιν
οἰνηροῖσιν ὀλίγοις,[190] μὴ ἄγαν ψυχροῖσι· ψῦχος γὰρ
272 ἐν τοῖσι | τοιούτοισι σπασμὸν ἐπικαλέεται. ἐπιτήδεια
δὲ καὶ σεύτλων φύλλα ἢ βηχίου ἢ ἄλλου τινὸς τῶν
τοιούτων ἐν οἴνῳ μέλανι αὐστηρῷ ἡμίεφθα ἐπιτιθέντα
ἰητρεύειν ἐπί τε τὸ ἕλκος ἐπί τε τὰ περιέχοντα· κη-
ρωτῇ δὲ χλιαρῇ ὑποχρίειν αὐτὸ τὸ ἕλκος. ἢν δὲ ἡ ὥρη

time has come for footwear to be discussed, the most suitable are the so-called "mud shoes," since this kind of boot yields least to the foot; indeed, the foot yields rather to it. The Cretan form of footwear is also suitable.

63. In cases where the bones of the leg are dislocated and cause a lesion, and they protrude right through the joints at the foot, whether inward or outward, you should not reduce the injury; but let any practitioner who chooses do so. For you may be certain that where there is permanent reduction, these patients will die, and their life will last only a few days: few live beyond seven days, and it will be a convulsion that causes their death. But gangrene of the leg and foot may also be a sequel. It should be well known that this will happen; and I do not believe that even hellebore, given on the day of the accident and drunk again, would do any good; if anything could help, something of this kind would come nearest; but I have no confidence even in it. But if there is no reduction or attempt at reduction to begin with, most patients will survive. The leg and foot should be positioned as the patient himself wishes, only avoiding their being unsupported or unstable. Treat with pitch-cerate and a few compresses steeped in wine, not too cold; for cold in such cases provokes convulsions. Other suitable applications are leaves of beet or colt's foot, or something similar, half-boiled in dark astringent wine, and applied both to the wound and the parts around it. Anoint the wound itself with warm cerate. If it

---

190 ὀλίγοις MV: om. B

χειμερινὴ ᾖ, καὶ ἔρια ῥυπαρὰ οἴνῳ καὶ ἐλαίῳ καταρραίνοντα χλιαροῖσιν ἄνωθεν ἐπιτέγγειν· καταδεῖν δὲ μηδέν, μηδὲ περιπλάσσειν μηδενί· εὖ γὰρ εἰδέναι χρὴ ὅτι πίεξις καὶ ἀχθοφορίη πᾶν κακὸν τοῖσι τοιούτοισίν ἐστιν. ἐπιτήδεια δὲ πρὸς τὰ τοιαῦτα[191] καὶ τῶν ἐναίμων μετεξέτερα, ὅσοισιν αὐτῶν συμφέρει· ἔρια δὲ ἐπιτιθέντα, οἴνῳ ἐπιτέγγοντα, πολὺν χρόνον ἐᾶν. τὰ δὲ ὀλιγημερώτατα τῶν ἐναίμων καὶ ὅσα ῥητίνῃ προσκαταλαμβάνεται οὐχ ὁμοίως ἐπιτήδεια ἐκείνοισίν ἐστιν. χρονίη γὰρ ἡ κάθαρσις τῶν ἑλκέων τούτων γίνεται· πολὺν γὰρ χρόνον πλαδαρὴ γίνεται· τινὰς δὲ τούτων χρηστὸν ἐπιδέειν. εἰδέναι μὲν δή που σάφα χρὴ ὅτι ἀνάγκη τὸν ἄνθρωπον χωλὸν αἰσχρῶς γενέσθαι· καὶ γὰρ ὁ ποὺς ἐς τὸ ἄνω ἀνέσπασται τῶν
274 τοιούτων, καὶ τὰ ὀστέα | τὰ διολισθήσαντα ἔξω ἐξέχοντα φαίνεται. οὔτε γὰρ ψιλοῦται τῶν τοιούτων ὀστέων οὐδὲν ὡς ἐπὶ τὸ πολύ, εἰ μὴ κατὰ βραχύ τι, οὔτε ἀφίσταται, ἀλλὰ περιωτειλοῦται λεπτῇσιν ὠτειλῇσι καὶ ἀσθενέσι, καὶ ταῦτα ἢν ἀτρεμίζωσι πολὺν χρόνον· εἰ δὲ μή, ἑλκύδριον ἐγκαταλειφθῆναι κίνδυνος ἀναλθές. ὅμως δέ, περὶ οὗ ὁ λόγος, οὕτω μὲν ἰητρευόμενοι σῴζονται, ἐμβληθέντος δὲ τοῦ ἄρθρου καὶ ἐμμείναντος, ἀποθνῄσκουσιν.

64. Ωὑτὸς δὲ λόγος οὗτος, ἢν καὶ τὰ τοῦ πήχεος ὀστέα παρὰ τὸν καρπὸν τῆς χειρὸς ἕλκος ποιήσαντα ἐξίσχῃ, ἤν τε ἐς τὸ εἴσω μέρος τῆς χειρός, ἤν τε ἐς τὸ ἔξω. σάφα γὰρ ἐπίστασθαι χρὴ ὅτι ἀποθανεῖται ἐν ὀλίγῃσιν ἡμέρῃσι τοιούτῳ θανάτῳ, οἵῳ περ καὶ

is winter, apply an upper moist dressing of crude wool, sprinkling it with tepid wine and oil; but avoid all bandaging and dressing with plasters, for you must keep well in mind that pressure and weight do nothing but harm in such cases. Some of the applications for fresh wounds are also suitable for these injuries, in cases where they are useful: cover with wool, sprinkle with wine, and leave the dressing on for a long time. Wound remedies that last a very short time, and that involve resin, are not so suitable for these patients, because the cleaning of their wounds then takes more time, since their moist stage is prolonged. Bandaging is good for some of these cases. Finally, you should bear clearly in mind that the patient will necessarily be lame and deformed; for his foot is drawn up, and the projection of the dislocated bones is obvious. There is no denudation of the bones as a rule, except to a slight extent, nor do they exfoliate; but they get scarred over with a thin and weak cicatrix: that is, if the patients remain at rest for a long time; otherwise there is the risk of a small ulcer being left. However, to return to our subject, those treated in this way are saved; but if the joint is reduced and remains in place, they die.

64. The same remarks apply to cases where the bones of the forearm cause a lesion at the wrist and protrude there, whether on the inner or outer side of the hand. For you must understand clearly that the patient will die in a few days in the way mentioned above, if the bones are

191 τὰ τοιαῦτα MV: ταῦτα B

πρόσθεν εἴρηται, ὅτῳ ἂν ἐμβληθέντα τὰ ὀστέα ἐμμείνῃ. οἷσι δ' ἂν μὴ ἐμβληθῇ μηδὲ πειρηθῇ ἐμβάλλεσθαι, οὗτοι πολὺ πλείονες περιγίνονται. ἰητρείη δὲ τοιαύτη τοῖσι τοιούτοισιν ἐπιτηδείη, οἵη περ εἴρηται· τὸ δὲ σχῆμα αἰσχρὸν τοῦ χωλώματος ἀνάγκη εἶναι, καὶ τοὺς δακτύλους τῆς χειρὸς ἀσθενέας καὶ ἀχρείους· ἢν μὲν γὰρ ἐς τὸ εἴσω μέρος ὀλίσθῃ τὰ ὀστέα, ξυγκάμπτειν οὐ δύνανται τοὺς δακτύλους· ἢν δὲ ἐς τὸ ἔξω μέρος, ἐκτανύειν οὐ δύνανται.

65. Οἷσι δ' ἂν κνήμης ὀστέον, ἕλκος ποιησάμενον
276 παρὰ τὸ | γόνυ, ἔξω ἐξίσχῃ, ἤν τε ἐς τὸ εἴσω μέρος, ἤν τε ἐς τὸ ἔξω, τούτοισιν ἢν μέν τις ἐμβάλῃ, ἔτι ἑτοιμότερος ὁ θάνατός ἐστιν ἤπερ τοῖσιν ἑτέροισιν, καίπερ καὶ ἐκείνοισιν ἕτοιμος ἐών. ἢν δὲ μὴ ἐμβαλὼν ἰητρεύῃς, ἐλπίδες μὲν σωτηρίης οὕτω μόνως εἰσίν· κινδυνωδέστερα δὲ ταῦτα τῶν ἑτέρων γίνεται καὶ ὅσῳ ἂν ἀνωτέρω καὶ ὅσῳ ἂν ἰσχυρότερα ᾖ καὶ ἀπὸ ἰσχυροτέρων ὠλισθήκῃ. ἢν δὲ τὸ ὀστέον τὸ τοῦ μηροῦ πρὸς τοῦ γόνατος ἕλκος ποιησάμενον ἐξολίσθῃ, ἐμβληθὲν μὲν καὶ ἐμμεῖναν, ἔτι βιαιότερον καὶ θᾶσσον τὸν θάνατον ποιεῖ ἢ τὰ πρόσθεν εἰρημένα·[192] μὴ ἐμβληθὲν δὲ πολὺ κινδυνωδέστερον ἢ τὰ πρόσθεν· ὅμως δὲ μούνη ἐλπὶς αὕτη σωτηρίης.

66. Ωὑτὸς δὲ λόγος καὶ περὶ τῶν κατὰ τὸν ἀγκῶνα ἄρθρων, καὶ περὶ τῶν τοῦ πήχεος καὶ βραχίονος· ὅσα γὰρ ἂν τούτων ἐξαρθρήσαντα ἐξίσχῃ ἕλκος ποιησάμενα, πάντα, ἢν ἐμβληθῇ, θάνατον φέρει, μὴ ἐμβαλλόμενα[193] δέ, ἐλπίδα σωτηρίης· χώλωσις δὲ ἑτοίμη

reduced and remain in place; but if there is no reduction or attempt at reduction, the great majority survive. The suitable treatment in such cases is just as was described, but the deformity due to the impairment will certainly be apparent, and the fingers will be weak and useless; for if the bones are displaced inward, patients cannot flex their fingers, if outward, they cannot extend them.

65. In cases where a bone of the (sc. lower) leg causes a lesion at the knee and projects either to the outer or inner side, death is more imminent if you reduce the dislocation, than in the other cases, although it is imminent in them too. If you treat it without reduction, this is the only hope of survival. These cases are the more dangerous, the higher they are located, and the stronger the dislocated parts and those from which they are dislocated happen to be. If the thigh at the knee causes a lesion and is dislocated as a result, on being reduced and kept in place it will cause still more violent and rapid death than the cases discussed above; when not reduced, there is far more danger than in the former cases, yet this is the only hope of survival.

66. The same remarks apply to the joints at the elbow, both of the forearm and of the upper arm; for if any one of them is dislocated and projects, causing a lesion, they all bring a fatal issue if reduced; but if not reduced, there is a hope of recovery, although those who survive are cer-

---

192 *ποιεῖ ἢ τὰ πρόσθεν εἰρημένα* B: *ποιήσει τῶν πρόσθεν εἰρημένων* MV

193 *ἐμβαλλόμενα* B: *ἐμβληθὲν* MV

τοῖσι περιγινομένοισι. θανατωδέστερα δὲ τοῖσιν[194] *ἐμβαλλομένοισίν ἐστι τὰ ἐπανωτέρω*[195] *τῶν ἄρθρων, ἀτὰρ καὶ τοῖσι μὴ*[196] *ἐμβαλλομένοισι κινδυνωδέστερα δὲ*[197] *ταῦτα. εἰ δέ τινι τὰ ἀνώτατα ἄρθρα ἐξαρθρή-*
278 *σαντα* | *ἕλκος ποιήσαντα ἔξω ἐξίσχοι, ταῦτα δ' ἂν ἔτι καὶ ἐμβαλλόμενα ταχυθανατώτατα ἂν εἴη καὶ μὴ ἐμβαλλόμενα κινδυνωδέστατα. ἰητρείη δὲ εἴρηται οἵη τις ἐμοὶ δοκέει ἐπιτηδειοτάτη τῶν τοιούτων εἶναι.*

67. *Οἷσι δὲ ἄρθρα δακτύλων, ἢ ποδὸς ἢ χειρός, ἐξαρθρήσαντα ἐξέσχεν ἕλκος ποιησάμενα, μὴ κατεηγότος τοῦ ὀστέου, ἀλλὰ κατ' αὐτὴν τὴν ξύμφυσιν ἀποσπασθέντος, τούτοισιν ἢν ἐμβληθέντα ἐμμείνῃ, ἔνι μέν τις κίνδυνος σπασμοῦ, ἢν μὴ χρηστῶς ἰητρεύωνται· ὅμως δὲ*[198] *ἄξιον ἐμβάλλειν, προειπόντα ὅτι φυλακῆς πολλῆς καὶ μελέτης δέεται. ἐμβάλλειν μέντοι ῥήϊστον καὶ δυνατώτατον καὶ τεχνικώτατόν ἐστι*[199] *τῷ μοχλίσκῳ, ὡς καὶ πρόσθεν εἴρηται ἐν τοῖσι καταγνυμένοισι καὶ ἐξίσχουσιν ὀστέοισιν. ἔπειτα ἀτρεμέειν ὡς μάλιστα χρή, καὶ κατακεῖσθαι καὶ ὀλιγοσιτέειν· ἄμεινον δὲ καὶ φαρμακεῦσαι ἄνω κούφῳ φαρμάκῳ. τὸ δὲ ἕλκος θεραπεύειν*[200] *ἢ ἐναίμοισι τοῖσιν ἐπιτέγκτοισι ἢ οἷσι κεφαλῆς*[201] *ὀστέα κατεηγότα ἰητρεύεται, κατάψυχρον δὲ κάρτα μηδὲν προσφέρειν. ἥκιστα μὲν οὖν τὰ πρῶτα ἄρθρα κινδυνώδεά ἐστι, τὰ δ' ἐπάνω*[202] *κινδυνωδέστερα. ἐμβάλλειν δὲ χρὴ αὐθή-*

[194] *θανατωδέστερα δὲ τοῖσιν* MV: *θανατωδες γὰρ ἅτε τοῖς* B [195] *ἐπανώτερα* B: *ἀνώτερα* MV [196] *μὴ* B: om. MV

tain to be maimed. More fatal when reduced are (sc. compound) dislocations of the more proximal joints; and they too involve greater danger even when unreduced. If anyone has the most proximal joints dislocated and projecting through the lesion made, it is there that reduction brings the swiftest death; and there too the danger is greatest, even without reduction. The kind of treatment which seems to me most suitable in such cases has already been described.

67. When the joints of the fingers or toes are dislocated and project through a lesion, without the bone being fractured, but simply torn away at its connection, in these cases reduction and fixation involve some danger of convulsions, if they are not competently carried out; still, it is worthwhile to reduce the dislocation, giving warning beforehand as to the need of great caution and care. The easiest and most effective reduction, and that most in accord with the art, is the one carried out with a small lever, as described before in relation to fractured and protruding bones. Afterward the patient should keep as quiet as possible, lie down, and take a little food. It is quite beneficial to give a mild emetic. Treat the lesion either with moist applications for fresh wounds, or remedies used for head fractures; but do not apply anything very cold. The distal joints, then, are least dangerous, and the higher ones more so. You should make the reduction on the first or the fol-

---

197 δὲ B: αὐτὰ MV 198 Add. τι MV
199 ἐστι MV: ἐπὶ B 200 θεραπ. B: ἰητρεύειν MV
201 κεφαλῆς B: ἐν κεφαλῃ MV
202 ἐνάνω Kühlewein: δὲ ἔτι ἄνω codd.

280 μερα ἢ τῇ ὑστεραίῃ, | τριταίῳ δὲ καὶ τεταρταίῳ ἥκιστα· τεταρταῖα γὰρ ἐόντα ἐπισημαίνει τῇσι παλιγκοτίῃσι μάλιστα. οἷσιν ἂν οὖν μὴ αὐτίκα ἐγγένηται[203] ἐμβάλλειν, ὑπερβαίνειν[204] χρὴ ταύτας τὰς εἰρημένας ἡμέρας· ὅ τι γὰρ ἂν εἴσω δέκα ἡμερέων ἐμβάλλῃς, πᾶν καταληπτέον. ἢν δὲ ἄρα ἐμβεβλημένῳ σπασμὸς ἐπιγένηται, ἐκβάλλειν[205] τὸ ἄρθρον χρὴ ταχύ, καὶ θερμῷ τέγγειν ὡς πλειστάκις, καὶ τὸ ὅλον σῶμα θερμῶς καὶ λιπαρῶς καὶ μαλθακῶς ἔχειν, καὶ μάλιστα κατὰ τὰ ἄρθρα. κεκάμφθαι δὲ μᾶλλον ἢ ἐκτετάσθαι πᾶν τὸ σῶμα χρή. προσδέχεσθαι μέντοι χρὴ[206] κατὰ τοὺς δακτύλους τὰ ἄρθρα τὰ ἐμβαλλόμενα ἀποστατικὰ ἔσεσθαι· τὰ γὰρ πλεῖστα οὕτω γίνεται, ἢν καὶ ὁτιοῦν φλεγμονῆς ὑπογένηται, ὡς, εἰ μὴ δι᾽ ἀμαθίην τῶν δημοτέων ἐν αἰτίῃ ἔμελλεν ὁ ἰητρὸς ἔσεσθαι, οὐδὲν ἂν πάντως οὐδ᾽ ἐμβάλλειν ἔδει.

68. τὰ μὲν οὖν κατὰ τὰ ἄρθρα ὀστέα ἐξίσχοντα ἐμβαλλόμενα οὕτω κινδυνώδεά ἐστιν, ὡς εἴρηται. |
282 ὅσα δὲ κατὰ τὰ ἄρθρα τὰ κατὰ τοὺς δακτύλους ἀποκόπτεται τελείως, ταῦτα ἀσινέα τὰ πλεῖστά ἐστιν, εἰ μή τις ἐν αὐτῇ τῇ τρώσει λειποθυμήσας βλαβείη· καὶ ἰητρείη φαύλη ἀρκέσει τῶν τοιούτων ἑλκέων. ἀτὰρ καὶ ὅσα μὴ κατὰ τὰ ἄρθρα, ἀλλὰ κατ᾽ ἄλλην τινὰ ἴξιν τῶν ὀστέων ἀποκόπτεται, καὶ ταῦτα ἀσινέα ἐστί, καὶ ἔτι εὐαλθέστερα τῶν ἑτέρων· καὶ ὅσα κατὰ τοὺς δακτύλους ὀστέα καταγέντα ἐξίσχει μὴ κατὰ τὸ ἄρθρον, καὶ ταῦτα ἀσινέα ἐμβαλλόμενα ἐστίν. ἀποκόψιες δὲ τέλειαι ὀστέων καὶ κατὰ τὰ ἄρθρα καὶ ἐν ποδὶ

lowing day, but not on the third or fourth, since the onset of complications occurs mostly on the fourth day. In cases, then, where immediate reduction fails, you should pass over the days named. Any case you reduce within ten days is liable to result in convulsions. If these supervene after reduction, it is necessary quickly to re-dislocate the joint, make frequent warm affusions, and keep the whole body warm, comfortable and peacefully at rest, especially at the joints. The whole body should rather be flexed than extended. In any case it is to be expected that the articular ends of the phalanges will exfoliate after reduction, since this happens in most cases, if there is any amount of inflammation. So, if the surgeon is not likely to incur blame owing to the ignorance of laymen, no reduction should be made at all.

68. The dangers, then, of reducing bones which project through the skin at the joints are such as have been described. Cases of complete amputation of fingers or toes at the joints are usually without danger, unless a patient suffers from syncope at the time of injury, and ordinary treatment will suffice for such wounds. Again, where the amputation is not at the joints, but somewhere along the length of the bones, these cases too are not dangerous, and heal even more readily than the former. Also, if the protrusion of fractured finger bones is not at a joint, reduction is without danger here too. Complete amputations even at the joints both of the foot and hand, or of the leg at the

---

203 ἐγγέν. B: γέν. MV
204 ὑπερβαίνειν MV: ὑπερβάλλειν B
205 ἐκβάλλειν HI: ἐμβ. BMV 206 χρὴ B: om. MV

καὶ ἐν χειρὶ καὶ ἐν κνήμῃ, τοῖσι παρὰ τὰ σφυρά, καὶ ἐν πήχει τοῖσι παρὰ τοὺς καρπούς, τοῖσι πλείστοισιν ἀποκοπτομένοισιν ἀσινέα γίνεται, ὅσα ἂν μὴ αὐτίκα λειποθυμίη ἀνατρέψῃ ἢ τεταρταίοισιν ἐοῦσι πυρετὸς ξυνεχὴς ἐπιγένηται.

69. Ἀποσφακελίσιες μέντοι σαρκῶν, καὶ ἐν τρώμασιν αἱμορρόοισι γενομένοισιν ἢ ἀποσφίγξεσιν ἰσχυραῖς, καὶ ἐν ὀστέων κατήγμασι γενομένοισι[207] πιεχθεῖσι μᾶλλον τοῦ καιροῦ, καὶ ἐν ἄλλοισι δεσμοῖσι βιαίοισιν, ἀπομελανθέντα[208] ἀποπίπτει πολλοῖσι· καὶ οἱ πολλοὶ περιγίνονται τῶν τοιούτων, καὶ οἷσι μηροῦ μέρος τι ἀποπίπτει καὶ τῶν σαρκῶν καὶ τοῦ ὀστέου, καὶ οἷσι βραχίονος, ἧσσόνως δέ. πήχεος δὲ καὶ κνήμης ἀποπεσούσης, καὶ ἔτι εὐφορωτέροις[209] περιγίνονται. οἷσι μὲν οὖν κατεαγέντων τῶν ὀστέων ἀποσφίγξιες αὐτίκα ἐγένοντο καὶ μελασμοί, τούτοισι μὲν
284 ταχεῖαι αἱ περιρρήξιες γίνονται τοῦ | σώματος, καὶ τὰ ἀποπίπτοντα ταχέως ἀποπίπτει, ἤδη τῶν ὀστέων προενδεδωκότων. οἷσι δὲ ὑγιέων ἐόντων τῶν ὀστέων οἱ μελασμοὶ γίνονται, αἱ μὲν σάρκες ταχέως θνῄσκουσι καὶ τούτοισι, τὰ δὲ ὀστέα βραδέως ἀφίσταται, ᾗ ἂν τὰ ὅρια τοῦ μελασμοῦ γένηται καὶ ἡ ψίλωσις τοῦ ὀστέου. χρὴ δέ, ὅσα ἂν κατωτέρω τοῦ σώματος τῶν ὁρίων τοῦ μελασμοῦ ἔῃ, ταῦτα, ὅταν ἤδη πάμπαν τεθνήκῃ καὶ ἀναλγέα ᾖ, ἀφαιρέειν κατὰ τὸ ἄρθρον, προμηθεόμενον ὅπως μή τιτρώσκῃς.[210] ἢν γὰρ ὀδυνηθῇ ἀποταμνόμενος καὶ μήπω κυρήσῃ τὸ σῶμα τεθνεὸς ταύτῃ ᾗ ἀποτέμνεται, κάρτα κίνδυνος ὑπὸ τῆς

ankle, and of the forearm at the wrist, are in most cases without danger, unless syncope overcomes the person at once, or continuous fever supervenes on the fourth day.

69. Gangrene of the tissues occurring in bleeding wounds or violent strangulations, in fractures of bones that are under more pressure than they should be, and in other cases of overly tight bandaging: the intercepted parts turn black, and in many cases exfoliate. The majority of such patients survive, even when a part of the thigh comes away with the soft parts and the bone, or also parts of the arm, although less frequently with these. When the forearm or leg comes away, patients survive even more certainly. Now, in cases of fractured bones, when strangulation arises at once in conjunction with blackening, lines of demarcation rapidly appear on the part, and what is about to come away does so quickly, since the bones are already dissolving; but in cases where the darkening comes on while the bones are sound, the flesh dies rapidly there too, but the bones separate only slowly along the border between the blackness and the denudation of the bone. Concerning parts of the limb that are below the limit of the mortification, when these are quite dead and painless, they should be taken off at the joint, taking care not to wound any living part. For if the patient suffers pain during the amputation, and the limb happens not to be dead yet at the place where it is being cut away, there is a great

---

207 γενομένοισι MV: om. B 208 ἀπομελανθέντα B: ἀποληφθέντα MV 209 εὐφορωτέροις B: εὐφόρως MV

210 τιτρώσκῃς Ermerins: τιτρώσῃ B: τι τρῶσις M: τιτρώσκησιν V

ὀδύνης λειποθυμῆσαι· αἱ δὲ τοιαῦται λειποθυμίαι πολλοὺς παραχρῆμα ἤδη ἀπώλεσαν. μηροῦ μὲν οὖν ὀστέον, ψιλωθὲν ἐκ τοιούτου τρόπου, ὀγδοηκοσταῖον εἶδον ἐγὼ[211] ἀποστάν· ἡ μέντοι κνήμη τούτῳ τῷ ἀνθρώπῳ κατὰ τὸ γόνυ ἀφῃρέθη εἰκοσταίη· ἐδόκε δέ μοι καὶ ἐγγυτέρω· οὐ γὰρ[212] ἀλλ' ἐπὶ τὸ προμηθέστερον ἔδοξέ τι ποιέειν. κνήμης δὲ ὀστέα ἐκ τοιούτου μελασμοῦ, μάλα κατὰ μέσην τὴν κνήμην ἐόντα, ἑξηκοσταῖά μοι ἀπέπεσεν, ὅσα ἐψιλώθη αὐτῶν. διενέγκοι μὲν γὰρ ἄν τι καὶ ἰητρείη ἰητρείης ἐς τὸ θᾶσσόν τε
286 καὶ βραδύτερον τὰ ὀστέα ψιλούμενα ἀποπίπτειν· | διενέγκοι δ' ἄν τι καὶ πίεξις πιέξιος καὶ ἐπὶ τὸ ἰσχυρότερόν τε καὶ ἀσθενέστερον, καὶ ἐς τὸ θᾶσσόν τε καὶ βραδύτερον ἀπομελανθέντα ἀποθανεῖν τὰ νεῦρα καὶ τὰς σάρκας καὶ τὰς ἀρτηρίας καὶ τὰς φλέβας. ἐπεὶ ὅσα μὴ ἰσχυρῶς ἀποληφθέντων θνῄσκει, ἔνια τῶν τοιούτων οὐκ ἀφικνέεται ἐς ὀστέων ψιλώματα, ἀλλ' ἐπιπολαιότερα ἐκπίπτει· ἔνια δὲ οὐδὲ ἐς νεύρων ψιλώματα ἀφικνεῖται, ἀλλ' ἐπιπολαιότερα ἐκπίπτει. διὰ[213] οὖν ταύτας τὰς εἰρημένας[214] προφάσιας οὐκ ἔστιν ἓν οὔνομα ἀριθμοῦ τῷ χρόνῳ θέσθαι, ἐν ὁπόσῳ ἕκαστα τούτων κρίνεται.

Προσδέχεσθαι δὲ χρὴ μάλα τὰ τοιαῦτα ἰήματα· ἐσιδέειν γὰρ φοβερώτερά ἐστιν ἢ ἰητρεύειν· καὶ ἰητρείη πραεῖα ἀρκέε τοιούτοισιν πᾶσι· αὐτὰ γὰρ ἑωυτὰ κρίνει μοῦνον. τῆς δὲ διαίτης ἐπιμελεῖσθαι χρὴ ὡς κατὰ δύναμιν ἀπύρετος ἔῃ, καὶ ἐν σχήμασι δικαίοισιν εὐθετίζειν τὸ σῶμα· δίκαια δὲ ταῦτα μηδὲ μετέωρον risk of syncope from the pain; and syncopes of this kind have brought sudden death to many. I have seen a bone of the thigh, denuded in this way, separate on the eightieth day; this patient's leg separated at the knee on the twentieth day, but I held this to be too early, and resolved to act rather on the careful side. The bones of the leg in a similar case of gangrene that I had, just in the middle of the leg, came away on the sixtieth day, as far as they were denuded. One kind of treatment or another would make a great difference in the rapidity or slowness with which the denuded bones come away. So too pressure, if stronger or weaker, would make a difference in the rapidity or slowness of the blackening and mortification of the cords, muscles, arteries, and veins. For where the parts perish without great strangulation, the denudation sometimes does not extend to the bones, but the more superficial tissues are sloughed off; sometimes the denudation does not even extend even to the cords, but only the more superficial tissues are sloughed off. For the reasons stated, then, it is impossible to fix on some definite time when each of these cases will be determined.

The treatment of such cases should be readily taken on, since they are more formidable to look at than to treat, and mild treatment is sufficient for them all, since they reach their crisis spontaneously. The diet must be carefully managed, so that the patient will be, as far as possible, without fever, and hold his body in the correct postures; correctly

211 ἔγω B: om. MV   212 ἅμα add. MV
213 διὰ MV: δεῖ B
214 εἰρημένας B: om. MV

ποιέειν, μηδὲ ἐς τὸ κάτω ῥέπον, ἀλλὰ μᾶλλον ἐς τὸ ἄνω, ποτὶ καὶ ἔστ' ἂν τελείως περιρραγῇ· αἱμορραγιέων γὰρ ἐν τούτῳ τῷ χρόνῳ κίνδυνος. διὰ τοῦτο οὖν οὐ χρὴ κατάρροπα τὰ τρώματα ποιέειν, ἀλλὰ τἀναντία. ἐπεὶ ὅταν γε ὁ χρόνος ἐγγένηται πλείων καὶ καθαρὰ τὰ ἕλκεα γένηται, οὐκ ἔτι ταῦτα τὰ σχήματά
288 ἐστιν ἐπιτήδεια, | ἀλλ' ἡ εὐθεῖα θέσις, καὶ ἐνίοτε ἐπὶ τὸ κατάρροπον ῥέποντα. ἀνὰ χρόνον γὰρ ἐνίοισι τούτων ἀποστάσιες πύου[215] γίνονται, καὶ ὑποδεσμίδων δέονται. προσδέχεσθαι δὲ χρὴ τοὺς τοιούτους ἀνὰ χρόνον ὑπὸ δυσεντερίης πιέζεσθαι· καὶ γὰρ ἐπὶ τοῖσι μελαινομένοισι, τοῖσι πλείστοισιν ἐπιγίνεται δυσεντερίη, καὶ ἐπὶ τοῖσιν αἱμορραγήσασιν ἐξ ἑλκέων. ἐπιγίνεται δὲ ὡς ἐπὶ τὸ πολὺ κεκριμένων ἤδη καὶ τῶν μελασμῶν καὶ τῆς αἱμορραγίης, καὶ ὁρμᾶται μὲν λάβρως καὶ ἰσχυρῶς· ἀτὰρ οὔτε πολυήμερος γίνεται οὔτε θανατώδης· οὔτε γὰρ μάλα ἀπόσιτοι γίνονται οἱ τοιοῦτοι, οὔτε ἄλλως ξυμφέρει κενεαγγέειν.

70. Μηροῦ δὲ ὀλίσθημα ἐς ἰσχίον ὧδε χρὴ ἐμβάλλειν, ἢν ἐς τὸ εἴσω μέρος ὠλισθήκῃ. ἀγαθὴ μὲν ἥδε καὶ δικαίη καὶ κατὰ φύσιν ἡ ἐμβολή, καὶ δή τι καὶ ἀγωνιστικὸν ἔχουσα, ὅστις γε καὶ τοῖσι τοιούτοισιν ἥδεται κομψευόμενος. κρεμάσαι χρὴ τὸν ἄνθρωπον τῶν ποδῶν πρὸς μεσόδμην δεσμῷ δυνατῷ μέν, μαλθακῷ δὲ καὶ πλάτος ἔχοντι. τοὺς δὲ πόδας διέχειν τι[216] χρὴ ὅσον τέσσαρας δακτύλους ἀπ' ἀλλήλων, ἢ καὶ ἔλασσον· χρὴ δὲ καὶ ἄνωθεν[217] τῶν ἐπιγουνίδων
290 προσπεριβεβλῆσθαι πλατεῖ ἱμάντι καὶ μαλθακῷ, |

placed limbs are neither elevated as a whole nor sloping downward, but rather sloping upward, especially before the line of separation is fully demarcated; for there is a danger of hemorrhage in this period. Consequently, you should do not let the injured part hang down, but the reverse. When a considerable time has elapsed, and the wounds have become clean, the suitable position is no longer the same as before, but rather the horizontal position, and sometimes one sloping downward. In time purulent abscesses form in some of these cases, and they require underbandages. It is to be expected that such patients will be troubled, after a time, with dysentery; for dysentery supervenes in most cases of mortification, and when there is hemorrhaging from wounds. It comes on, as a rule, when the mortification or hemorrhage has already reached its crisis: in the beginning, it is copious and violent, but it neither lasts long nor threatens life. The patients in such cases do not lose their appetite much, nor is there any advantage in a restricted diet.

70. Dislocation of the thigh at the hip should be reduced as follows, if it is dislocated inward. This is a good and correct method, and in accordance with nature, and one too that has something daring about it, which would please the dilettante in these matters. You should suspend the patient by his feet from a crossbeam with a band that is strong, but soft, and quite wide. The feet should be about four inches apart, or even less. The patient should also be bound around above the kneecaps with a broad,

215 πύου Galen: ὀστέου BMV
216 τι B: om. MV
217 ἄνωθεν B: ἐπάνωθεν MV

ἀνατείνοντι πρὸς τὴν μεσόδμην· τὸ δὲ σκέλος τὸ σιναρὸν ἐντετάσθαι ὡς δύο δακτύλοις μᾶλλον τοῦ ἑτέρου χρή. ἀπὸ τῆς γῆς τὴν κεφαλὴν ἀπεχέτω ὡς δύο πήχεας, ἢ ὀλίγῳ πλέον ἢ ἔλασσον. τὰς δὲ χεῖρας παρατεταμένας παρὰ τὰς πλευρὰς προσδεδεμένος[218] ἔστω μαλθακῷ τινί. ταῦτα δὲ πάντα ὑπτίῳ κατακειμένῳ κατασκευασθήτω, ὡς ὅτι ἐλάχιστον χρόνον κρέμηται. ὅταν δὲ κρεμασθῇ, ἄνδρα χρὴ εὐπαίδευτον καὶ μὴ ἀσθενέα, ἐνείραντα τὸν πῆχυν μεσηγὺ τῶν μηρῶν, ἔπειτα θέσθαι τὸν πῆχυν μεσηγὺ τοῦ περινέου καὶ τῆς κεφαλῆς τοῦ μηροῦ τῆς ἐξεστηκυίης. ἔπειτα ξυνάψαντα τὴν ἑτέρην χεῖρα πρὸς τὴν διῃρμένην, παραστάντα ὀρθὸν παρὰ τὸ σῶμα τοῦ κρεμαμένου, ἐξαπίνης ἐκκρεμασθέντα μετέωρον αἰωρηθῆναι ὡς ἰσορροπώτατον. αὕτη ἡ ἐμβολὴ παρέχεται πάντα ὅσα χρὴ κατὰ φύσιν· αὐτό τε γὰρ τὸ σῶμα κρεμάμενον τῷ ἑωυτοῦ βάρει κατάτασιν ποιέεται, ὅ τε ἐκκρεμασθεὶς ἅμα μὲν τῇ κατατάσει ἀναγκάζει ὑπε-
292 ραιωρέεσθαι | τὴν κεφαλὴν τοῦ μηροῦ ὑπὲρ τῆς κοτύλης, ἅμα δὲ τῷ ὀστέῳ τοῦ πήχεος ἀπομοχλεύει καὶ ἀναγκάζει ἐς τὴν ἀρχαίην φύσιν ὀλισθάνειν. χρὴ δὲ παγκάλως μὲν τοῖσι δεσμοῖσιν ἐσκευάσθαι, φρονέοντα δὲ καὶ ὡς ἰσχυρότατον τὸν ἐξαιωρούμενον εἶναι.

71. Ὡς μὲν οὖν καὶ πρόσθεν εἴρηται, μέγα τὸ διαφέρον ἐστὶ τῶν φυσίων τοῖσι ἀνθρώποισιν ἐς τὸ εὐέμβλητα εἶναι καὶ δυσέμβλητα· καὶ διότι μέγα διαφέρει, εἴρηται πρόσθεν ἐν τοῖσι περὶ ὤμου. ἐνίοισι γὰρ ὁ μηρὸς ἐμπίπτει ἀπ᾽ οὐδεμιῆς παρασκευῆς, ἀλλ᾽

soft band stretching up to the beam; and the injured leg should be stretched about two inches further than the other one. The patient's head should hang about two cubits, more or less, from the ground, and he should hold his arms extended along his sides, and fastened with something soft. Let all these preparations be made while he is lying on his back, in order that the period of suspension can be as short as possible. When he is suspended, let an assistant who is well trained and not weak insert his forearm between the patient's thighs, and bring it down between the perineum and the dislocated head of the thigh. Then, clasping the inserted hand with the other hand, while standing erect beside the body of the suspended patient, let him suddenly hang himself on him, and hold himself in the air as evenly balanced as possible. This mode of reduction provides everything requisite according to nature, for the body itself when suspended exerts extension by its own weight; the assistant who is suspended while making the extension, forces the head of the thigh to be lifted up above its socket, and at the same time levers it out with the bone of his forearm, and forces it to slip into its normal place. But the bands must be perfectly prepared, and attention paid that the suspended assistant is the strongest man available.

71. Now, as was said before, there is a great difference in the constitution of individuals in the ease and difficulty with which their dislocated joints can be reduced; and the reason for this great difference was given before in the part about the shoulder. Thus, in some, the thigh can be

218 προσδεδεμένος MV: προσλελαμμένος B

ὀλίγης μὲν κατατάσιος, ὅσον τῇσι χερσὶ κατιθῦναι, βραχείης δὲ κιγκλίσιος. πολλοῖσι δὲ συγκάμψασι τὸ σκέλος κατὰ τὸ ἄρθρον ἐνέπεσεν, ἤδη ἀμφίσφαλσιν ποιησάμενον. ἀλλὰ γὰρ τὰ πολὺ πλείω οὐκ ἐνακούει τῆς τυχούσης παρασκευῆς. διὰ τοῦτο ἐπίστασθαι[219] μὲν χρὴ τὰ κράτιστα περὶ ἑκάστου ἐν πάσῃ τῇ τέχνῃ· χρῆσθαι δὲ οἷσιν ἂν δόξῃ ἑκάστοτε. εἴρηνται μὲν οὖν τρόποι κατατασίων καὶ ἐν τοῖσιν ἔμπροσθεν γεγραμμένοισιν, ὥστε χρῆσθαι τούτων ὅστις ἂν παρατύχῃ. δεῖ γὰρ ἀντικατατετάσθαι ἰσχυρῶς, ἐπὶ θάτερα μὲν
294 τοῦ σκέλεος, ἐπὶ θάτερα δὲ τοῦ | σώματος. ἢν γὰρ εὖ καταταθῇ, ὑπεραιωρηθήσεται ἡ κεφαλὴ τοῦ μηροῦ ὑπὲρ τῆς ἀρχαίης ἕδρης· καὶ ἢν μὲν ὑπεραιωρηθῇ οὕτως, οὐδὲ κωλῦσαι ἔτι ῥηΐδιον ἵζεσθαι αὐτὴν ἐς τὴν ἑωυτῆς ἕδρην, ὥστε ἤδη πᾶσα ἀρκέει μόχλευσίς τε καὶ κατόρθωσις. ἀλλὰ γὰρ ἐλλείπουσιν ἐν τῇ κατατάσει· διὰ τοῦτο ὄχλον πλείω παρέχει ἡ ἐμβολή. χρὴ δὲ οὐ μοῦνον παρὰ τὸν πόδα τὰ δεσμὰ ἐξηρτῆσθαι, ἀλλὰ καὶ ἄνωθεν τοῦ γούνατος, ἵνα μὴ κατὰ τὸ τοῦ γούνατος ἄρθρον ἐν τῇ τανύσει ἡ ἐπίδοσις ᾖ μᾶλλον ἢ κατὰ τὸ τοῦ ἰσχίου ἄρθρον. οὕτω μὲν οὖν χρὴ τὴν κατάτασιν τὴν πρὸς τὸ τοῦ ποδὸς μέρος ἐσκευάσθαι· ἀτὰρ καὶ τὴν ἐπὶ θάτερα κατάτασιν, μὴ μοῦνον ἐκ τῆς περὶ τὸ στῆθος κατὰ[220] τὰς μασχάλας περιβολῆς ἀντιτείνεσθαι, ἀλλὰ καὶ ἱμάντι μακρῷ, διπτύχῳ, ἰσχυρῷ, προσηνεῖ, παρὰ τὸν περίνεον περιβεβλημένῳ,[221] παρατεταμένῳ, ἐπὶ μὲν τὰ ὄπισθεν παρὰ τὴν ῥάχιν, ἐπὶ δὲ τὰ ἔμπροσθεν παρὰ τὴν κληῗδα, προσ-

reduced without any apparatus, by the aid of slight extension, such as can be applied with the hands, and a slight movement; while in many, flexion of the leg at the joint and making a movement of circumduction reduces it. But much more frequently, they do not yield to any ordinary procedure; for this reason, you must be acquainted with the most effective methods which the whole range of the art has available for each case, and employ them where each seems appropriate. Now methods of extension have been described in previous chapters, so that you may use any one of them which happens to be available. Strong extension must be applied both ways, of the leg in one direction, and of the body in the other; for if good extension is made, the head of the thigh will be lifted over its original seat, and when it is brought up like this, nothing can prevent it from easily settling into its normal position, so that any kind of leverage and adjustment is sufficient. But it is in the extension that operators fail, and that is why the reduction causes so much trouble. It is necessary to attach the bands not only at the foot, but also above the knee, so that the force of the extension will not be exerted on the knee joint rather than on the hip joint. This then is how the extension toward the foot end should be arranged. But there should also be traction in the other direction, not only by means of a band around the chest and under the axillae, but also by a long double band, strong and soft, passed around the perineum and stretched behind along the spine and in front along the collarbone, and attached

219 ἐπίστασθαι MV: ἐπισπασθαι B
220 κατὰ B: καὶ MV 221 περιβεβ. B: βεβ. MV

ἠρτημένῳ πρὸς τὴν ἀρχὴν τὴν ἀντικατατείνουσαν. οὕτω διαναγκάζεσθαι, τοῖσι μὲν ἔνθα διατεινομένοισι, τοῖσι δὲ ἔνθα, ὅκως δὲ ὁ ἱμὰς ὁ παρὰ τὸν περίνεον μὴ περὶ τὴν κεφαλὴν[222] τοῦ μηροῦ παρατεταμένος ἔσται, ἀλλὰ μεσηγὺ τῆς κεφαλῆς καὶ τοῦ περινέου· ἐν δὲ τῇ κατατάσει κατὰ μὲν τὴν κεφαλὴν τοῦ μηροῦ ἐρείσας
296 | τὴν πυγμὴν ἐς τὸ ἔξω ὠθείτω. ἢν δὲ μετεωρίζηται ἑλκόμενος, διέρσας τὴν χεῖρα καὶ ἐπισυνάψας τῇ ἑτέρῃ χειρὶ ἅμα μὲν συγκατατεινέτω, ἅμα δὲ ἐς τὸ ἔξω ξυναναγκαζέτω·[223] ἄλλος δέ τις τὸ παρὰ τὸ γόνυ τοῦ μηροῦ ἡσύχως ἐς τὸ ἔσω μέρος κατορθούτω.

72. Εἴρηκα δὲ καὶ πρόσθεν ἤδη[224] ὅτι ἐπάξιον, ὅστις ἐν πόλει πολυανθρώπῳ ἰητρεύει, ξύλον κεκτῆσθαι τετράγωνον ὡς ἑξάπηχυ,[225] ἢ ὀλίγῳ μέζον, εὖρος δὲ ὡς δίπηχυ, πάχος δὲ ἀρκέει σπιθαμιαῖον· ἔπειτα κατὰ μῆκος μὲν ἔνθεν καὶ ἔνθεν ἐντομὴν ἔχειν χρή, ὡς μὴ ὑψηλοτέρη τοῦ καιροῦ ἡ μηχάνησις εἴη· ἔπειτα φλιὰς βραχείας, ἰσχυρὰς καὶ ἰσχυρῶς ἐνηρμοσμένας, ὀνίσκον ἔχειν ἑκατέρωθεν· ἔπειτα ἀρκέει μὲν ἐν
298 τῷ ἡμίσει τοῦ ξύλου | —οὐδὲν δὲ κωλύει καὶ[226] διὰ παντός—ἐντετμῆσθαι ὡς καπέτους μακρὰς πέντε ἢ ἕξ, διαλειπούσας ἀπ᾽ ἀλλήλων ὡς τέσσαρας δακτύλους, αὐτὰς δὲ ἀρκέει εὖρος τριδακτύλους εἶναι καὶ βάθος οὕτως. ἔχειν δὲ κατὰ μέσον τὸ ξύλον καὶ καταγλυφὴν χρὴ βαθυτέρην, ἐπὶ τετράγωνον, ὡς τριῶν δακτύλων· καὶ ἐς μὲν τὴν καταγλυφὴν ταύτην, ὅταν δοκῇ προσδεῖν, ξύλον ἐμπηγνύναι ἐναρμόζον τῇ καταγλυφῇ, τὸ δὲ ἄνω στρογγύλον· ἐμπηγνύναι δέ,

to the anchor of the counterextension. With the cords arranged this way, apply forces stretching some of them in one direction, and the others in the opposite direction, taking care that the band at the perineum is not stretched over the head of the thigh, but between it and the perineum. During the extension, press your fist against the head of the thigh, and thrust it outward. If the pulling lifts the patient up, insert one hand between the thighs and, clasping it with the other hand, combine extension with pressure outward. Have someone else adjust the knee end of the bone gently inward.

72. I have already said before that it is worthwhile for someone who practices in a populous city to acquire a quadrangular plank, six cubits long or a little more, and about two cubits wide, while for thickness a span is sufficient. Next, it should have a cut made at either end of the long sides, in order that the mechanism will not be higher than is suitable. Then provide strong short supports, firmly fitted in, and having windlasses at each end. It suffices, next, to cut out five or six long grooves about four inches apart; it will be enough if they are three inches broad and the same in depth, occupying half (sc. the length of) the plank, though there is no objection to their extending its whole length. The plank should also have a deeper opening cut out in the middle, about three inches square; and into this hole insert, when requisite, a post, fitted to it, but rounded in the upper part. Insert it, whenever it seems

222 περὶ τὴν κεφαλὴν B: ἐπὶ τῇ κεφαλῇ MV
223 ξυναναγκ. MV: συνκαταγκ. B
224 ἤδη B: om. MV
225 χρὴ add. B
226 καὶ B: om. MV

ἐπήν ποτε δοκέῃ ξυμφέρειν, μεσηγὺ τοῦ περινέου καὶ τῆς κεφαλῆς τοῦ μηροῦ. τοῦτο τὸ ξύλον ἑστεὸς κωλύει τὴν ἐπίδοσιν ἐπιδιδόναι τὸ σῶμα τοῖσι πρὸς ποδῶν ἕλκουσιν· ἐνίοτε γὰρ ἀρκέει αὐτὸ τὸ ξύλον τοῦτο ἀντὶ τῆς ἄνωθεν ἀντικατατάσιος· ἐνίοτε δὲ καὶ κατατεινομένου τοῦ σκέλεος ἔνθεν καὶ ἔνθεν, αὐτὸ τὸ ξύλον τοῦτο, χαλαρὸν ἐγκείμενον ᾗ τῇ ᾗ τῇ, ἐκμοχλεύειν ἐπιτήδειον ἂν εἴη τὴν κεφαλὴν τοῦ μηροῦ ἐς τὸ ἔξω μέρος. διὰ τοῦτο γὰρ καὶ αἱ κάπετοι ἐντετμέαται, ὡς καθ᾽ ὁκοίην ἂν αὐτῶν ἁρμόσῃ, ἐμβαλλόμενος ξύλινος μοχλὸς μοχλεύοι, ἢ παρὰ τὰς κεφαλὰς τῶν ἄρθρων, ἢ κατὰ τὰς κεφαλὰς τελέως ἐρειδόμενος ἅμα τῇ κατατάσει, ἤν τε ἐς τὸ ἔξω μέρος ξυμφέρῃ ἐκμοχλεύεσθαι, ἤν τε ἐς τὸ εἴσω, καὶ ἤν τε στρογγύλον τὸν μοχλὸν ξυμφέρῃ εἶναι, ἤν τε πλάτος ἔχοντα· ἄλλος γὰρ ἄλλῳ τῶν ἄρθρων ἁρμόζει. εὔχρηστος δέ ἐστιν ἐπὶ πάντων
300 *ἄρθρων ἐμβολῆς*[227] *τῶν κατὰ τὰ σκέλεα αὕτη ἡ | μόχλευσις ξὺν τῇ κατατάσει. περὶ οὗ οὖν ὁ λόγος ἐστί, στρογγύλος ἁρμόζει ὁ μοχλὸς εἶναι· τῷ μέντοι ἔξω ἐκπεπτωκότι ἄρθρῳ πλατὺς ἁρμόσει εἶναι. ἀπὸ τούτων τῶν μηχανέων καὶ ἀναγκέων οὐδὲν ἄρθρον μοι δοκέει οἷόν τε εἶναι ἀπορηθῆναι ἐμπεσεῖν.*

73. *Εὕροι δ᾽ ἄν τις καὶ ἄλλους τρόπους τούτου τοῦ ἄρθρου ἐμβολῆς· εἰ γὰρ τὸ ξύλον τὸ μέγα τοῦτο ἔχοι κατὰ μέσον καὶ ἐκ πλαγίων φλιὰς δύο ὡς ποδιαίας, ὕψος δὲ ὅκως ἂν δοκέῃ ξυμφέρειν, τὴν μὲν ἔνθεν, τὴν δὲ ἔνθεν· ἔπειτα ξύλον πλάγιον ἐνείη ἐν τῇσι φλιῇσιν ὡς κλιμακτήρ, ἔπειτα εἰ διέρσειεν*[228] *τὸ ὑγιὲς σκέλος*

useful, between the perineum and the head of the thigh. This post, when fixed, prevents the body from sliding when traction is made toward the feet; in fact, sometimes the post by itself is a substitute for counterextension upward. Sometimes also, when the leg is pulled in one direction or the other, this same post, being so placed as to have free play to either side, will be suitable for levering the head of the thigh outward. It is for this purpose, too, that the grooves are cut, so that a wooden lever can be inserted into whichever one of them is appropriate, and brought to bear either at the side of the heads of the joints, or directly upon them, exerting pressure simultaneously with the extension, whether the leverage required is outward or inward, and whether the lever needs to be rounded or flat, for one shape suits one joint, and another another. This leverage, combined with extension, is very effective in the reductions of all the joints of the legs. As regards our present subject, it is proper for the lever be rounded; but for an external dislocation of the joint, a flat one will be suitable. It seems to me that no joint will be impossible to reduce by these machines and procedures.

73. Someone might find other ways of reducing this joint. For if this large plank had two props in the middle and to the sides, about a foot long and their height as seems suitable, one on one side, the other on the other, and then a crossbar of wood was inserted into the props like a ladder step, you could insert the sound leg between

227 ἐμβολῆς recc. Littré: ἐμβολὴν B: ἐμβολὴ M: ἐμβολῇ V
228 εἰ διέρσειεν B: διέρσαι MV

μεσηγὺ τῶν φλιέων, τὸ δὲ σιναρὸν ἄνωθεν τοῦ κλιμακτῆρος ἔχοι ἐναρμόζον ἀπαρτὶ[229] πρὸς τὸ ὕψος καὶ πρὸς τὸ ἄρθρον, ᾗ ἐκπέπτωκεν· ῥηΐδιον δὲ ἁρμόζειν·[230] τὸν γὰρ κλιμακτῆρα ὑψηλότερόν τι χρὴ ποιέειν τοῦ μετρίου, καὶ ἱμάτιον πολύπτυχον, ὡς ἂν ἁρμόσῃ, ὑποτείνειν ὑπὸ τὸ σῶμα.[231] ἔπειτα χρὴ ξύλον ἔχον |
302 πλάτος μέτριον, καὶ μῆκος ἄχρι τοῦ σφυροῦ ὑποτεταμένον, ὑπὸ τὸ σκέλος εἶναι, ἱκνεύμενον ἐπέκεινα τῆς κεφαλῆς τοῦ μηροῦ ὡς οἷόν τε·[232] προσκαταδεδέσθαι δὲ δεῖ πρὸς τὸ σκέλος, ὅκως ἂν μετρίως ἔχῃ. κἄπειτα κατατεινομένου τοῦ σκέλεος, εἴτε ξύλῳ ὑπεροειδεῖ, εἴτε τούτων τινὶ τῶν κατατασίων, ὁμοῦ χρὴ καταναγκάζεσθαι τὸ σκέλος περὶ τὸν κλιμακτῆρα ἐς τὸ κάτω μέρος ξὺν τῷ ξύλῳ τῷ προσδεδεμένῳ· τὸν δέ τινα κατέχειν τὸν ἄνθρωπον ἀνωτέρω τοῦ ἄρθρου κατὰ τὸ ἰσχίον. καὶ γὰρ οὕτως ἅμα μὲν ἡ κατάτασις ὑπεραίροιτο τὴν κεφαλὴν τοῦ μηροῦ ὑπὲρ τῆς κοτύλης, ἅμα δὲ ἡ μόχλευσις ἀπωθέοι τὴν κεφαλὴν τοῦ μηροῦ ἐς τὴν ἀρχαίην φύσιν. αὗται πᾶσαι αἱ εἰρημέναι ἀνάγκαι ἰσχυραὶ καὶ πᾶσαι κρέσσους τῆς ξυμφορῆς, ἤν τις ὀρθῶς καὶ καλῶς σκευάζῃ. ὥσπερ δὲ καὶ πρόσθεν ἤδη εἴρηται, πολύ τι ἀπὸ ἀσθενεστέρων κατατασίων καὶ φαυλοτέρης κατασκευῆς τοῖσι πλείστοισιν ἐμπίπτει.

74. Ἢν δὲ ἐς τὸ ἔξω ἡ κεφαλὴ τοῦ μηροῦ ὀλίσθῃ,
304 τὰς μὲν κατατάσιας | ἔνθα καὶ ἔνθα οὕτω[233] χρὴ ποιέεσθαι ὥσπερ εἴρηται τοιουτοτρόπως· τὴν δὲ μόχλευσιν πλάτος ἔχοντι μοχλῷ μοχλεύειν χρὴ ἅμα τῇ

the props, and have the injured one on the top of the bar, fitting exactly to its height and to the joint where it was dislocated. This is easy to arrange; for the crossbar should be put somewhat higher than the middle, and a folded garment spread under the patient's body, so that it fits. Then a piece of wood of suitable width, and of a length sufficient to reach to the ankle, should be extended under the leg, going up as far as possible beyond the head of the thigh; it should be bound to the leg in a suitable manner. Then, while the leg is being extended either by a pestle-shaped piece of wood or by any of the methods of extension mentioned above, you must simultaneously force the leg downward over the crossbar with the piece of wood attached to it, while an assistant holds the patient down at the hip above the joint. In this way the extension will raise the head of the thigh over its socket at the same time as the leverage pushes it back into its natural position. All the forcible methods of reduction mentioned are effective, and all are able to correct the injury, if they are properly and skillfully applied. But, as was said before, in the majority of cases the joint is put in with much weaker extensions and a more ordinary apparatus.

74. When the head of the thigh slips outward, extension should be made in both directions as described, or in a similar fashion. The leverage should be carried out with

---

229 ἀπαρὶ B: ἀτάρ τι MV
230 ἅρμόζειν B: νομίζειν MV
231 Add. ἁπλοῦν B
232 οἶόν τε MV: οἴονται B
233 οὕτω B: om. MV

κατατάσει, ἐκ τοῦ ἔξω μέρους ἐς τὸ εἴσω ἀναγκάζοντα, κατά τε αὐτὸν τὸν γλουτὸν τιθέμενον τὸν μοχλὸν καὶ ὀλίγῳ ἀνωτέρω· ἐπὶ δὲ τὸ ὑγιὲς ἰσχίον κατὰ τὸν γλουτὸν ἀντιστηριζέτω τις τῇσι χερσὶν ὡς μὴ ὑπείκῃ τὸ σῶμα, ἢ ἑτέρῳ τινὶ τοιούτῳ μοχλῷ ὑποβάλλων καὶ ἐρείσας, ἐκ τῶν καπέτων τὴν ἁρμόζουσαν ἀντικατεχέτω· τοῦ δὲ μηροῦ τοῦ ἐξηρθρηκότος τὸ παρὰ τὸ γόνυ εἴσωθεν ἔξω παραγέτω ἡσύχως. ἡ δὲ κρέμασις οὐχ ἁρμόσει τούτῳ τῷ τρόπῳ τῆς ὀλισθήσιος τοῦ ἄρθρου· ὁ γὰρ πῆχυς τοῦ ἐκκρεμαμένου[234] ἀπωθοίη ἂν τὴν κεφαλὴν τοῦ μηροῦ ἀπὸ τῆς κοτύλης. τὴν μέντοι ξὺν τῷ ξύλῳ τῷ ὑποτεινομένῳ μόχλευσιν μηχανήσαιτ᾽ ἄν τις ὥστε ἁρμόζειν καὶ τούτῳ τῷ τρόπῳ τοῦ ὀλισθήματος, ἔξωθεν προσαρτέων. ἀλλὰ τί καὶ δεῖ; ἢν γὰρ ὀρθῶς μὲν καὶ εὖ κατατείνηται, ὀρθῶς δὲ μοχλεύηται, τί οὐκ ἂν ἐμπέσοι ἄρθρον οὕτως ἐκπεπτωκός;

75. Ἢν δὲ ἐς τοὔπισθεν μέρος ἐκπεπτώκῃ ὁ μηρός,
306 τὰς | μὲν κατατάσιας καὶ ἀντιτάσιας οὕτω χρὴ ποιέεσθαι, ὡς εἴρηται. ἐπιστορέσαντα δὲ ἐπὶ τὸ ξύλον ἱμάτιον πολύπτυχον, ὡς μαλακώτατον ἔῃ,[235] πρηνέα κατακλίναντα τὸν ἄνθρωπον, οὕτω κατατείνειν· ἅμα δὲ τῇ κατατάσει χρὴ τῇ σανίδι καταναγκάζειν τὸν αὐτὸν τρόπον ὡς τὰ ὑβώματα, κατ᾽ ἴξιν τοῦ πυγαίου ποιησάμενον τὴν σανίδα, καὶ μᾶλλον ἐς τὸ κάτω μέρος ἢ ἐς τὸ ἄνω τῶν ἰσχίων· καὶ ἡ ἐντομὴ ἡ ἐν τῷ τοίχῳ τῇ σανίδι μὴ εὐθεῖα ἔστω, ἀλλ᾽ ὀλίγον καταφερὴς πρὸς τὸ τῶν ποδῶν μέρος. αὕτη ἡ ἐμβολὴ κατὰ

a broad lever at the same time as the extension, forcing it from without inward, the lever being applied to the buttock itself and a little above it. Have someone provide countersupport to the hip on the sound side at the buttock with his hands, so that the body does not move; or provide counterpressure by slipping a similar lever under the joint, using a suitable groove as fulcrum. Let the bone of the dislocated thigh be gently brought from within outward at the knee. The suspension method is not suitable for this form of dislocation, since the forearm of the person who hangs himself on to (the patient) would push the head of the thigh away from its socket. But you could arrange the leverage with the board attached in such a way as to suit this form of dislocation as well, by fitting it to the outside. But what else is needed? For if the extension is correct and well executed, and the leverage correct, what joint dislocated in this way would not fall back into place?

75. If the thigh is dislocated backward, extension and counterextension should be made in the way described. Spread a cloak folded many times on to the plank, so that it will be as soft as possible; with the patient lying prone, provide extension this way, and at the same time as the extension you must provide downward pressure with the plank, as in cases of humpback, putting the plank in a line with the buttock, and rather below than above the hips. The groove in the wall for the board should not be level, but sloping a little downward in the direction of the feet. This mode of reduction is most naturally in accord with

234 ἐκκρε. MV: ἐγκρε. B
235 ἔῃ MV: om. B

φύσιν τε μάλιστα τῷ τρόπῳ τούτῳ τοῦ ὀλισθήματός ἐστι καὶ ἅμα ἰσχυροτάτη. ἀρκέσειε δ' ἂν ἴσως ἀντὶ τῆς σανίδος καὶ ἐφεζόμενόν τινα, ἢ τῇσι χερσὶν ἐρεισάμενον ἢ ἐπίβαντα ἐξαπίνης ὁμοίως ἐπαιωρηθῆναι ἅμα τῇ κατατάσει. ἄλλη δὲ οὐδεμία ἐμβολὴ τῶν πρόσθεν εἰρημένων κατὰ φύσιν ἐστὶ τῷ τρόπῳ τούτῳ τοῦ ὀλισθήματος.

76. Ἢν δὲ ἐς τοὔμπροσθεν ὀλίσθῃ, τῶν μὲν κατατασίων ὁ αὐτὸς τρόπος ποιητέος· ἄνδρα δὲ χρὴ ὡς ἰσχυρότατον ἀπὸ χειρῶν καὶ ὡς εὐπαιδευτότατον, ἐνερείσαντα τὸ θέναρ τῆς ἑτέρης χειρὸς παρὰ τὸν βουβῶνα, καὶ τῇ χειρὶ τῇ ἑτέρῃ τὴν ἑωυτοῦ χεῖρα προσκαταλαβόντα, ἅμα μὲν ἐς τὸ κάτω μέρος ὠθέειν
308 τὸ | ὀλίσθημα, ἅμα δὲ ἐς τοὔμπροσθεν τοῦ γούνατος μέρος. οὗτος γὰρ ὁ τρόπος τῆς ἐμβολῆς κατὰ φύσιν μάλιστα τούτῳ τῷ ὀλισθήματί ἐστιν. ἀτὰρ καὶ ὁ κρεμασμὸς ἐγγύς τι τοῦ κατὰ φύσιν· δεῖ μέντοι τὸν ἐγκριμνάμενον ἔμπειρον εἶναι, ὡς μὴ ἐκμοχλεύῃ τῷ πήχει τὸ ἄρθρον, ἀλλὰ περὶ[236] μέσον τὸν περίνεον καὶ κατὰ τὸ ἱερὸν ὀστέον τὴν ἐγκρέμασιν[237] ποιέηται.

77. Εὐδοκιμέει δὲ δὴ καὶ ἀσκῷ τοῦτο τὸ ἄρθρον ἐμβάλλεσθαι· καὶ ἤδη μέν τινας εἶδον οἵτινες ὑπὸ φλαυρότητος[238] καὶ τὰ ἔξω ἐκκεκλιμένα καὶ τὰ ὄπισθεν ἀσκῷ ἐπειρῶντο ἐμβάλλειν, οὐ γινώσκοντες ὅτι ἐξέβαλλον αὐτὸ μᾶλλον ἢ ἐνέβαλλον. ὁ μέντοι πρῶτος ἐπινοήσας δῆλον ὅτι πρὸς τὰ ἔσω ὠλισθηκότα ἀσκῷ ἐμβάλλειν ἐπειρήσατο. ἐπίστασθαι μὲν οὖν χρὴ ὡς χρηστέον ἀσκῷ, εἰ δέοι χρῆσθαι· διαγινώ-

this form of dislocation, and at the same time very effective. It would probably be sufficient, in place of the board, for someone to sit on the part, or to apply pressure with his hands, or to step on the patient, in each case bringing his weight suddenly to bear at the moment of extension. None of the other modes of reduction described above is in natural conformity with this dislocation.

76. If the dislocation is forward, the same kinds of extension are to be used. A well-trained man with very strong hands should apply pressure on the patient's groin with the palm of one hand, and grasp this one with the other hand, while at the same time he presses the dislocated part downward as well as toward the forward part of the knee. This mode of reduction is in most natural accord with this particular dislocation. But suspension also has a certain proximity with nature; however, the man who hangs himself on (sc. to the patient) must be experienced, in order to avoid levering the joint out with his arm, but make the suspension exert its effect at the middle of the perineum and over the sacrum.

77. There is also a well-regarded method of reducing this joint with a bladder; and I have even seen some practitioners who, through incompetence, kept trying to reduce even external and posterior dislocations with a bladder, not knowing that they were putting it out rather than putting it in. The first inventor of the method, however, obviously used the bladder in trying to reduce inward dislocations. Now, it must be understood how to use the bladder, if required, but still be borne in mind that many other

236 περὶ MV: κατὰ B 237 ἐγκρ. B: ἐκκρ. MV
238 φλαυρ. B: φαυλ. MV

σκειν δὲ δεῖ[239] ὅτι ἕτερα πολλὰ ἀσκοῦ κρέσσω ἐστίν. χρὴ δὲ τὸν μὲν ἀσκὸν ἐνθεῖναι[240] ἐς τοὺς μηροὺς ἀφύσητον ἐόντα, ὡς ἂν δύναιτο ἀνωτάτω πρὸς τὸν περίνεον ἀνάγοντα. ἀπὸ δὲ τῶν ἐπιγουνίδων ἀρξάμενον, ταινίῃ πρὸς ἀλλήλους τοὺς μηροὺς καταδῆσαι ἄχρι τοῦ ἡμίσεος τῶν μηρῶν· ἔπειτα ἐς ἕνα τῶν ποδῶν,[241] τὸν λελυμένον, ἐνθέντα αὐλὸν ἐκ χαλκίου, φῦσαν ἐσαναγκάζειν ἐς τὸν ἀσκόν· τὸν δὲ ἄνθρωπον |
310 πλάγιον κατακέεσθαι, τὸ σιναρὸν σκέλος ἐπιπολῆς ἔχοντα. ἡ μὲν οὖν παρασκευὴ αὕτη ἐστί· σκευάζονται δὲ κάκιον οἱ πλεῖστοι ἢ ὡς ἐγὼ εἴρηκα· οὐ γὰρ καταδέουσι τοὺς μηροὺς ἐπὶ συχνόν, ἀλλὰ μοῦνον τὰ γούνατα, οὐ δὲ κατατείνουσι·[242] χρὴ δὲ καὶ προσκατατείνειν. ὅμως δὲ ἤδη τινὲς ἐνέβαλλον ῥηϊδίου τοῦ πρήγματος ἐπιτυχόντες. εὐφόρως δὲ οὐ πάνυ ἔχει διαναγκάζεσθαι οὕτως· ὅ τε γὰρ ἀσκὸς ἐμφυσώμενος[243] οὐ τὰ ὀγκηρότατα αὐτοῦ ἔχει πρὸς τῷ ἄρθρῳ τῆς κεφαλῆς, ἣν δεῖ μάλιστα ἐκμοχλεύσασθαι, ἀλλὰ κατὰ ἑωυτὸν αὐτὸς μέσος καὶ τῶν μηρῶν ἴσως ἢ κατὰ τὸ μέσον ἢ ἔτι κατώτερον. οἵ τε αὖ μηροὶ φύσει γαυσοὶ πεφύκασιν, ἄνωθεν γὰρ σαρκώδεές τε καὶ ξύμμηροι, ἐς δὲ τὸ κάτω ὑπόξηροι, ὥστε καὶ ἡ τῶν μηρῶν φύσις ἀπαναγκάζει[244] τὸν ἀσκὸν ἀπὸ τοῦ ἐπικαιροτάτου χωρίου. εἴ τε οὖν τις σμικρὸν ἐνθήσει τὸν ἀσκόν, σμικρὴ ἡ ἰσχὺς ἐοῦσα ἀδύνατος[245] ἔσται ἀναγκάζειν τὸ ἄρθρον. εἰ δὲ δεῖ ἀσκῷ χρῆσθαι, ἐπὶ

[239] δεῖ B: χρὴ MV

[240] ἐνθεῖναι B: καταθεῖναι MV

methods are more effective. The bladder must be applied in the uninflated state between the thighs, and brought up as close as possible to the perineum. Bind the thighs to one another with a band beginning from above the knee-caps and continuing halfway up the thighs; then, inserting a brass tube into one of the feet of the bladder which has been untied, force air into it. The patient should lie on his side with the injured leg on top. This, then, is the arrangement; but most operators make less adequate preparation than I have described. They do not fasten the thighs together over most of their length, but only at the knees; nor do they extend the leg, although extension is necessary. Still, some operators have succeeded with the reduction, because they chanced upon an easy case. But a forcible reduction is by no means easy to achieve this way: for the inflated bladder does not direct its greatest force against the articular extremity of the head, which is the place that must in particular be levered out, but at its own middle, which presumably corresponds with middle of the thighs, or is even lower down; furthermore, the thighs themselves have a natural curve, being at the top muscular and close together, but lower down leaner, so that the natural form of the thighs also forces the bladder away from the most opportune place. If someone inserts a small bladder, its power being small, it will be unable to reduce the joint. So, if you must use a bladder, the thighs are to be bound

---

241 ποδῶν B: πολλῶν MV

242 κατατ. B: προσκατατ. MV

243 ἐμφυσῶ. MV: φυσῶ B   244 ἄπαν. Littré: ἔπαν. BMV

245 ἀδύνατος BM: δυνατὸν V (M add. as alternative in marg.)

πολὺ οἱ μηροὶ ξυνδετέοι πρὸς ἀλλήλους, καὶ ἅμα τῇ κατατάσει τοῦ σώματος ὁ ἀσκὸς φυσητέος· τὰ δὲ σκέλεα ἀμφότερα ὁμοῦ καὶ καταδεῖν ἐν τούτῳ τῷ τρόπῳ τῆς ἐμβολῆς ἐπὶ τὴν τελευτήν.[246] |

312 78. Χρὴ δὲ περὶ πλείστου μὲν ποιέεσθαι ἐν πάσῃ τῇ τέχνῃ ὅκως ὑγιᾶ ποιήσεις τὸν νοσέοντα· εἰ δὲ πολλοῖσι τρόποισι οἷόν τε εἴη ὑγιᾶ ποιεῖν, τὸν ἀοχλότατον[247] χρὴ αἱρέεσθαι· καὶ γὰρ ἀνδραγαθικώτερον[248] τοῦτο καὶ τεχνικώτατον, ὅστις μὴ ἐπιθυμεῖ δημοειδέος κιβδηλίης. περὶ οὗ οὖν ὁ λόγος ἐστί, τοιαίδε ἄν τινες κατοικίδιοι κατατάσιες τοῦ σώματος εἶεν, ὥστε ἐκ τῶν παρεόντων τὸ εὔπορον εὑρίσκειν. τοῦτο μέν, εἰ τὰ δεσμὰ τὰ ἱμάντινα μὴ παρείη τὰ μαλθακὰ καὶ προσηνέα, ἀλλὰ ἢ σιδήρεα[249] ἢ ὅπλα ἢ σχοινία, ἢ ταινίῃσι χρὴ ἢ ἐκρήγμασι[250] τρυχίων ἐρινέων περιελίσσειν ταύτῃ μάλιστα ᾗ μέλλει τὰ δεσμὰ καθέξειν, καὶ ἔτι ἐπὶ πλέον· ἔπειτα οὕτω δεῖν τοῖσι δεσμοῖσι· τοῦτο δέ, ἐπὶ κλίνης χρή, ἥτις ἰσχυροτάτη καὶ μεγίστη τῶν παρεουσέων κατατετάσθαι καλῶς[251] τὸν ἄνθρωπον· τῆς δὲ κλίνης τοὺς πόδας, ἢ τοὺς πρὸς κεφαλῆς ἢ τοὺς πρὸς ποδῶν, ἐρηρεῖσθαι πρὸς τὸν οὐδόν, εἴτε ἔξωθεν συμφέρει, εἴτε ἔσωθεν· παρὰ δὲ τοὺς ἑτέρους πόδας παραβεβλῆσθαι ξύλον τετράγωνον πλάγιον, διῆκον ἀπὸ τοῦ ποδὸς πρὸς τὸν πόδα· καὶ ἢν μὲν λεπτὸν ᾖ τὸ ξύλον, προσδεδέσθω πρὸς τοὺς πόδας τῆς κλίνης, ἢν δὲ παχὺ ᾖ, οὐ δεῖ.[252] ἔπειτα τὰς ἀρχὰς χρὴ τῶν δεσμῶν καὶ τῶν πρὸς κεφαλῆς καὶ τῶν πρὸς
314 | ποδῶν προσδῆσαι ἑκατέρας πρὸς ὕπερον ἢ πρὸς

together for a good length, and the bladder inflated simultaneously with the extension of the body; also tie both legs together at their extremities in this form of reduction.

78. The most important thing to do in all the practice of your art is to make the ill person well; and if it is possible to make him well in many ways, you should choose the least troublesome way. This is more honorable and more in accord with the art for anyone who is not merely striving after vulgar success. To return to our subject, there are certain domestic means of exerting extension that can readily be found among whatever is at hand. First, if no soft supple leather straps are available, but there are iron chains, ship's tackle, or ropes, you should wrap these in cloths or torn woolen rags, especially the part of them where they are to be fastened on, and somewhat further, and then proceed to bind them on as straps. Another possibility is to use a bed, the strongest and largest available, for making effective extension; the legs of the bed, either at the head or at the foot, should press against a threshold, outside or inside, as is opportune; and a quadrangular plank should be laid crosswise against the other legs, reaching from one to the other. If the plank is thin, fasten it to the legs of the bed, but if thick, this is unnecessary. Next, you must tie the ends of the bands, both those at the head and those at the feet respectively, to a pestle, or some

246 ἐπὶ τὴν τελευτήν MV: om. B 247 ἀοχλότ. MV: ὀχλωτ. B 248 -κώτερον MV: -κώτατον B
249 σιδήρεα MV: σειραὶ Kühlewein: σιρᾶ B
250 ἐκρή. Littré: ἐρρή. BMV 251 καλῶς MV: om. B
252 οὐ δεῖ Kühlewein: οὐδέν MV: μηδέν B

ἄλλο τι τοιοῦτο ξύλον. ὁ δὲ δεσμὸς ἐχέτω ἰθυωρίην κατὰ τὸ σῶμα ἢ καὶ ὀλίγον ἀνωτέρω, ξυμμέτρως δὲ ἐκτετάσθω[253] πρὸς τὰ ὕπερα, ὡς, ὀρθὰ ἑστεῶτα, τὸ μὲν παρὰ τὸν οὐδὸν ἐρείδηται, τὸ δὲ παρὰ τὸ ξύλον τὸ παραβεβλημένον· κἄπειτα οὕτω τὰ ὕπερα ἀνακλῶντα χρὴ τὴν κατάτασιν ποιέεσθαι.[254] ἀρκέει δὲ καὶ κλῖμαξ ἰσχυροὺς ἔχουσα τοὺς κλιμακτῆρας, ὑποτεταμένη ὑπὸ τὴν κλίνην, ἀντὶ τοῦ οὐδοῦ τε καὶ τοῦ ξύλου τοῦ παρατεταμένου, ὡς τὰ ὕπερα, πρὸς τῶν κλιμακτήρων τοὺς ἁρμόζοντας ἔνθεν καὶ ἔνθεν προσερηρεισμένα, ἀνακλώμενα,[255] οὕτω τὴν κατάτασιν ποιέηται τῶν δεσμῶν.

Ἐμβάλλεται δὲ μηροῦ ἄρθρον καὶ τόνδε τὸν τρόπον, ἢν ἐς τὸ εἴσω ὀλισθήκῃ καὶ ἐς τὸ ἔμπροσθεν· κλίμακα χρὴ κατορύξαντα ἐπικαθίσαι τὸν ἄνθρωπον, ἔπειτα τὸ μὲν ὑγιὲς σκέλος ἡσύχως κατατείναντα προσδῆσαι, ὅκου ἂν ἁρμόσῃ· ἐκ δὲ τοῦ σιναροῦ ἐς κεράμιον ὕδωρ ἐγχέαντας ἐκκρεμάσαι, ἢ σφυρίδα λίθους ἐμβαλόντας. ἕτερος τρόπος ἐμβολῆς· ἢν ἐς τὸ
316 εἴσω | ὀλισθῇ· στρωτῆρα χρὴ διαδῆσαι μεταξὺ δύω στύλων, ὕψος ἔχοντα συμμέτρως· προεχέτω δὲ τοῦ στρωτῆρος ὁ νοσέων[256] κατὰ τὸ ἓν μέρος, ὁκόσον τὸ πυγαῖον· περιδήσας δὲ περὶ τὸ στῆθος τοῦ ἀνθρώπου ἱμάτιον, ἐπικαθίσαι τὸν ἄνθρωπον ἐπὶ τὸ προέχον τοῦ στρωτῆρος· εἶτα προσλαβεῖν τὸ στῆθος πρὸς τὸν στῦλον πλατεῖ τινι· ἔπειτα τὸ μὲν ὑγιὲς σκέλος κατεχέτω τις, ὡς μὴ περισφάλληται· ἐκ δὲ τοῦ σιναροῦ

other such piece of wood. Let each band be in line with the body, or slanting a little upward, and evenly stretched to the pestles, so that, when they are vertical, one will rest against the threshold, and the other against the crossbeam; then you should make the extension by drawing back the pestles arranged in this way. It is also adequate to stretch a ladder with strong crossbars under the bed as a substitute for the threshold and crossbeam, so that the pestles can be fastened to the steps at either end, and, when drawn back, can in this way exert extension on the bands.

The joint of the thigh can also be reduced in the following manner, if it is dislocated inward or forward. You should fix a ladder in the ground, and seat the patient on top of it; then, gently extend the sound leg and fasten it at a suitable point. From the injured limb suspend a jar and pour water into it, or a basket and put stones into it. Another method of reducing the joint, if dislocated inward: fasten a crossbar between two supports at a moderate height, with one end of it projecting a buttock's length. After passing a cloak round the patient's chest, seat him on the projecting crossbar, and then fasten his chest to the upright with a broad band. Then have an assistant hold the healthy leg, to prevent the patient from slipping around,

---

253 ἐκτετ. M: ἐντετ. BV
254 ποιέεσθαι B: ποιέειν MV
255 ἀνακλώμενα B: om. MV
256 ὁ νοσέων trans. after μέρος MV

κρεμάσαι[257] *βάρος, ὅσον ἂν ἁρμόζῃ, ὥσπερ καὶ πρόσθεν ἤδη*[258] *εἴρηται.*

79. Πρῶτον *μὲν οὖν δεῖ εἰδέναι ὅτι πάντων τῶν ὀστέων αἱ ξυμβολαί εἰσιν ὡς ἐπὶ πολὺ ἡ κεφαλὴ καὶ ἡ κοτύλη· ἐφ' ὧν δὲ καὶ ἡ χώρα κοτυλοειδὴς καὶ ἐπίμακρος·*[259] *ἔνιαι δὲ τῶν χωρέων γληνοειδέες εἰσίν. ἀεὶ δὲ ἐμβαλέειν δεῖ πάντα τὰ ἐκπίπτοντα ἄρθρα, μάλιστα μὲν εὐθὺς παραχρῆμα ἔτι θερμῶν ἐόντων· εἰ δὲ μή, ὡς τάχιστα· καὶ γὰρ τῷ ἐμβάλλοντι ῥηΐτερον καὶ θᾶσσόν ἐστιν ἐμβαλέειν, καὶ τῷ ἀσθενέοντι πολὺ ἀπονωτέρη ἡ ἐμβολὴ ἡ πρὶν διοιδεῖν ἐστίν. δεῖ δὲ ἀεὶ παντα τὰ ἄρθρα, ὁκόταν μέλλῃς ἐμβαλέειν, προαναμαλάξαι καὶ διακιγκλίσαι· ῥᾷον γὰρ ἐθέλει* | *ἐμβάλλεσθαι. παρὰ πάσας δὲ τὰς τῶν ἄρθρων ἐμβολὰς ἰσχναίνειν δεῖ τοὺς ἀνθρώπους, μάλιστα μὲν περὶ τὰ μέγιστα ἄρθρα καὶ χαλεπώτατα ἐμβάλλεσθαι, ἥκιστα δὲ περὶ τὰ ἐλάχιστα καὶ ῥηΐδια.*

318

80. *Δακτύλων δὲ ἢν ἐκπέσῃ ἄρθρον τι τῶν τῆς χειρός, ἤν τε τὸ πρῶτον ἤν τε τὸ δεύτερον, ἤν τε τὸ τρίτον, ωὑτὸς τρόπος τῆς ἐμβολῆς· χαλεπώτερα μέντοι ἀεὶ τὰ μέγιστα τῶν ἄρθρων ἐμβάλλειν. ἐκπίπτει δὲ τέσσαρας τρόπους, ἢ ἄνω ἢ κάτω ἢ ἐς τὸ πλάγιον ἑκατέρωθεν, μάλιστα μὲν ἐς τὸ ἄνω, ἥκιστα δὲ ἐς τὰ πλάγια, ἐν τῷ σφόδρα κινέεσθαι. ἑκατέρωθεν δὲ τῆς χώρης, οὗ ἐκβέβηκεν, ὥσπερ ἄμβη ἐστίν. ἢν μὲν οὖν ἐς τὸ ἄνω ἐκπέσῃ ἢ ἐς τὸ κάτω, διὰ τὸ λειοτέρην εἶναι ταύτην τὴν χώρην ἢ ἐκ τῶν πλαγίων, καὶ ἅμα μικρῆς ἐούσης τῆς ὑπερβάσιος, ἢν μεταστῇ τὸ ἄρθρον, ῥηΐ-*

and hang a suitable weight from the injured leg, as has already been described.

79. First, it must be understood that the connections between all the bones generally consist of a head and a socket. In some, the socket is large, cup-shaped, and oblong; but in others, the sockets are shallowly concave. It is imperative to reduce all dislocated joints, preferably at once and while the parts are still warm, failing that, as soon as possible, since reduction, before swelling sets in, is accomplished much more easily and quickly by the operator, and is much less painful for the patient. When you are about to reduce any joint, you must always first make it supple and move it around; for in this way it will be more easily reduced. In all cases of reduction, the patient must be put on a restricted diet, especially when the joints are very large and difficult to reduce, but least when they are very small and easy.

80. If any joint of the fingers, whether the first, second, or third, is dislocated, the mode of reduction is identical, although the largest joints are always the hardest to reduce. Dislocation takes place in four ways, up or down or to either side: chiefly it is upward, most rarely to the sides, and then as the result of some violent movement. On each side of the articular socket there is a sort of rim: thus, if a displacement is upward or downward, it is easier to reduce when it happens, because the part is flatter (sc. in these directions) than at the sides, making the obstacle to get

---

257 *κρεμάσαι* B: *ἐκκρεμάσαι* MV
258 *ἤδη* B: om. MV
259 *ἐπίμακρος* MV: *ὑπόμακρος* B

διόν ἐστιν ἐμβάλλειν. τρόπος δὲ τῆς ἐμβολῆς ὅδε·
περιελίξαι τὸν δάκτυλον ἄκρον ἢ ἐπιδέσματί τινι ἢ
ἄλλῳ τρόπῳ τοιούτῳ τινί, ὅκως, ὁκόταν κατατείνῃς
ἄκρου λαβόμενος, μὴ ἀπολισθάνῃ· ὅταν δὲ περιελί-
ξῃς, τὸν μέν τινα διαλαβέσθαι ἄνωθεν τοῦ καρποῦ
τῆς χειρός, τὸν δὲ τοῦ κατειλημμένου· ἔπειτα κατατεί-
νειν πρὸς ἑωυτὸν ἀμφοτέρους εὖ μάλα, καὶ ἅμα ἀπῶ-
σαι τὸ ἐξεστηκὸς ἄρθρον[260] ἐς τὴν χώρην. ἢν δὲ ἐς τὰ
πλάγια, τῆς μὲν κατατάσιος ωὑτὸς τρόπος. ὅταν δὲ
δὴ δοκέῃ σοι ὑπερβεβηκέναι τὴν ἄμβην,[261] ἅμα χρὴ
κατατείναντας ἀπῶσαι ἐς τὴν χώρην εὐθύς, ἕτερον δέ
τινα ἐκ τοῦ ἑτέρου μέρεος τοῦ δακτύλου φυλάσσειν
καὶ ἀνωθέειν, ὅκως μὴ πάλιν ἐκεῖθεν ἀπολίσθῃ. ἐμ-
320 βάλλουσι δὲ ἐπιεικέως | καὶ αἱ σαῦραι αἱ ἐκ τῶν φοι-
νίκων πλεκόμεναι, ἢν κατατείνῃς ἔνθεν καὶ ἔνθεν τὸν
δάκτυλον, λαβόμενος τῇ μὲν ἑτέρῃ τῆς σαύρης,[262] τῇ
δὲ ἑτέρῃ τοῦ καρποῦ τῆς χειρός. ὁκόταν δὲ ἐμβάλλῃς,
ἐπιδεῖν δεῖ ὀθονίοισιν ὡς τάχιστα, λεπτοτάτοισι κε-
κηρωμένοισι κηρωτῇ μήτε λίην μαλακῇ μήτε λίην
σκληρῇ, ἀλλὰ μετρίως ἐχούσῃ. ἡ μὲν γὰρ σκληρὴ
ἀφέστηκεν ἀπὸ τοῦ δακτύλου· ἢν δὲ ὑγρὴ ἔῃ, θερμαι-
νομένου τοῦ δακτύλου διατήκεται καὶ ἀπόλλυται.
λύειν δὲ ἄρθρον δακτύλου τριταῖον ἢ τεταρταῖον· τὸ
δὲ ὅλον, ἢν μὲν φλεγμήνῃ, πυκνότερον λύειν· εἰ δὲ μή,
ἀραιότερον. κατὰ πάντων δὲ τῶν ἄρθρων ταῦτα λέγω.
καθίσταται δὲ τοῦ δακτύλου τὸ ἄρθρον τεσσαρεσκαι-
δεκαταῖον. ὁ αὐτὸς δέ ἐστι τρόπος χειρὸς δακτύλων
τε καὶ ποδός θεραπείης.

over small. The mode of reduction is as follows. Wrap a bandage or something of the kind around the end of the finger, in such a way that it will not slip off when you grasp the end and make the extension. When the bandage has been applied, one person should take hold of the wrist from above, and another person of the wrapped finger. Next, each of these is to pull forcefully in his own direction, and at the same time push the projecting joint back into place. In the case of lateral dislocation, the mode of extension is the same. When you think the bone has passed over the rim of the joint, push it at once into place, while keeping up the extension; an assistant should keep guard over the other side of the finger and make counterpressure, to prevent another dislocation to that side. The "lizards" woven from palm tissue are also effective means of reduction, if you extend the finger both ways, grasping the "lizard" with one hand and the wrist with the other hand. After reduction you must at once apply very light bandages soaked in cerate, that are neither too soft nor too hard, but of medium consistency; for a hard bandage gets detached from the finger; but if it is moist, as the finger becomes warm, the bandage melts and disintegrates. Change the dressing of a finger joint on the third or fourth day; in general, if there is inflammation, change it oftener; if not, more rarely: I apply this rule to all joints. A finger joint is healed in fourteen days. The mode of treatment is the same for both fingers and toes.

---

260 *ἄρθρον* MV: om. B
261 *ἄμβην* Kühlewein: *γραμμήν* BMV
262 *σαύρης* B: *αὔρης* MV

81. Παρὰ πάσας δὲ τὰς τῶν ἄρθρων ἐμβολὰς δεῖ ἰσχναίνειν καὶ λιμαγχονέειν ἄχρι ἑβδομαίου· καὶ εἰ μὲν φλεγμαίνοι, πυκνότερον λύειν, εἰ δὲ μή, ἀραιότερον· ἡσυχίην δὲ δεῖ ἔχειν ἀεὶ τὸ πόνεον ἄρθρον, καὶ ὡς κάλλιστα ἐσχηματισμένον κέεσθαι.

82. Γόνυ δὲ εὐηθέστερον ἀγκῶνος διὰ τὴν εὐσταλίην καὶ τὴν εὐφυΐην, διὸ καὶ ἐκπίπτει καὶ ἐμπίπτει[263] ῥᾷον· ἐκπίπτει δὲ πλειστάκις εἴσω, ἀτὰρ καὶ ἔξω καὶ
322 ὄπισθεν. ἐμβολαὶ δὲ καὶ ἐκ | τοῦ συγκεκάμφθαι ἢ ἐκλακτίσαι ὀξέως, ἢ ξυνελίξας ταινίης ὄγκον, ἐν τῇ ἰγνύῃ θείς, ἀμφὶ τοῦ τοῦτον ἐξαίφνης ἐς ὄκλασιν ἀφιέναι τὸ σῶμα. δύναται δὲ καὶ κατατεινόμενον μετρίως, ὥσπερ ἀγκών, ἐμπίπτειν[264] τὰ ὄπισθεν· τὰ δὲ ἔνθα ἢ ἔνθα, ἐκ τοῦ συγκεκάμφθαι ἢ ἐκλακτίσαι, ἀτὰρ καὶ ἐκ κατατάσιος μετρίης. ἡ διόρθωσις ἅπασι κοινή. ἢν δὲ μὴ ἐμπέσῃ τοῖσι μὲν ὄπισθεν, ξυγκάμπτειν οὐ δύνανται, ἀτὰρ οὐδὲ τοῖσιν ἄλλοισι πάνυ. μινύθει δὲ μηροῦ καὶ κνήμης ἔμπροσθεν· ἢν δὲ ἐς τὸ εἴσω, βλαισότεροι, μινύθει δὲ τὰ ἔξω. ἢν δὲ ἐς τὸ ἔξω, γαυσότεροι, χωλοὶ δὲ ἧσσον· κατὰ δὲ τὸ παχύτερον ὀστέον ὀχέει, μινύθει δὲ τὰ ἔσω. ἐκ γενεῆς δὲ ἢ ἐν αὐξήσει κατὰ λόγον τὸν πρόσθεν.

263 καὶ ἔμπιπτει B: om. MV
264 ἐμπίπτειν MV: ἐκπίπτει B

81. In all reductions of joints, the patient should have an attenuating and fasting diet up to the seventh day; if there is inflammation, change the dressing oftener; if not, more rarely. The injured joint should always be kept at rest, and be placed in the best possible position.

82.[7] A dislocation of the knee is easier to repair than one of the elbow, because of its simple, regular form, which allows it to be dislocated and reduced more easily. Most often the knee is dislocated inward, but sometimes also outward or backward. Methods of reduction: by flexion of the knee on itself, or a sharp kick outward, or by placing a mass of rolled-up bandage in the patient's ham, onto which he suddenly crouches down bringing the full weight of his body to bear. It is also possible by moderate extension to reduce dislocations of the knee toward the back, the same way this works for the elbow. Dislocations of the knee to one side or the other are treated by flexion, by kicking outward—generally the reduction backward—or by moderate extension. Adjustment is the same for all these. If the bone does not fall back into place, patients with dislocation toward the back are unable to flex the knee, but in the other dislocations this is hardly possible either: the front of the thigh and the lower leg suffer atrophy. If dislocation is inward, patients are more knock-kneed, and there is atrophy of the muscles on the outer side; if outward, they are more bowlegged, but less lame, since the weight rests on the larger bone (tibia); the inner side atrophies. Cases that are congenital or occur during adolescence follow the rule given above.

[7] The text of chapters 82–87 appears to be taken from *Instruments of Reduction* 26–31.

83. Τὰ δὲ κατὰ τὰ σφυρὰ κατατάσιος ἰσχυρῆς δέεται, ἢ τῇσι χερσὶν ἢ ἄλλοισι τοῖσι, κατορθώσιος δὲ ἅμα ἀμφότερα ποιεούσης· κοινὸν δὲ τοῦτο ἅπασιν.

84. Τὰ δὲ ἐν ποδὶ ὡς καὶ τὰ ἐν χειρὶ ὑγιές. |

324 85. Τὰ δὲ τῆς κνήμης ξυγκοινωνέοντα καὶ μὴ ἐκπεσόντα ἐκ γενεῆς ἢ καὶ ἐν αὐξήσει ἐξαρθρήσαντα, ταὐτὰ ἃ καὶ ἐν χειρί.

86. Ὁκόσοι δὲ πηδήσαντες ἄνωθεν ἐστηρίξαντο τῇ πτέρνῃ, ὥστε διαστῆναι τὰ ὀστέα καὶ φλέβας ἐκχυμωθῆναι καὶ νεῦρα ἀμφιφλασθῆναι, ὁκόταν γένηται οἷα τὰ δεινά, κίνδυνος μὲν σφακελίσαντα τὸν αἰῶνα πρήγματα παρασχεῖν· ῥοιώδη μὲν γὰρ τὰ ὀστέα, τὰ δὲ νεῦρα ἀλλήλοισι κοινωνέοντα. ἐπεὶ καὶ οἷσιν ἂν μάλιστα κατεηγεῖσιν ἢ ὑπὸ τρώματος ἢ ἐν κνήμῃ ἢ ἐν μηρῷ, ἢ νεύρων ἀπολισθέντων ἃ κοινωνέει τούτων, ἢ ἐκ κατακλίσιος ἀμελέος, ἐμελάνθη πτέρνη, καὶ τούτοισι τὰ παλιγκοτέοντα δ' ἐκ τῶν τοιούτων. ἔστιν ὅτε καὶ πρὸς τῷ σφακελισμῷ γίνονται πυρετοὶ ὀξέες, λυγμώδέες, γνώμης ἁπτόμενοι, ταχυθάνατοι, καὶ ἐμφλέβων αἱμορροιέων πελιώσιες καὶ γαγγραινώσιες. σημεῖα δὲ τῶν παλιγκοτησάντων, ἢν τὰ ἐκχυμώματα καὶ τὰ μελάσματα καὶ τὰ περὶ ταῦτα ὑπόσκληρα καὶ ὑπέρυθρα ‹ᾖ›.[265] ἤ τε ξὺν σκληρύσματι πελιδνωθῇ,
326 κίνδυνος μελανθῆναι· | ἢν δὲ ὑποπέλια ᾖ,[266] ἢ καὶ πέλια μάλα καὶ ἐκκεχυμωμένα, ἢ ὑπόχλωρα καὶ μαλακά, ταῦτα ἐπὶ πᾶσι τοῖσι τοιούτοισιν ἀγαθά. ἴησις,

[265] Add. ᾖ Kühlewein [266] ᾖ V: om. BM

83. Dislocations at the ankle require forceful extension with the hands or other means, making the adjustment simultaneously. This is common to all.

84. Dislocations in the foot heal the same way as those in the hand.

85. The bones connecting (sc. the foot) with the leg, when not dislocated from birth or in adolescence, follow the same course as those of the hand.

86. Patients that have fallen from a height and landed on their heel so as to break the bones apart will have ecchymosis from the vessels and contusion of the ligaments: when such serious injuries are sustained, there is a danger, as necrosis supervenes, of a chronic disability, since the bones are unstable and the ligaments are interconnected. When there is also a fracture or a wound in the calf or the thigh, or the ligaments holding the bones become loose, or from heedless positioning in bed the heel has become necrotic, these things will cause such patients to have complications. Sometimes with the necrosis come very acute fevers accompanied by hiccups, trembling, and disturbances of the mind; these are quickly fatal; also, lividity of the blood vessels, and gangrenes. Signs of these complications are ecchymoses and dark bruises, with a degree of induration and redness in the surrounding parts; for if with the indurations lividity develops, there is a danger of necrosis; but if the part is only slightly livid, or even very livid after there has been ecchymosis, or it is greenish in color and soft, these are favorable signs in all such cases. Treatment: if patients remain free of fever, give them a

ἢν μὲν ἀπύρετος ᾖ, ἐλλέβορον· εἰ δὲ μή, μή· ἀλλὰ ποτὸν ὀξύγλυκυ, εἰ δέοι. ἐπίδεσις δὲ ἄρθρων· ἐπὶ δὲ πάντα, μᾶλλον τοῖσι φλάσμασιν, ὀθονίοισι πλείοσι καὶ μαλθακωτέροισι· πίεξις ἧσσον· προσπεριβάλλειν δὲ τὰ πλεῖστα τῇ πτέρνῃ. τὸ σχῆμα, ὅπερ ἡ ἐπίδεσις, ὡς μὴ ἐς τὴν πτέρνην ἀποπιέζηται· νάρθηξι δὲ μὴ χρῆσθαι.

87. Οἷσι δ᾽ ἂν ἐκβῇ ὁ ποὺς ἢ αὐτὸς ἢ ξὺν τῇ ἐπιφύσει, ἐκπίπτει μὲν μᾶλλον ἐς τὸ εἴσω· ἢν δὲ μὴ ἐμπέσῃ, λεπτύνεται ἀνὰ χρόνον ἰσχίον καὶ μηρὸς καὶ κνήμης τὸ ἀντίον τοῦ ὀλισθήματος. ἐμβολὴ δὲ ἄλλη, ὥσπερ καρποῦ, κατάτασις δὲ ἰσχυρή. ἴησις δέ, νόμος ἄρθρων. παλιγκοτέει, ἧσσον δὲ καρποῦ, ἢν ἡσυχάσωσι. δίαιτα μείων, ἐλινύουσι. τὰ δὲ ἐκ γενεῆς ἢ ἐν αὐξήσει, κατὰ λόγον τῶν πρότερον.

dose of hellebore, but not otherwise, in which case you can give a drink of oxymel, if it is necessary. Bandaging should be as is used for joints, and suited in particular to contusions; and with plenty of soft bandages, the pressure quite restrained, more water, and additional bandages, in particular around the heel; the position chosen should be as it was for bandaging, to avoid pressure on the heel: do not employ splints.

87. When the foot is dislocated, either by itself or together with its (tibiofibular) epiphysis, it usually moves toward the inside. If it is not reduced, over time the hip will atrophy, and also the thigh and calf on the side opposite to the dislocation. Reduction is as with the wrist, but requires more forceful extension; treatment follows the rule for joints. Relapse does occur, but less often than with the wrist, if the patient remains at rest. Diet should be reduced, since the patient is at rest. In congenital and adolescent cases follow the method given above.

INSTRUMENTS OF REDUCTION

INTRODUCTION

The title Μοχλικόν (Mochlikon) = *Instruments of Reduction* is included among the surgical works in Erotian's Hippocratic census, and a dozen terms from the text are glossed in his *Collection*;[1] among these explanations, many references are made to Bacchius' *Words* (3 BC), including specific attributions to all three of Bacchius' *syntaxes*.[2] In his Hippocratic *Glossary*, Galen draws on the text of *Instruments of Reduction* for three terms that occur only in this treatise and includes five more that also appear elsewhere in the Collection.[3] In his other works, in particular his Hippocratic commentaries, references to the treatise and/or its title also occasionally occur.[4] Otherwise, the treatise remains virtually unknown until it appears in the manuscript M in the tenth century (today its sole extant independent Greek witness), leaving only the slightest trace of its existence in medieval Islamic or Latin medical literature.[5]

1 See Nachmanson, pp. 347–58; Herophilus, pp. 486–88.

2 Cf. Erotian: A73 (p. 21.10); fr. 40 (p. 111.4); fr. 45 (pp. 113.1f.).

3 See Perilli, p. 412, for references.

4 Cf. *Testimonien*, vol. II,1, pp. 327f.

5 Cf. Sezgin, p. 28.

Instruments of Reduction is an account of the bones, their injuries, and applicable treatments arranged in forty-three chapters according to the following general sections and chapters.

A. Osteology

1	Hands and feet
	Lower limb including the hip bone
	Vertebral column together with the ribs and sternum
	Upper limb including the shoulder blade and collarbone

B. Disorders of the head

2	Fracture of the nose
3	Injury of the external ear
4	Dislocation of the mandible
39	Cases of deliquescence

C. Disorders of the upper limb

5	Dislocations of the shoulder
6	Avulsion of the acromion
7–15	Subluxation and dislocations of the elbow
16,18	Dislocation of the wrist
17	Dislocation of the hand
19	Luxation of the finger joints

D. Disorders of the lower limb

20–25	Dislocations of the thigh
26	Dislocations of the knee
27	Dislocations of the ankle
28–33	Dislocations of the foot; clubfoot

E. Abnormal spinal curvatures

36	Spinal curvatures and fractured ribs
37	Curvatures resulting from falling

F. General surgical phenomena

34 Amputations

35 Gangrene

38 Instruments of reduction

40–41 Subluxations and dislocations: signs and treatment

42 Open fractures

43 Severity of dislocations

As has been explained above,[6] most of *Instruments of Reduction* represents an abbreviated form of texts found in *Joints* and *Fractures*. Only the accounts of the bones, of fractures of the nose, and of injury of the external ear, contained in the first three chapters, are fuller than their counterparts in other treatises.

Critical editions of the text with varying amount of commentary are those of Littré, Petrequin, Kühlewein, and Withington, and English translations have been provided by Adams and Withington. In preparing the present volume, I have collated the text from the manuscript M and availed myself of the work of my predecessors, in particular Adams and Withington.

[6] P. x.

ΜΟΧΛΙΚΟΝ

IV 340 Littré 1. Ὀστέων φύσις· δακτύλων μὲν ἁπλᾶ καὶ ὀστέα καὶ ἄρθρα, χειρὸς δὲ καὶ ποδὸς πολλά, ἄλλα ἀλλοίως ξυνηρθρωμένα· μέγιστα δὲ τὰ ἀνωτάτω. πτέρνης δὲ ἕν, οἷον ἔξω φαίνεται, πρὸς δὲ αὐτὴν οἱ ὀπίσθιοι τένοντες τείνουσι. κνήμης δὲ δύο, ἄνωθεν καὶ κάτωθεν ξυνεχόμενα, κατὰ μέσον δὲ διέχοντα σμικρόν· τὸ ἔξωθεν, κατὰ τὸν σμικρὸν δάκτυλον λεπτότερον βραχεῖ, πλεῖστον δὲ ταύτῃ διεχούσῃ καὶ σμικροτέρῃ ῥοπῇ κατὰ γόνυ, καὶ ὁ τένων ἐξ αὐτοῦ πέφυκεν, ὁ παρὰ τὴν ἰγνύην ἔξω. ἔχουσι δὲ κάτωθεν κοινὴν ἐπίφυσιν πρὸς ἣν ὁ ποὺς κινέεται· ἄλλην δὲ ἄνωθεν ἔχουσιν ἐπίφυσιν, ἐν ᾗ τὸ τοῦ μηροῦ ἄρθρον κινέεται, ἁπλόον καὶ εὐσταλὲς ὡς[1] ἐπὶ μήκει· εἶδος κονδυλῶδες, ἔχον ἐπιμυλίδα· αὐτὸς δὲ ἔγκυρτος ἔξω καὶ ἔμπροσθεν· ἡ δὲ κεφαλὴ ἐπίφυσίς ἐστι στρογγύλη, ἐξ ἧς τὸ νεῦρον τὸ ἐν τῇ κοτύλῃ τοῦ ἰσχίου πέφυκεν· ὑποπλάγιον δὲ καὶ

[1] εὐ. ὡς Littré: ἐυσταλεέως M

INSTRUMENTS OF REDUCTION

1. Arrangement of the bones. In the digits the bones and joints are simple, while in the hand and foot many bones are fitted together in various different ways; the largest bones are the ones highest up. In the heel there is only one bone, which appears as a projection; to it the posterior tendons (Achilles tendons) are attached. In the lower leg there are two bones (fibula and tibia); these are attached to one another at the top and bottom, but slightly separated in the middle. The one facing outward (fibula), being on the same side as the little toe, is somewhat narrower than the other one, especially in the area where they are separated; it is more slanted there, but at the knee has less inclination; a tendon (of the biceps femoris) passes upward from the fibula and runs along the outside of the ham. These two bones share a common epiphysis below, in which the foot moves. They also have an epiphysis at the top, in which the end of the femur is articulated and moves; the knee joint is simple and compact, considering the length of bone; it has the form of a knuckle and on it is a kneecap. The thigh bone (femur) itself curves outward toward the side and front; its head has the form of a spherical epiphysis, from which the cord (ligamentum teres) passes to the socket of the hip bone (acetabulum) where

τοῦτο προσήρτηται, ἧσσον δὲ βραχίονος. τὸ δὲ ἰσχίον προσίσχεται πρὸς τῷ μεγάλῳ σφονδύλῳ τῷ παρὰ τὸ ἱερὸν ὀστέον χονδρονευρώδει δεσμῷ.

Ῥάχις δὲ ἀπὸ μὲν τοῦ ἱεροῦ ὀστέου μέχρι τοῦ μεγάλου σφονδύλου κυφή· κύστις τε καὶ γονὴ καὶ
342 ἀρχοῦ τὸ ἐγκεκλιμένον ἐν τούτῳ. | ἀπὸ δὲ τούτου ἄχρι φρενῶν ἦλθεν ἡ ἰθύλορδος, καὶ αἱ ψόαι κατὰ τοῦτο· ἐντεῦθεν δὲ ἄχρι τοῦ μεγάλου σφονδύλου τοῦ ὑπὲρ τῶν ἐπωμίδων ἰθυκυφής· ἔτι δὲ μᾶλλον δοκέει ἤ ἐστιν· αἱ γὰρ ὄπισθεν τῶν σπονδύλων ἀποφύσιες ταύτῃ ὑψηλόταται· τὸ δὲ τοῦ αὐχένος ἄρθρον λορδόν. σφόνδυλοι δὲ ἔσωθεν ἄρτιοι πρὸς ἀλλήλους, ἀπὸ δὲ τῶν ἔξωθεν χόνδρων νεύρῳ[2] ξυνεχόμενοι· ἡ δὲ ξυνάρθρωσις αὐτῶν ἐν τῷ ὄπισθεν τοῦ νωτιαίου· ὄπισθεν δὲ ἔχουσιν ἔκφυσιν ὀξείην ἔχουσαν ἐπίφυσιν χονδρώδεα· ἔνθεν νεύρων ἀπόφυσις καταφερής, ὥσπερ καὶ οἱ μῦες παραπεφύκασιν ἀπὸ αὐχένος ἐς ὀσφῦν, πληρεῦντες δὲ πλευρέων καὶ ἀκάνθης τὸ μέσον. πλευραὶ δὲ κατὰ τὰς διαφύσιας τῶν σφονδύλων νευρίῳ προσπεφύκασιν ἀπ᾽ αὐχένος ἐς ὀσφῦν ἔσωθεν, ἐπίπροσθεν δὲ κατὰ τὸ στῆθος χαῦνον καὶ μαλθακὸν τὸ ἄκρον ἔχουσαι· εἶδος ῥαιβοειδέστατον[3] τῶν ζώων· στενότατος γὰρ ταύτῃ ὁ ἄνθρωπος ἐπ᾽ ὄγκον·[4] ᾗ δὲ μὴ πλευραί εἰσιν, ἔκφυσις πλαγίη, βραχείη καὶ πλατείη· ἐφ᾽ ἑκάστῳ σφονδύλῳ νευρίῳ προσπεφύκασιν.

[2] *νεύρῳ* Littré: *νεύρων* M
[3] *ῥαιβο-* Erotian: *ῥομβο-* M
[4] *ἐπ᾽ ὄγκον* H Littré: *ἐπόγκων* M

it is inserted. This attachment is at an angle, but less so than the one at the shoulder (long head of the biceps brachii muscle). At the spine the hip bone is attached to the great vertebra (fifth lumbar) next the sacrum by a cartilaginous ligament.

The spine from the (coccyx of the) sacrum to the great vertebra is convex backward, and the bladder, reproductive parts, and the inclined part of the rectum are inside it. Ascending from there the spine is convex forward as far as the diaphragm, running along with the psoas muscles; then from the diaphragm up to the great vertebra above the shoulders (seventh cervical), it is convex backward, seeming even more convex than it is because the posterior processes of the vertebrae there are most prominent; in the neck the spine is convex forward. On the inside, the vertebrae fit into one another, and they are attached to one another by a ligament from the outside cartilages. The joints connecting the spinal vertebrae are posterior to the spinal marrow (i.e., spinal cord), and on their posterior surface they have a pointed process with a cartilaginous epiphysis. From there ligaments run down, passing in the same way the muscles do from the side of the neck to the loin, occupying the space between the ribs and the spinal ridge. The ribs are attached in the intervals between the vertebrae from the neck down to the loins, by a small ligament on the inside, while at their anterior end they have a spongy, soft termination against the sternum. In their shape, the ribs of humans are the most bent of all animals', since in that area the human is the species narrowest (in the anterior-posterior direction) in relation to its mass. Where there are no longer any ribs, there is a short, broad process attached to each vertebra by a small ligament.

Στῆθος δὲ ξυνεχὲς αὐτὸ ἑωυτῷ, διαφύσιας ἔχον πλαγίας, ᾗ πλευραὶ προσήρτηνται, χαῦνον δὲ καὶ
344 χονδρῶδες. κληῖδες | δὲ περιφερέες ἐς τοὔμπροσθεν, ἔχουσαι πρὸς μὲν τὸ στῆθος βραχείας κινήσιας, πρὸς δὲ τὸ ἀκρώμιον συχνοτέρας. ἀκρώμιον δὲ ἐξ ὠμοπλατέων πέφυκεν, ἀνομοίως δὲ τοῖσι πλείστοισι. ὠμοπλάτη δὲ χονδρώδης τὸ πρὸς ῥάχιν, τὸ δ' ἄλλο χαύνη, τὸ ἀνώμαλον ἔξω ἔχουσα, αὐχένα δὲ καὶ κοτύλην ἔχουσα χονδρώδεα, ἐξ ἧς αἱ πλευραὶ κίνησιν ἔχουσι, εὐαπόλυτος ἐοῦσα ὀστέων, πλὴν βραχίονος. τούτου δὲ ἐκ τῆς κοτύλης νευρίῳ ἡ κεφαλὴ ἐξήρτηται, χόνδρου χαύνου περιφερῆ ἐπίφυσιν ἔχουσα· αὐτὸς δ' ἔγκυρτος ἔξω καὶ ἔμπροσθεν πλάγιος, οὐκ ὀρθὸς πρὸς κοτύλην· τὸ δὲ πρὸς ἀγκῶνα αὐτοῦ πλατὺ καὶ κονδυλῶδες καὶ βαλβιδῶδες· καὶ στερεόν, ἔγκοιλον ὄπισθεν, ἐν ᾧ ἡ κορώνη ἡ ἐκ τοῦ πήχεος, ὅταν ἐκταθῇ ἡ χείρ, ἔνεστιν· ἐς τοῦτο καὶ τὸ ναρκῶδες νεῦρον, ὃ ἐκ τῆς διαφύσιος τῶν τοῦ πήχεος ὀστέων, ἐκ μέσων ἐκπέφυκε καὶ περαίνεται.

2. Ῥὶς δὲ κατεαγεῖσα ἀναπλάσσεσθαι οἵη τε αὐθω-
346 ρόν· τοῦτο κοινόν· κὴν | μὲν οὖν ὁ χόνδρος, ἐντίθεσθαι[5] ἄχνην ὀθονίου, ἐναποδέοντα λοπῷ Καρχηδονίῳ, ἢ ἄλλῳ ὃ μὴ ἐρεθιεῖ· τῷ λοπῷ δὲ τὰς παραλλάξιας παρακολλᾶν καὶ ἀναλαμβάνειν· ταῦτα δὲ ἐπίδεσις κακοποιέει. ἴησις ἄλλη· ἅμα δὲ τῷ ξυμβαλεῖν, ξὺν μάννῃ[6] ἢ θείῳ ξὺν κηρωτῇ· αὐτίκα ἀναπλάσσεις,

[5] -τίθεσθαι Foes in note, Petrequin: -τίθεται M

[6] ἅμα δὲ . . . μάννῃ conj. Petrequin: τῷ ξυμμανῇ M

The sternum is a continuous bone with angular interstices into which the ribs are inserted; it is spongy and cartilaginous. The collarbones are curved in front; they have a slight mobility at the sternum, but more movement at the acromion. The acromion is an outgrowth of the shoulder blade, arranged differently than in most (sc. animals). The shoulder blade is cartilaginous next to the spine, but otherwise spongy; it is irregular along its outer side, and its neck and articular cavity (glenoid) are cartilaginous. It allows the ribs their mobility, being independent of the skeleton except where it is attached to the arm. The head of this bone (humerus) is attached from its articular cavity by a small ligament (tendon of the biceps) and has a roundish epiphysis of spongy cartilage; the humerus is convex toward the outside, and at the front it is placed on an angle rather than meeting its socket straight on. The end toward the elbow is broad, shaped like a knuckle, and grooved; it is solid, and on its back side it has a hollow into which the coronoid process of the ulna (olecranon) enters when the arm is extended. Here there is a cord that can cause a loss of sensation, coming out from between the bones of the forearm and ending there.

2. The nose on being fractured can be adjusted at once. This is done in all cases, and if it is the cartilage that is affected, insert pieces of lint and bind them with thin Carthaginian leather or some other nonirritating material; use the leather to elevate and secure the distorted parts: a bandage would be harmful in such a case. Another treatment: simultaneously approximate the parts and apply frankincense powder, or sulfur in a cerate. You will adjust this at once and then hold the parts in place, by putting

ἔπειτα ἀνακωχήσεις, τοῖσι δακτύλοισι ἐσματευόμενον καὶ παραστρέφοντα· καὶ τὸ Καρχηδόνιον· πωροῖτο δ' ἂν καὶ ἢν ἕλκος ἐνῇ· καὶ ἢν ὀστέα ἀπιέναι μέλλῃ—οὐ γὰρ παλιγκοτώτατα—οὕτω ποιητέα.

3. Οὖς κατεαγὲν μὴ ἐπιδεῖν, μηδὲ καταπλάσσειν· ἢν δέ τι δέῃ, ὡς κουφότατον, ἡ κηρωτή· καὶ θείῳ κατακολλᾷν. ὧν δὲ ἔμπυα διὰ παχέος εὑρίσκεται, πάντα δὲ τὰ ὑπόμυξα καὶ τὰ ὑγρῇ σαρκὶ πλήρεα ἐξαπατᾷ· οὐ μὴ βλάβῃ στομωθὲν τὸ τοιοῦτο· ἔστι γὰρ ἄσαρκα καὶ ὑδατώδεα, μύξης πλέα· ὅπου δὲ καὶ οἷα ἐόντα θανατώδεά ἐστι, παρεῖται. ὤτων καῦσις πέρην, τάχιστα ὑγιάζει· κυλλὸν δὲ καὶ μεῖον γίνεται τὸ οὖς, ἢν πέρην καυθῇ. ἢν δὲ στομωθῇ, κούφῳ ἐναίμῳ δεήσει χρῆσθαι. |

348 4. Γνάθοι δὲ κατασπῶνται μὲν πολλάκις καὶ καθίστανται· ἐκπίπτουσι δὲ ὀλιγάκις, μάλιστα μὲν χασμωμένοισιν· οὐ γὰρ ἐκπίπτει, εἰ μή τις χανὼν μέγα παραγάγοι· ἐκπίπτει δὲ μᾶλλον, ὅτι τὰ νεῦρα ἐν πλαγίῳ καὶ λελυγισμένα συνδιδοῖ. σημεῖα· προέχει ἡ κάτω γνάθος καὶ παρέστραπται τἀναντία τοῦ ἐκπτώματος· συμβάλλειν οὐ δύνανται· ἢν δὲ ἀμφότεραι, προΐσχουσι μᾶλλον, συμβάλλουσιν ἧσσον, ἀστραβέες· δηλοῖ δὲ τὰ ὅρια τῶν ὀδόντων τὰ ἄνω τοῖσι κάτω κατ' ἴξιν. ἢν οὖν ἀμφότεραι ἐκπεσοῦσαι μὴ αὐτίκα ἐμπέσωσι, θνήσκουσι δεκαταῖοι οὗτοι μάλιστα πυρετῷ συνεχεῖ νωθρῇ τε καρώσει· οἱ γὰρ μύες οὗτοι τοιοῦτοι. γαστὴρ ἐπιταράσσεται ὀλίγα ἄκρητα·

your finger into the nostrils, feeling the dislocation, and reducing it; use Carthaginian leather here too. The break will consolidate even if a lesion is present; and even if the bones are destined to break away, this is the treatment to use, since there is no chance of a complication.

3. A fractured ear you should neither bandage nor plaster. If it does require some kind of support, this should be very light, e.g., a cerate; also fasten it down with sulfur. In cases where the ears are suppurating, the pus will only be found in their depth, since all tissues full of mucus and moisture are deceptive. To incise such parts will not cause any harm, since they lack flesh and consist of water and mucus (I am omitting where and which cases are likely to cause death). Cautery through the ears brings prompt recovery, but the ear becomes contracted and deformed if it is cauterized by this method. If the ear is incised, it will require the use of a weak styptic.

4. The lower jaw is often draw downward and then falls back up into its normal place, but only rarely is it truly dislocated, and then usually from yawning; for it does not dislocate unless someone in the midst of a wide yawn draws it to one side. It dislocates mainly because the ligaments, being oblique and twisted, give way. Signs: the lower jaw projects and is drawn to the side opposite to the dislocation; such patients cannot close their mouth. If both sides are dislocated, the projection is greater, as is the disability to close the mouth, and there is no deviation to one side, as is shown by the upper and lower teeth being aligned. Now if there is bilateral dislocation and it is not immediately reduced, such patients die around the tenth day, generally with continuous fever, stupor and somnolence, so powerful are these muscles. The belly is disturbed by

καὶ ἣν ἐμέωσιν, τοιαῦτα ἐμέουσιν· ἡ δ᾽ ἑτέρη ἀσινεστέρη. ἐμβολὴ δὲ ἡ αὐτὴ ἀμφοτέρων· κατακειμένου ἢ καθημένου, τῆς κεφαλῆς ἐχόμενον, περιλαβόντα τὰς γνάθους ἀμφοτέρας ἀμφοτέρῃσι χερσὶν ἔσωθεν καὶ ἔξωθεν, τρία ἅμα[7] ποιῆσαι· ὦσαι ἐς ὀρθὸν καὶ ἐς τοὐπίσω, καὶ συσχεῖν τὸ στόμα. ἴησις· μαλάγμασι καὶ σχήμασι καὶ ἀναλήψει γενείου· ποιοῦσι ταῦτα τῇ ἐμβολῇ.

5. Ὦμος δὲ ἐκπίπτει κάτω· ἄλλῃ δὲ οὔπω ἤκουσα. δοκέει μὲν γὰρ ἐς τοὔμπροσθεν ἐκπίπτειν, ὧν αἱ σάρκες αἱ περὶ τὸ ἄρθρον μεμινυθήκασι διὰ τὴν φθίσιν,[8]
350 οἷον καὶ βουσὶ χειμῶνος φαίνεται διὰ | λεπτότητα. καὶ ἐκπίπτει μᾶλλον τοῖσι δὲ λεπτοῖσιν ἢ ἰσχνοῖσιν ἢ ξηροῖσι καὶ τοῖς ὑγράσματα περὶ τὰ ἄρθρα ἔχουσιν ἄνευ φλεγμονῆς· αὕτη γὰρ συνδεῖ. οἱ δὲ καὶ βουσὶν ἐμβάλλοντες καὶ ἀποπερονῶντες ἐξαμαρτάνουσι, καὶ ὅτι διὰ τὴν χρῆσιν, ὡς χρῆται βοῦς σκέλει, λήθει, καὶ ὅτι κοινὸν καὶ ἀνθρώπῳ οὕτως ἔχοντι τὸ σχῆμα τοῦτο· τό τε Ὁμήρειον· καὶ διότι λεπτότατοι βόες τηνικαῦτα. ὅσα τε τὸν πῆχυν πλάγιον ἀπὸ πλευρέων ἄραντες δρῶσιν, οὐ πάνυ δύνανται δρᾶν, οἷσιν ἂν μὴ ἐμπέσῃ. οἷσι μὲν οὖν ἐκπίπτει μάλιστα, καὶ ὡς ἔχουσιν, εἴρηται. οἷσι δὲ ἐκ γενεῆς, τὰ ἐγγύτατα μᾶλλον βραχύνεται ὀστέα, οἷον ἐν τούτῳ οἱ γαλιάγκωνες·

[7] τρία ἅμα Foes in note: τρίαμμα M
[8] φθίσιν Littré: φύσιν M

scanty, unmixed stools, and if there is vomiting, it is the same kind of material. Dislocation on one side is less harmful. Reduction is carried out the same way in both cases. With the person lying down or sitting, his head is held in place, and you take hold of the lower jaw on both sides with your two hands, placing (sc. your fingers) both inside and outside his mouth, and make three movements at the same time: draw the lower jaw straight, press it backward, and close the mouth. Treatments are emollients, positioning, and elevation of the chin: all these contribute to the reduction.

5. The shoulder dislocates downward, and I have never heard of any other way. It appears, however, to be dislocated anteriorly in patients whose muscles around the joint are diminished due to a wasting disease, the way that dislocations are also seen in cattle in winter due to emaciation. Dislocation occurs more often in subjects who are of slight build, lean and dry, or also in those who have moisture around their joints without any inflammation, since inflammation holds joints together. People who try to reduce dislocations in cattle and fix them with a pin are mistaken in not noticing that the ox has this appearance due to the way it employs its leg, and that human beings too suffer the identical condition if they have the same deformity—they are also disregarding the Homeric reference (*Iliad* 14, 180) and the reason why cattle are very lean in that season. Movements involving lateral elevation of the elbow away from the ribs are quite impossible for patients whose joint has not been reduced. This, then, is an account of the persons most liable to the dislocation, and how it occurs. In those who have the condition from birth, the proximal bones are most shortened, as, e.g., in the case of the "weasel-armed," the forearm less so, the

πῆχυς δὲ ἧσσον, χεὶρ δὲ ἔτι ἧσσον, τὰ δ᾽ ἄνωθεν οὐδέν. καὶ ἀσαρκότατα ἐγγύς· μινύθει δὲ μάλιστα τὰ ἐναντία τῶν ὀλισθημάτων, καὶ τὰ ἐν αὐξήσει, ἧσσον δέ τινι τῶν ἐκ γενεῆς.

Καὶ τὰ παραπυήματα,[9] τὰ κατ᾽ ἄρθρον βαθέα, νεογενέσι μάλιστα παρ᾽ ὦμον γίνεται, καὶ τούτοισιν ὥσπερ τὰ ἐξαρθρήσαντα ποιέει. ἢν δὲ ηὐξημένοισι, τὰ μὲν ὀστέα οὐ μειοῦται, οὐδὲ γὰρ ἔχει ᾗ[10] ἄλλα οὐ ξυναύξεται ὁμοίως, αἱ δὲ μινυθήσιες τῶν σαρκῶν·
352 τοῦτο γὰρ καθ᾽ ἡμέρην καὶ αὔξεται καὶ | μειοῦται, καὶ καθ᾽ ἡλικίας. † καὶ ἃ δύναται σχήματα, καὶ ἃ σημαίνει[11] τὸ παρὰ τὸ ἀκρώμιον κατεσπασμένον καὶ κοῖλον, διότι ὅταν τὸ ἀκρώμιον ἀποσπασθῇ καὶ κοῖλον ᾖ, οἴονται τὸν βραχίονα ἐκπεπτωκέναι· † κεφαλὴ δὲ τοῦ βραχίονος ἐν τῇ μασχάλῃ φαίνεται· αἴρειν οὐ δύνανται, οὐδὲ παράγειν ἔνθα ἢ ἔνθα ὁμοίως· ὁ ἕτερος ὦμος μηνύει.

Ἐμβολαί· αὐτὸς μὲν τὴν πυγμὴν ὑπὸ μασχάλην ὑποθεὶς τὴν κεφαλὴν ἀνωθέειν, τὴν δὲ χεῖρα ἐπιπαράγειν ἐπὶ τὸ στῆθος. ἄλλη· ἐς τοὐπίσω περιαναγκάσαι, ὡς ἀμφισφαλῇ. ἄλλη· κεφαλὴ μὲν πρὸς ἀκρώμιον, χερσὶ δὲ ὑπὸ μασχάλην, κεφαλὴν ἀπάγειν βραχίονος, γούνασι δὲ ἀγκῶνα ἀπωθεῖν, ἢ ἀντὶ τῶν γουνάτων ἕτερον τὸν ἀγκῶνα παράγειν ὡς τὸ πρότερον· ἢ κατ᾽

[9] παραπυήματα recc.: παρὰ ποιήματα M

[10] ᾗ Littré: οἱ M [11] ἃ σημαίνει Kühlewein: αὖ σημεῖον M: ἃ σημαῖει M in marg.

hand least, and the bones above the joint not at all. The greatest loss of flesh is near the joint, and the greatest atrophy is on the side opposite to the dislocation, and when the disorder appears in adolescence, although less than in congenital cases.

In neonates concomitant suppurations occur deep inside next to the joint, most often beside the shoulder, and such cases have the same effect as dislocations. In adults, the bones are not shortened, since there is no occasion for one bone to have less growth than another, although there is a loss of muscle substance, since these increase and decrease daily, and according to age. † And what positions are able . . . , and what the point of the shoulder being avulsed and hollow indicates, for which reason when the acromion is torn away and leaves a hollowness, people think that the arm has been dislocated. †[1] The head of the arm becomes evident in the axilla, and patients are unable to raise the arm or move it equally to one side or the other: the opposite shoulder makes this clear.

Reductions: have the patient put his fist into his axilla and press the head upward while bringing the arm toward his chest. Another method: force the arm backward so as to rotate it. Another: with your head against the acromion and hands under the axilla, lift the head of the arm and push the elbow back with your knees, or, instead of using the knees, have another person bring the elbow to the side as just described. Or suspend the patient on your shoulder,

[1] The text of this passage is difficult to understand and may contain a lacuna.

ὤμου ἵζεσθαι, ὑποθεὶς τῇ μασχάλῃ τὸν ὦμον· ἢ τῇ πτέρνῃ ἐνθέντα ἐκπληρώματα τῇ μασχάλῃ, δεξιῇ δεξιόν· ἢ περὶ ὕπερον· ἢ περὶ κλιμακτῆρα· ἢ περίοδος ξὺν τῷ ξύλῳ τῷ ὑπὸ χεῖρα τεινομένῳ. ἴησις· τὸ σχῆμα, πρὸς πλευρῇσι βραχίων, χεὶρ ἄκρη ἄνω, ὦμος ἄνω· οὕτως ἐπίδεσις, ἀνάληψις. ἢν δὲ μὴ ἐμπέσῃ, ἀκρώμιον προσλεπτύνεται.

6. Ἀκρώμιον ἀποσπασθέν, τὸ μὲν εἶδος φαίνεται οἷόν περ ὤμου ἐκπεσόντος, στερίσκεται δὲ οὐδενός, ἐς δὲ τὸ αὐτὸ οὐ καθίσταται. σχῆμα τὸ αὐτὸ ὃ καὶ ἐκπεσόντι, ἐν ἐπιδέσει καὶ ἀναλήψει· ἐπίδεσις καὶ ὡς νόμος.

7. Ἀγκῶνος ἄρθρον παραλλάξαν μὲν ἢ πρὸς πλευρὴν ἢ ἔξω, μένοντος τοῦ ὀξέος αὐτοῦ[12] ἐν τῷ κοίλῳ τοῦ βραχίονος, ἐς ἰθὺ κατατείνοντα, τὰ ἐξέχοντα ἀνωθέειν ὀπίσω καὶ ἐς τὸ πλάγιον.

8. Τὰ δὲ τελέως ἐκβάντα ἢ ἔνθα ἢ ἔνθα· κατάτασις
354 μὲν ἐν ᾗ | ὁ βραχίων ἐπιδεῖται· οὕτω γὰρ τὸ καμπύλον τοῦ ἀγκῶνος οὐ κωλύσει. ἐκπίπτει δὲ μάλιστα ἐς τὸ πρὸς πλευρέων μέρος. τὰς δὲ κατορθώσιας, ἀπάγοντα ὅτι πλεῖστον, ὡς μὴ ψαύῃ τῆς κορώνης ἡ κεφαλή, μετέωρον περιάγειν καὶ περικάμψαι, καὶ μὴ ἐς ἰθὺ

[12] αὐτοῦ Potter: τοῦ codd.

placing the support under his axilla, or with your heel, interposing packing material in the axilla—use your right heel for the right shoulder—or with a pestle-shaped piece of wood, or a ladder, or make a circular movement with a wooden lever fixed under the arm. Treatment: position the arm against the ribs, raise the extremities of the hand, and elevate the shoulder: bandage and suspend in this attitude. If not reduced, the acromion will atrophy too.

6. Avulsion of the acromion: this presents the same appearance as a dislocation of the shoulder, except that there is no loss of function; on being reduced, however, the bone does not remain in its normal position. Position for bandaging and suspension are the same as for dislocation of the humerus; bandaging is according to the normal rules.

7.[2] When there is a displacement of the elbow joint either toward the side or outward with its point (olecranon) remaining in the cavity (olecranon fossa) of the arm, hold the joint extended in a straight line and force the projecting part obliquely backward.

8. Complete dislocations of the elbow in either direction require extension in the position in which the arm is bandaged: this will prevent the curved part of the elbow (coronoid process of the ulna) from getting in the way; generally, the dislocation is in the direction of the ribs. For adjustment, separate the bones as far as possible so that the head (of the arm) will not come into contact with the coronoid process; with the arm raised, employ the movements of circumduction and flexion without forcing the

[2] Chapters 7–19 of *Instruments of Reduction* are introduced in slightly altered form into the text of *Joints* as chapters 17–29.

βιάζεσθαι, ἅμα δὲ ὠθέειν τἀναντία ἐφ᾽ ἑκάτερα, καὶ παρωθέειν ἐς χώρην. συνωφελοίη δ᾽ ἂν καὶ ἐπίστρεψις[13] ἀγκῶνος ἐν τούτοισιν, ἐν τῷ μὲν ἐς τὸ ὕπτιον, ἐν τῷ δὲ ἐς τὸ πρηνές. ἴησις δέ, σχήματος μὲν ὀλίγον ἀνωτέρω ἄκρην χεῖρα ἀγκῶνος ἔχειν, βραχίονα δὲ κατὰ πλευράς οὕτω δὲ καὶ ἀνάληψις. εὔφορον καὶ φύσις καὶ χρῆσις ἐν τῷ κοινῷ· ἢν ἄρα μὴ κακῶς πωρωθῇ, πωροῦται δὲ ταχέως. ἴησις· ὀθονίοισι κατὰ τὸν νόμον τὸν ἀρθριτικόν, καὶ τὸ ὀξὺ προσεπιδεῖν.

9. Παλιγκοτώτατον δὲ ἀγκὼν πυρετοῖσιν, ὀδύνῃ, ἀσώδει, ἀκρητοχόλῳ· ἀγκῶνος δὲ μάλιστα ὀπίσω διὰ τὸ ναρκῶδες, δεύτερον τὸ ἔμπροσθεν. ἴησις δ᾽ ἡ αὐτή· ἐμβολαὶ δὲ τοῦ μὲν ὀπίσω ἐκτείνοντα κατατεῖναι. σημεῖον δέ· οὐ γὰρ δύνανται ἐκτείνειν· τοῦ δὲ ἔμπροσθεν οὐ δύνανται ξυγκάμπτειν. τούτῳ δὲ ἐνθέντα τι σκληρὸν συνειλεγμένον, περὶ τοῦτο συγκάμψαι ἐξ ἐκτάσιος ἐξαίφνης.

10. Διαστάσιος δὲ ὀστέων σημεῖον κατὰ τὴν φλέβα τὴν κατὰ βραχίονα σχιζομένην διαψαύοντι. |

356 11. Ταῦτα δὲ ταχέως διαπωροῦται· ἐκ γενεῆς δὲ βραχύτερα τὰ κάτω ὀστέα τοῦ σίνεος, πλεῖστον τὰ δ᾽ ἐγγύτατα πήχεος, δεύτερον χειρός, τρίτον δακτύλων. βραχίων δὲ καὶ ὦμος ἐγκρατέστερα διὰ τὴν τροφήν· ἡ δ᾽ ἑτέρη χεὶρ διὰ τὰ ἔργα πλείω ἔτι ἐγκρατεστέρη.

[13] -στρεψις Froben: -στρεψας M

arm back in a straight line, and at the same time press the two bones in opposite directions to bring them around into place. In cases like this it is helpful to turn the elbow alternately between supination and pronation. For treatment, keep the hand positioned somewhat higher than the elbow, and the arm against the ribs, also during suspension. The position is comfortable, natural and practicable altogether; if there is no unfavorable ankylosis, the callus will form quickly. Treat by applying bandages according to the method regularly employed for joints, and include the point of the elbow inside them.

9. The elbow is very liable to exacerbation with fevers, pain, nausea, and (sc. the vomiting of) unmixed bile, especially a dislocation backward, owing to the numbness, but secondly a dislocation forward. Treatment is the same. Modes of reduction for backward dislocation: extension and counterextension; a sign for this dislocation is that patients cannot extend the arm; for forward dislocation, a sign is that they cannot flex it. In this case treatment is to insert some firm rolled-up material into the bend of the elbow, and, starting in the extended position, suddenly to flex the arm over it.

10. Separation of the bones (sc. of the forearm) is recognized by palpation of the blood vessel of the arm where it bifurcates.

11. In these cases (unreduced dislocations of the elbow) the callus is rapidly formed; if the dislocation is congenital, the bones below the injury will be shortened, those of the forearm nearest the injury most; secondly those of the hand, thirdly those of the fingers. The upper arm and the shoulder are stronger because they receive more nutriment. The opposite arm is stronger still be-

μινύθησις δὲ σαρκῶν, εἰ μὲν ἔξω ἐξέπεσεν, εἴσω· εἰ δὲ μή, ἐς τοὐναντίον ᾗ ἐξέπεσεν.

12. Ἀγκὼν δὲ ἢν μὲν ἔσω ἢ ἔξω ἐκβῇ, κατάτασις μὲν ἐν σχήματι ἐγγωνίῳ, τῷ πήχει κοινῷ πρὸς βραχίονα· καὶ μασχάλην ἀναλαβὼν ταινίῃ ἀνακρεμάσαι, ἀγκῶνι δὲ ἄκρῳ ὑποθείς παρὰ δὲ τὸ ἄρθρον βάρος ἐκκρεμάσαι, ἢ χερσὶ καταναγκάσαι. ὑπεραιωρηθέντος δὲ τοῦ ἄρθρου, αἱ παραγωγαὶ τοῖσι θέναρσιν, ὡς τὰ ἐν χερσίν. ἐπίδεσις ἐν τούτῳ τῷ σχήματι, καὶ ἀνάληψις καὶ θέσις.

13. Τὰ δὲ ὄπισθεν, ἐξαίφνης ἐκτείνοντα διορθοῦν τοῖσι θέναρσιν· ἅμα δὲ δεῖ ἐν τῇ διορθώσει, καὶ τοῖσιν ἑτέροισιν. ἢν δὲ πρόσθεν, ἀμφὶ ὀθόνιον συνειλεγμένον, εὔογκον, συγκάμπτοντα ἅμα διορθοῦσθαι.

14. Ἢν δὲ ἑτεροκλινὲς ᾖ, ἐν τῇ διορθώσει ἀμφότερα ἅμα χρὴ ποιέειν· τῆς δὲ μελέτης κοινὸν καὶ τὸ σχῆμα καὶ ἡ ἐπίδεσις· δύναται δὲ καὶ ἐκ τῆς διατάσιος κοινῇ ξυμπίπτειν πάντα. |

358 15. Τῶν δὲ ἐμβολέων αἱ μὲν ἐξ ὑπεραιωρήσιος ἐμβάλλονται, αἱ δὲ ἐκ κατατάσιος, αἱ δὲ ἐκ περισφάλσιος· αὗται δὲ ἐκ τῶν ὑπερβολέων τῶν σχημάτων ἢ τῇ ἢ τῇ σὺν τῷ τάχει.

16. Χειρὸς δὲ ἄρθρον ὀλισθάνει ἢ ἔσω ἢ ἔξω, ἔσω δὲ τὰ πλεῖστα. σημεῖα δ᾽ εὔσημα· ἢν μὲν ἔσω, ξυγκάμπτειν τοὺς δακτύλους οὐ δύνανται· ἢν δὲ ἔξω,

cause of its work. Atrophy of the muscles occurs on the inner side, if the dislocation is outward, and otherwise on the side opposite to the dislocation.

12. If the elbow is dislocated inward or outward, extension should be made in an angled position with the forearm held normally in relationship to the upper arm. And elevating the axilla, suspend it with a band; put a weight under the point of the elbow near the joint and hang this down, or press (sc. the arm) down with your hands. With the articular end raised, make the adjustments with the palms of your hands, the same way as for dislocations of the hands. Bandaging is applied in this position, together with suspension and fixation.

13. Dislocations backward: while holding the joint extended, suddenly put it in with the palms of your hands; the two actions must be executed simultaneously, as in the other cases. If the dislocation is forward, make a combined flexion and adjustment around a large rolled bandage.

14. If there is a deviation to one side, in the reduction both movements should be combined (extension/flexion and adjustment). Positioning and bandaging follow the common rule of management. It is also possible to reduce all these cases by the common method of double extension.

15. Some reductions are performed by lifting over, others by extension, and others by circumduction. All these actions operate by overshooting the natural position in one direction or another with a rapid movement.

16. The joint of the hand is dislocated either inward or outward, but usually inward. The signs are easily recognized: if inward patients cannot flex their fingers, if outward they cannot extend them. Reduction: placing the

ἐκτείνειν. ἐμβολὴ δέ· ὑπὲρ τραπέζης τοὺς δακτύλους ἔχων, τοὺς μὲν τείνειν, τοὺς δὲ ἀντιτείνειν· τὸ δὲ ἐξέχον ἢ θέναρι ἢ πτέρνῃ ἅμα ἀπωθέειν πρόσω καὶ κάτωθεν, κατὰ τὸ ἕτερον ὀστέον ὄγκον μαλθακὸν ὑποθείς, κἢν μὲν ἄνω, καταστρέψας τὴν χεῖρα, ἢν δὲ κάτω, ὑπτίην. ἴησις, ὀθονίοισιν.

17. Ὅλη δὲ ἡ χεὶρ ὀλισθάνει ἢ ἔσω ἢ ἔξω ἔνθα ἢ ἔνθα· μάλιστα δὲ ἔσω· ἔστι δ᾿ ὅτε ἡ ἐπίφυσις ἐκινήθη· ἔστι δ᾿ ὅτε τὸ ἕτερον τῶν ὀστέων διέστη. τούτοισι κατάτασις ἰσχυρὴ ποιητέη, καὶ τὸ μὲν ἐξέχον ἀπωθέειν, τὸ δὲ ἕτερον ἀντωθέειν, δύο εἴδεα ἅμα καὶ ἐς τοὐπίσω καὶ ἐς τὸ πλάγιον, ἢ χερσὶν ἐπὶ τραπέζης ἢ πτέρνῃ. παλίγκοτα δὲ καὶ ἀσχήμονα, τῷ χρόνῳ δὲ κρατύνεται ἐς χρῆσιν. ἴησις, ὀθονίοισι ξὺν τῇ χειρὶ καὶ τῷ πήχει· καὶ νάρθηκας μέχρι δακτύλων τιθέναι· ἐν νάρθηξι δὲ τεθέντα ταῦτα πυκνότερον λύειν ἢ τὰ κατήγματα, καὶ καταχύσει πλέονι χρῆσθαι.

18. Ἐκ γενεῆς δέ, βραχυτέρη ἡ χεὶρ γίνεται, καὶ ἡ μινύθησις σαρκῶν μάλιστα τἀναντία ἢ ὡς τὸ ἔκπτωμα· ηὐξημένῳ δὲ ὀστέα μένει. |

360 19. Δακτύλου δὲ ἄρθρον ὀλισθὸν μὲν εὔσημον· οὐ δεῖ γράφειν.[14] ἐμβολὴ δέ· κατατείναντα ἐς ἰθὺ τὸ μὲν ἐξέχον ἀπωθέειν, τὸ δὲ ἐναντίον ἀντωθέειν. ἴησις, ἢ σταιτὶ ἢ ὀθονίοισι.[15] μὴ ἐμπεσὸν δὲ ἐπιπωροῦται ἔξω-

[14] οὐ δεῖ γράφειν M: del. Kühlewein Withington

[15] ἢ σταιτὶ ἢ ὀθονίοισι Kühlewein: ἧς γε τι ἡ ὀθονίοισι M

patient's fingers on a table, have one assistant extend the joint and another make counterextension, while the operator with his palm or heel presses the projecting part away both forward and downward, having placed some thick and soft material under the other bone. The hand should be prone if the dislocation is upward, and supine if it is downward. Treatment is with bandages.

17. The entire hand can dislocate either inward or outward, or laterally toward one side or the other, but mainly inward. So sometimes the (sc. inferior radioulnar) epiphysis is displaced, and sometimes one of the two bones separates. In these cases, forceful extension must be made, and the projecting bone pressed back, while the other bone is forced in the opposite direction—the two movements, backward and on an angle, being made simultaneously on a table, with your hands or a heel. These injuries give rise to serious consequences and cause deformity, but with time the joint becomes strong and regains its function. Treatment is with bandages around the hand and forearm, and also the application of splints reaching to the fingers. While the splints are attached to the limb, loosen them more frequently than you would for fractures, and employ more copious affusions.

18. When the dislocation is congenital, the hand becomes shortened, and there is atrophy of the muscles, especially on the side opposite to the displacement. But in an adult the bones are unaltered.

19. Dislocation of a finger joint is easy to recognize, and there is no need to describe it. Reduction: while holding the joint extended in a straight line, press the projecting part back, and make counterpressure on the opposite side; treatment is with either a tape or bandages. If the joint is

θεν. ἐκ γενεῆς δὲ ἢ ἐν αὐξήσει ἐξαρθρήσαντα τὰ ὀστέα βραχύνεται κάτω τοῦ ὀλισθήματος· καὶ σάρκες μινύθουσι ἀντία μάλιστα ἢ ὡς τὸ ἔκπτωμα· ηὐξημένῳ δὲ τὰ ὀστέα μένει.

20. Μηροῦ ἄρθρον ἐκπίπτει τέσσαρας τρόπους· ἔσω πλεῖστα, ἔξω δεύτερον, τὰ δὲ ἄλλα ὁμοίως. σημεῖα· κοινὸν μὲν τὸ ἕτερον σκέλος· ἴδιον δὲ τοῦ μὲν ἔσω. παρὰ περίναιον ψαύεται ἡ κεφαλή· ξυγκάμπτουσι οὐχ ὁμοίως μακρότερον τὸ σκέλος, δοκέει δὲ καὶ πολύ, ἢν μὴ ἐς μέσον ἀμφότερα ἄγων παρατείνῃς· καὶ γὰρ οὖν ἔξω ὁ ποὺς καὶ τὸ γόνυ ῥέπει. ἢν μὲν οὖν ἐκ γενεῆς ἢ ἐν αὐξήσει ἐκπέσῃ, βραχύτερος ὁ μηρός, ἧσσον δὲ κνήμη, κατὰ λόγον δὲ ἆλλα· μινύθουσι δὲ σάρκες, μάλιστα δὲ ἔξω. οὗτοι κατοκνέουσιν ὀρθοῦσθαι, καὶ[16] εἱλέονται ἐπὶ τὸ ὑγιές· ἢν δὲ ἀναγκάζωνται, σκίπωνι ἢ ἑνὶ ἢ δυσὶν ὁδοιπορέουσι, τὸ δὲ σκέλος αἴρουσιν· ὅσῳ γὰρ μεῖον, τόσῳ ῥᾷον. ἢν δὲ ηὐξημένοισι, τὰ μὲν ὀστέα μένει, αἱ δὲ σάρκες μινύθουσι, ὡς προείρηται. ὁδοιπορέουσι δὲ περιφοράδην, ὡς βόες, ἐν δὲ τῷ κενεῶνι καμπύλοι, ἐπὶ τὸ ὑγιὲς
362 ἐξίσχιοι ἐόντες· | τὸ μὲν γὰρ ἀνάγκη ὑποβαίνειν ὡς ὀχέῃ, τὸ δὲ ἀποβαίνειν (οὐ γὰρ δύναται ὀχέειν), ὥσπερ οἱ ἐν ποδὶ ἕλκος ἔχοντες. κατὰ δὲ τὸ ὑγιές, πλάγιοι τῷ ξύλῳ τῷ σώματι ἀντικοντέουσι, τὸ δὲ σιναρὸν τῇ χειρὶ ὑπὲρ τοῦ γόνατος καταναγκάζουσιν ὡς ὀχέειν ἐν τῇ μεταβάσει τὸ σῶμα· ἰσχίῳ κατωτέρω

[16] Add. ἀλλ' M

not reduced, callus will be laid down on the outside. When the dislocation is congenital, or while the person is still growing, the bones below the dislocation become shortened, and the muscles atrophy, especially on the contralateral side. But in an adult the bones are unaltered.

20. The thigh joint becomes dislocated in four ways: most often, inward, second outward, and in the other two directions equally. Signs (as compared with the other leg) specific for inward dislocation are: the head (sc. of the thigh bone) is felt against the perineum; patients are unable to flex their thigh very well; this leg seems to be longer, and greatly so unless you draw the legs together at the midline, since the foot and the knee turn to the outside. Now if the dislocation is present from birth or occurs in adolescence, the thigh is shortened, the lower leg less so, and the rest in proportion; the muscles will atrophy, especially on the outer side. Such patients are unable to stand up straight, and hobble along on their healthy leg. If they are compelled to stand up, they walk with one or two crutches, and hold their (sc. affected) leg raised, which is the easier the shorter the leg is. If the patients are adults, their bones remain unaltered, but the muscles atrophy as has been described. They walk the way cattle do by swinging their legs, bent in at the flank, and bent at the hip on the healthy side, since they must bring the (sc. healthy) leg underneath to support their body, but hold the (sc. injured) one out of the way, since it is unable to walk, the way people with a wound on their foot do. On the healthy side they give support to the body with a piece of wood held at an angle, and hold the injured limb down with their hand above the knee in order to support their

ἢν χρῆται, κάτω τε ἧσσον μινύθει καὶ τὰ ὀστέα, μᾶλλον δὲ σάρκες.

21. Τοῦ δὲ ἔξω τἀναντία καὶ τὰ σημεῖα καὶ αἱ στάσιες· καὶ τὸ γόνυ καὶ ὁ ποὺς ἔσω ῥέπει βραχύ. τοῖσι δὲ ἐν αὐξήσει ἢ ἐκ γενεῆς παθοῦσιν οὐχ ὁμοίως
364 ξυναύξεται κατὰ τὸν αὐτὸν λόγον· | ἰσχίον ἀνωτέρω τινί, οὐχ ὁμοίως. οἷσι δὲ πυκινὰ ἐκπίπτει ἐς τὸ ἔξω ἄνευ φλεγμονῆς, ὑγροτέρῳ τῷ σκέλει χρῶνται, ὥσπερ ὁ μέγας τῆς χειρὸς δάκτυλος· μάλιστα δὲ οὗτος ἐκπίπτει φύσει· οἷς μὲν ἐκπίπτει μᾶλλον ἢ ἧσσον, καὶ οἷσιν ἐκπίπτει χαλεπώτερον ἢ ῥᾷον, καὶ οἷσιν ἐλπὶς θᾶσσον ἐμπεσεῖν, καὶ οἷσιν οὐκ ἀκὴ τούτου, καὶ οἷσι πολλάκις ἐκπίπτει, ἴησις τούτου. ἐκ γενεῆς δὲ ἢ ἐπ' αὐξήσει ἢ ἐν νούσῳ—μάλιστα γὰρ ἐκ νούσου—ἔστι μὲν οἷσιν ἐπισφακελίζει τὸ ὀστέον, ἀτὰρ καὶ οἷσι μή, πάσχει μὲν πάντα, ἧσσον δὲ ἢ τὸ ἔσω, ἢν χρηστῶς ἐπιμεληθῶσιν, ὥστε καὶ ὅλῳ βαίνοντας τῷ ποδὶ διαρρίπτειν· διὰ μελέτης πλείστης τοῖσι νηπιωτάτοισιν· ἐαθέντα κακοῦται, ἐπιμεληθέντα δὲ ὠφελεῖται· τοῖς ὅλοις, ἧσσον δέ τι, μινύθουσι.

22. Οἷσι δ' ἂν ἀμφότερα οὕτως ἐκπέσῃ, τῶν ὀστέων ταὐτὰ παθήματα· εὔσαρκοι δὲ πλὴν ἔσωθεν, ἐξεχέ-

body when they take a step. If the part below the hip is used, the bones atrophy less there, but the muscles more.

21. The signs and attitudes of outward (dislocation of the hip) are the opposite: both the knee and the foot turn slightly toward the inside. In adolescents and if present from birth, the limb does not grow in equal proportion to the other one; the hip bone is elevated to a certain degree, and does not develop equally. In cases of frequent dislocation toward the outside without inflammation, the leg tends to take on more fluid, just as the thumb does (the joint of the digits by nature most likely to dislocate). In some cases the dislocation is more or less complete, and in others it is more or less difficult; in others you can expect it to be reduced more promptly, whereas in others there is no hope of cure. For patients who have the dislocation often, treatment is available. In cases where the dislocation has existed since birth or arisen in adolescence, or where it has occurred subsequent to a disease—for disease is the principal cause—there is sometimes necrosis of the bone; but even when this complication is not present, such patients suffer all the same things, although less severely than those with dislocation inward, if they are properly cared for, so that in walking they can place the whole foot on the ground and balance on it. Most treatment must be afforded to the youngest patients: if neglected they will deteriorate, but if looked after they will improve. A degree of atrophy develops in the whole limb, but somewhat less than with dislocation toward the inside.

22. In cases where both thighs are dislocated (sc. outward) in this way, both bones suffer in the same way. Such patients retain their muscles in a good state, except on the inner side, their buttocks are prominent, and their thighs

γλουτοι, ῥοικοὶ μηροί, ἢν μὴ ἐπισφακελίσῃ. εἰ κυφοὶ τὰ ἄνωθεν ἰσχίων γένοιντο, ὑγιηροὶ μέν, ἀναυξέες δὲ τὸ σῶμα, πλὴν κεφαλῆς.

23. Οἷσι δὲ ὄπισθεν, σημεῖα· ἔμπροσθεν λαπαρώτερον, ὄπισθεν ἐξέχον, ποὺς ὀρθός· ξυγκάμπτειν οὐ δύνανται, εἰ μὴ μετ᾽ ὀδύνης, ἐκτείνειν ἥκιστα· τούτοισι σκέλος βραχύτερον. ἀτὰρ οὐδ᾽ ἐκτανύειν δύνανται κατ᾽ ἰγνύην ἢ[17] κατὰ βουβῶνα, ἢν μὴ πάνυ αἴρωσιν, οὐδὲ ξυγκάμπτειν. ἡγεῖται ἐν τοῖσι πλείστοισι τὸ ἄνω
366 ἄρθρον τὸ | πρῶτον· κοινὸν τοῦτο ἄρθροισι, νεύροισι, μυσίν, ἐντέροισιν, ὑστέρῃσιν, ἄλλοισιν· τούτοις τοῦ ἰσχίου τὸ ὀστέον καταφέρεται ἐς τὸν γλουτόν· διὰ τοῦτο βραχύ, καὶ ὅτι ἐκτείνειν οὐ δύνανται. σάρκες παντὸς τοῦ σκέλεος ἐν πᾶσι μινύθουσιν· ἐφ᾽ οἷσι δὲ μάλιστα, καὶ ᾗ, εἴρηται· τὰ ἔργα τὰ ἑωυτοῦ ἕκαστον τοῦ σώματος ἐργαζόμενον μὲν ἰσχύει, ἄργεον δὲ κακοῦται, πλὴν κόπου, πυρετοῦ, φλεγμονῆς. καὶ τὸ ἔξω, ὅτι ἐς σάρκα ὑπείκουσαν, βραχύτερον· τὸ δὲ ἔσω, ὅτι ἐπ᾽ ὀστέον προέχον, μακρότερον. ἢν μὲν οὖν ηὐξημένοισι μὴ ἐμπέσῃ, ἐπὶ βουβῶσι καμπύλοι ὁδοιποπορέουσι, καὶ ἡ ἑτέρη ἰγνύη κάμπτεται· στήθεσι μόγις καθικνέεται· χειρὶ τὸ σκέλος καταλαμβάνει, ἄνευ ξύλου, ἢν ἐθέλωσιν· ἢν μὲν γὰρ μακρότερον ᾖ, οὐ βήσεται· ἢν δὲ βαίνῃ, βραχύ. μινύθησις δὲ σαρκῶν, οἷσι πόνοι, καὶ ἡ ἴξις ἔμπροσθεν, καὶ τῷ ὑγιεῖ κατὰ λόγον· οἷσι δὲ ἐκ γενεῆς ἢ αὐξομένοισι ἢ ὑπὸ νούσου

[17] ἢ Withington: μὴ M

are curved, unless necrosis develops. If their body above the hips becomes humpbacked, they may recover, but they fail to grow, except in the head.

23. Signs of dislocation (of the thigh) in the backward direction. The front of the body becomes hollower, and the back more prominent; the foot is held straight. Patients cannot flex their thigh without pain, nor extend it at all; the leg is shortened in such cases. They cannot extend the leg at either the knee or the groin, unless they lift it very far up, nor can they flex it. In most cases, it is the proximal joint that initiates functions, a trait shared by the joints, ligaments, muscles, intestines, uterus, etc. In these dislocations, the hip bone is carried down to the buttock, causing the limb to be shorter and also be impossible to extend. In all cases the muscles of the whole leg atrophy; in which cases particularly, and in what manner have been mentioned. Each part of the body that carries out its proper activities gains in strength, but lack of activity—except due to fatigue, fever, or inflammation—causes deterioration. External dislocation, because the movement is toward tissue that yields, makes the leg appear shortened, while in a dislocation to the inside, it seems to be lengthened due to the fact that the movement is against a projecting bone. Now if in adults the bone is not reduced, they will walk bending forward at the groins and holding their opposite knee flexed; they hardly reach the ground with the ball of the foot, and if they wish to walk without a crutch, they hold their leg down with a hand. The crutch for walking should be short, because if it is too long the patient will not use the foot. There is loss of flesh down the front in painful cases, and also in proportion on the healthy side. In congenital dislocations

ἐνόσησε καὶ ἔξαρθρα ἐγένετο—ἐν αἷς, εἰρήσεται—οὗτοι μάλιστα κακοῦνται διὰ τὴν τῶν νεύρων καὶ ἄρθρων ἀργίην· καὶ τὸ γόνυ διὰ τὰ εἰρημένα συγκακοῦνται. ξυγκεκαμμένον οὗτοι ἔχοντες ὁδοιπορέουσιν ἐπὶ ξύλου, ἑνὸς ἢ δύο· τὸ δὲ ὑγιές, εὔσαρκον διὰ χρῆσιν.

24. Οἷσι δὲ ἐς τοὔμπροσθεν, σημεῖα τἀναντία· ὄπισθεν λαπαρόν, ἔμπροσθεν ἐξέχον· ἥκιστα συγκάμπτουσιν οὗτοι τὸ σκέλος, μάλιστα δὲ ἐκτείνουσι· ὀρθὸς πούς, σκέλος ἴσον, πτέρνα· βραχεῖ ἄκρως ἀνέσταλται. ᾗ πονέουσι μάλιστα οὗτοι αὐτίκα, καὶ οὗρον
368 ἴσχεται μάλιστα | ἐν τούτοισι τοῖσιν ἐξαρθρήμασιν· ἐν γὰρ τόνοις ἔγκειται τοῖσιν ἐπικαίροισιν. τὰ ἔμπροσθεν κατατέταται—ἀναυξέα, νοσώδεα, ταχύγηρα[18]—τὰ ὄπισθεν στολιδώδεις· οἷσιν ηὐξημένοισιν, ὁδοιπορέουσι ὀρθοί, πτέρνῃ μᾶλλον βαίνοντες· εἰ δὲ ἠδύναντο μέγα προβαίνειν, κἂν πάνυ· σύρουσι δέ. μινύθει δὲ ἥκιστα, τούτοισι δὲ ἡ χρῆσις αἰτία· μάλιστα δὲ ὄπισθεν. διὰ παντὸς τοῦ σκέλεος, ὀρθότεροι τοῦ μετρίου, ξύλου δέονται κατὰ τὸ σιναρόν. οἷσι δὲ ἐκ γενεῆς ἢ αὐξομένοισι, χρηστῶς μὲν ἐπιμεληθεῖσιν ἡ χρῆσις, ὥσπερ τοῖσιν ηὐξημένοισιν· ἀμεληθεῖσι δὲ βραχύ, ἐκτεταμένον· πωροῦται γὰρ τούτοισι, μάλιστα δὲ ἐς ἰθὺ τὰ ἄρθρα· αἱ δὲ τῶν ὀστέων μειώσιες καὶ αἱ τῶν σαρκῶν μινυθήσιες κατὰ λόγον.

25. Μηροῦ δὲ κατάτασις μὲν ἰσχυρή· καὶ ἡ διόρ-

[18] ἀναυξέα . . . ταχύγηρα M: del. Withington

and those in adolescents, or when the dislocation follows a disease—which diseases will be explained later—there is particular harm due to inactivity of the ligaments and the joints, and the knee too shares in this deterioration from the causes named. Such patients walk with the joint flexed, on one or two crutches, but the healthy limb remains strong due to its continued use.

24. In cases of dislocation (of the thigh) in the forward direction, the signs are reversed: the back of the body becomes hollower, and the front more prominent. Such patients can hardly flex the leg at all, but they are usually able to extend it; the foot is held straight, the leg is the same length as the other one as far as the heel, and at the end the foot is turned slightly up. Now with this dislocation pain is generally felt at once, and the urine is generally retained due to pressure (sc. of the head of the femur) on critical (sc. nervous) cords. The parts in front are stretched [fail to grow, are diseased, and age rapidly], while the parts at the back are wrinkled. Adults with this dislocation walk erect, but put more weight on their heel, and if they could take long strides they would lean completely on it; but they drag the leg. There is little atrophy in such cases, because of continued exercise, and what there is appears mainly in the back part of the leg. Because the whole leg is straighter than normal, a crutch is needed on the injured side. In congenital and adolescent cases, if exercise is well managed they will fare as adults do, but in those neglected, the leg will be shortened and held extended—ankylosis occurs in these cases with the leg in an extended position. Shortening of the bones and atrophy of the muscles are proportionate.

25. For treatment of dislocations of the thigh, strong extension is necessary, and adjustment is made in every

θωσις κοινή, ἢ χερσὶν ἢ σανίδι ἢ μοχλῷ, τὰ μὲν ἔσω στρογγύλῳ, τὰ δὲ ἔξω πλατεῖ, μάλιστα δὲ τὰ ἔξω. καὶ τὰ μὲν ἔσω ἀσκοῖσιν ἀκεσάμενον ἐς τὸ ὑπόξηρον τοῦ μηροῦ, κατατάσιος δὲ καὶ ξυνδέσιος σκελέων· κρεμάσαι διαλείποντα σμικρὸν τοὺς πόδας, ἔπειτα πλέξαντα ἐκκρεμασθῆναί τινα, ἐν τῇ διορθώσει ἀμφότερα ἅμα ποιεῦντα. καὶ τῷ ἔμπροσθεν τοῦτο ἱκανὸν καὶ τοῖσιν ἑτέροις, ἥκιστα δὲ τῷ ἔξω. ἡ τοῦ ξύλου ὑπόστασις, ὥσπερ ὤμῳ, ὑπὸ τὴν χεῖρα, οἷς ἔσω· τοῖσι γὰρ ἄλλοισιν ἧσσον· καταναγκάσεις δὲ μετὰ διατάσιος, μάλιστα τῶν ἔμπροσθεν ἢ ὄπισθεν, ἢ ποδὶ ἢ χειρὶ ἐφίζεσθαι ἢ τῇ σανίδι. |

370 26. Γόνυ δὲ εὐηθέστερον ἀγκῶνος διὰ τὴν εὐσταλίαν καὶ εὐφυΐαν, διὸ καὶ ἐκπίπτει καὶ ἐμπίπτει ῥᾷον. ἐκπίπτει δὲ πλειστάκις ἔσω, ἀτὰρ καὶ ἔξω καὶ ὄπισθεν. ἐμβολαὶ δέ· ἢ ἐκ τοῦ ξυγκεκάμφθαι, ἢ ἐκλακτίσαι ὀξέως, ἢ ξυνελίξας ταινίης ὄγκον, ἐν ἰγνύῃ θείς, ἀμφὶ τοῦτον ἐξαίφνης ἐς ὄκλασιν ἀφεῖναι τὸ σῶμα—μάλιστα αὐτὴ τῶν ὄπισθε. δύναται δὲ καὶ κατατεινόμενα μετρίως, ὥσπερ ἀγκών, ἐμπίπτειν τὰ ὄπισθεν· τὰ δὲ ἔνθα ἢ ἔνθα, ἐκ τοῦ ξυγκεκάμφθαι ἢ ἐκλακτίσαι κατατάσει—μάλιστα δὲ αὐτὴ τὸ ὄπισθεν.

case with the hands, a board, or a lever—rounded for inward dislocations, and flat for outward ones. It is outward dislocations that require extension most; for internal dislocations there is a treatment with wineskins applied to the thigh at the point where it becomes narrow, which involves holding it extended and binding the two legs together. You can also suspend the patient with his feet slightly parted, and then have an assistant hang down from his body by winding around it; the two acts of the adjustment should be simultaneous (i.e., extension and leverage outward). This suffices for dislocations toward the front and otherwise, but least of all for those toward the outside. The method with a piece of wood placed under the thigh, as under the arm in dislocations of the shoulder, is suitable for internal dislocations, but less so for the others. Especially for forward and backward dislocations, you must combine pressure downward with extension, by using a foot and arm, or a plank to exert force from above.

26. A dislocation of the knee is easier to reduce than one of the elbow, because of its simple, regular form, which allows it to be dislocated and reduced more easily. Most often the knee is dislocated inward, but sometimes also outward or backward. Methods of reduction: either by flexion of the knee on itself, or by a sharp kick outward, or by placing a mass of rolled-up bandage in the patient's ham, onto which he suddenly crouches down bringing the full weight of his body to bear—generally the reduction of backward dislocations. It is also possible by moderate extension to reduce dislocations (sc. of the knee) toward the back, the way this works for the elbow. Dislocations to one side or the other are treated by flexion, by kicking outward—especially dislocation backward—but also by mod-

ἀτὰρ καὶ ἐκ κατατάσιος μετρίης, ἡ διόρθωσις ἅπασι κοινή. ἢν δὲ μὴ ἐμπέσῃ, τοῖσι μὲν ὄπισθεν ξυγκάμπτειν οὐ δύνανται, ἀτὰρ οὐδὲ τοῖσιν ἄλλοισι πάνυ τι. μινύθει δὲ μηροῦ καὶ κνήμης [ὀστέον][19] τὸ ἔμπροσθεν. ἢν δὲ ἐς τὸ ἔσω, βλαισότεροι, μινύθει δὲ τὰ ἔξω· ἢν δὲ ἐς τὸ ἔξω, γαυσότεροι, χωλοὶ δὲ ἧσσον· κατὰ γὰρ τὸ παχύτερον ὀστέον ὀχέει· μινύθει δὲ τὰ ἔσω. ἐκ γενεῆς δὲ ἢ ἐν αὐξήσει, κατὰ λόγον τὸν προσθεν.

27. Τὰ δὲ κατὰ τὰ σφυρὰ κατατάσιος ἰσχυρῆς δεῖται τῇσι χερσί, κατορθώσιος δὲ ἅμα ἀμφότερα ποιεύσης· κοινὸν δὲ πᾶσιν. |

372 28. Τὰ δὲ ἐν ποδί, ὡς τὰ ἐν χειρί, ὑγιῆ.

29. Τὰ δὲ ἐν τῇ κνήμῃ συγκοινωνέοντα καὶ μὴ ἐμπεσόντα, ἐκ γενεῆς καὶ ἐν αὐξήσει ἐξαρθρήσαντα, ταὐτὰ ἃ καὶ ἐν χειρί.

30. Ὅσοι δὲ πηδήσαντες ἄνωθεν ἐστηρίξαντο τῇ πτέρνῃ, ὥστε διαστῆναι τὰ ὀστέα καὶ φλέβας ἐκχυμωθῆναι καὶ νεῦρα ἀμφιφλασθῆναι, ὅταν γένηται οἷα τὰ δεινότατα, κίνδυνος μὲν σφακελίσαντα τὸν αἰῶνα πρήγματα παρασχεῖν· ῥοιώδη μὲν γὰρ τὰ ὀστέα, τὰ δὲ νεῦρα ἀλλήλοισι κοινωνέοντα. ἐπεὶ καὶ οἷσιν ἂν κατεαγεῖσιν ἢ ὑπὸ τρώματος, οἷα ἐν κνήμῃ, ἢ μηρῷ, νεύρων ἀπολυθέντων ἃ κοινωνέει τούτοισιν, ἢ ἐξ ἄλλης κατακλίσιος ἀμελέος ἐμελάνθη ἡ πτέρνη, καὶ τούτοισι παλίγκοτα ἐκ τοιούτων. ἔστιν ὅτε πρὸς σφακελισμῷ γίνονται πυρετοὶ ὑπερόξεες, λυγγώδεες, τρο-

[19] ὀστέον M: del. Littré

erate extension. Adjustment is the same for all these. If the bone does not fall back into place, patients with a dislocation toward the back are unable to flex the knee, but in the other dislocations this is hardly possible either; the front of the thigh and the lower leg suffer atrophy. If dislocation is inward, patients are more knock-kneed, and there is atrophy of muscles of the outer side; if outward, they are more bowlegged, but less lame, since their weight rests on the larger bone (tibia); the inner side atrophies. Cases that are congenital or occur during adolescence follow the rule given above.

27. Dislocations at the ankle require forceful extension with the hands and adjustment, applying both in combination. This is common to all.

28. Dislocations in the foot, like those in the hand, heal the same way.

29. The bones in the lower leg, when dislocated from birth or in adolescence, follow the same course as those of the hand.

30. Patients that have fallen from a height and landed on their heel so as to break the bones apart will have ecchymosis from the vessels and contusion of the ligaments: when such serious injuries are sustained, there is a danger that necrosis will result, along with a chronic disability, since the bones are unstable and the ligaments are interconnected. When there is also a fracture, or a wound in the calf or the thigh, or the ligaments holding the bones are torn away, or from heedless positioning in bed the heel has become necrotic, these things will cause such patients to deteriorate. Sometimes with the necrosis come very acute fevers accompanied by hiccups and trembling,

μώδεες, γνώμης ἁπτόμενοι, ταχυθάνατοι, καὶ ἔτι φλεβῶν αἱμορρόων πελιώσιες καὶ γαγγραινώσιες. σημεῖα τῶν παλιγκοτησάντων· ἢν τὰ ἐκχυμώματα καὶ τὰ μελάσματα καὶ τὰ περὶ ταῦτα ὑπόσκληρα καὶ ὑπέρυθρα ᾖ· ἢν γὰρ ξὺν σκληρύσματι πελιωθῇ, κίνδυνος μελανθῆναι· ἢν δὲ ὑποπέλια ᾖ, καὶ πελιὰ μάλα καὶ κεχυμένα, ἢ ὑπόχλωρα καὶ μαλθακά, ταῦτα ἐν πᾶσι τοῖσι τοιούτοισιν ἀγαθά. ἴησις· ἢν μὲν ἀπύρετοι ἔωσιν, ἑλλεβορίζειν· ἢν δὲ μή, μή· ἀλλὰ ποτὸν ὀξύγλυκυ, εἰ δέοι. ἐπίδεσις δὲ ἡ ἄρθρων· ἔτι δὲ πάντα
374 μᾶλλον τοῖσι φλάσμασι· καὶ | ὀθονίοισι πλέοσι καὶ μαλθακωτέροισι χρῆσθαι· πίεξις ἧσσον· ὕδωρ πλέον· προσπεριβάλλειν τὰ πλεῖστα πτέρνῃ· τὸ σχῆμα ὅπερ ἡ ἐπίδεσις, ὡς μὴ ἐς τὴν πτέρνην ἀποπιέζηται· ἀνωτέρω γούνατος ἔστω εὔθετος· νάρθηξι μὴ χρῆσθαι.

31. Ὅταν δὲ ἐκστῇ ὁ πούς, ἢ αὐτὸς ἢ σὺν τῇ ἐπιφύσει, ἐκπίπτει μᾶλλον ἐς τὸ ἔσω· ἢν δὲ μὴ ἐμπέσῃ, λεπτύνεται ἀνὰ χρόνον ἰσχίου καὶ μηροῦ καὶ κνήμης τὸ ἀντίον τοῦ ὀλισθήματος. ἐμβολή, ὡς καρποῦ, κατάτασις δὲ ἰσχυροτέρη. ἴησις, νόμος ἄρθρων· παλιγκοτεῖ ἧσσον καρποῦ, ἢν ἡσυχάσῃ. δίαιτα μείων, ἐλινύουσι γάρ. τὰ δὲ ἐκ γενεῆς μὲν ἢ ἐν αὐξήσει, κατὰ λόγον τῶν πρότερον.

32. Ἐπεὶ τὰ[20] σμικρὸν ὀλισθηκότα ἐκ γενεῆς, ἔνια οἷά τε διορθοῦσθαι. μάλιστα δὲ ποδὸς κύλλωσις· κυλλώσιος γὰρ οὐχ εἷς τρόπος. ἡ δὲ ἴησις τούτου, κηρο-

[20] ἐπεὶ τὰ Littré: ἔπειτα M

which cause disturbance of the mind and are quickly fatal; also, lividity from bleeding vessels and gangrenes. Signs of deterioration are ecchymoses and dark bruises with a degree of induration and redness in the surrounding parts; for if with the indurations lividity develops, there is a danger of mortification; but if the part is only slightly livid, or even very livid after there has been ecchymosis, or it is greenish in color and soft, these are favorable signs in all such cases. Treatment: if patients remain free of fever, give them a dose of hellebore, but not otherwise, in which case a drink of oxymel, if it is necessary. Bandaging as it is used for joints, and suited in particular to contusions, *viz.* plenty of soft bandages; pressure quite restrained, more water, additional bandages in particular around the heel. Position chosen as it was for bandaging, to avoid pressure on the heel, elevate (sc. the foot) above the knee, and position it properly. Do not employ splints.

31. In cases where the foot is dislocated, either by itself or together with the epiphysis (tibiofibular), it usually moves toward the inside. If it is not reduced, over time there will be atrophy of the hip bone, thigh and calf on the side opposite to the dislocation. Reduction is otherwise as with the wrist, but requires more forceful extension; treatment follows the rule for joints. Relapse does occur, but less often than with the wrist if the patient remains at rest. Diet should be reduced, since the patient remains at rest. In congenital and adolescent cases follow the plan given above.

32. When the bones (sc. of the foot) are slightly dislocated from birth, some can be corrected, in particular a clubfoot, of which there is more than one variety. This is treated by molding as one would a model; apply a wax

πλαστέειν κηρωτῇ ῥητινώδει, ὀθόνια συχνά, ἢ πέλμα ἢ μολύβδιον προσεπιδέειν, μὴ χρωτί· ἀνάληψις, τά τε σχήματα ὁμολογείτω.

33. *Ἢν δὲ ἐξαρθρήσαντα ἕλκος ποιησάμενα ἐξ-
ίσχῃ, ἐώμενα ἀμείνω, ὥστε δὴ μὴ ἀπαιωρέεσθαι μηδ'
ἀπαναγκάζεσθαι. ἴησις δέ· πισσηρῇ ἢ σπλήνεσιν
οἰνηροῖσι θερμοῖσιν—ἅπασι γὰρ τούτοισι τὸ ψυχρὸν
κακόν—καὶ φύλλοισιν· χειμῶνος δέ, εἰρίοισι ῥερυπω-
μένοισι τῆς σκέπης εἵνεκα· μὴ καταπλάσσειν, μηδ'
ἐπιδεῖν· δίαιτα λεπτή· ψῦχος, ἄχθος πολύ, πίεξις,
ἀνάγκη, σχήματος τάξις· εἰδέναι μὲν οὖν ὀλέθρια
ταῦτα πάντα. μετρίως δὲ θεραπευθέντες, χωλοὶ αἰ-
σχρῶς· ἢν γὰρ παρὰ πόδας γένηται, ποὺς ἀνασπᾶται,
καὶ ἤν πῃ ἄλλῃ, κατὰ λόγον. ὀστέα οὐ μάλα ἀφίστα-*
376 *ται· μικρὰ γὰρ ψιλοῦται, | περιωτειλοῦται λεπτῶς.
τούτων τὰ μέγιστα κινδυνωδέστατα, καὶ τὰ ἀνωτάτω.
ἐλπὶς δὲ μούνη σωτηρίης, ἐὰν μὴ ἐμβάλῃ, πλὴν τὰ
κατὰ δακτύλους καὶ χεῖρα ἄκρην· ταῦτα δὲ προει-
πόντα*[21] *τοὺς κινδύνους. ἐγχειρέειν ἐμβάλλειν ἢ τῇ
πρώτῃ ἢ τῇ δευτέρῃ, εἰ δὲ μή, πρὸς τὰς δέκα· ἥκιστα
τεταρταῖα. ἐμβολή· οἱ μοχλίσκοι. ἴησις δέ, ὡς κε-
φαλῆς*[22] *ὀστέων, καὶ θερμή· ἐλλεβόρῳ δὲ καὶ αὐτίκα
καὶ ἔπειτα τοῖσιν ἐμβαλλομένοισι βέλτιον χρῆσθαι.
τὰ δ' ἄλλα εὖ εἰδέναι δεῖ ὅτι*[23] *ἐμβαλλομένων θάνατοι·
τὰ μέγιστα καὶ τὰ ἀνωτάτω μάλιστα καὶ τάχιστα.*

[21] -*ειπόντα* Kühlewein: *εἶπεν τὰ* M
[22] -*λῆς* Littré: *λὴ* M
[23] *δ. ὅ.* Littré: *διότι* M

plaster of cerate made with resin: bind this on with several bandages, or a sandal or a sheet of lead, but not against the skin. Support in the appropriate attitude with a sling.

33. If the dislocated bones (of the foot) make a wound and project, they are better left alone provided they are not left unsupported or exposed to violence. Treatment is with pitch cerate or compresses soaked in warm wine—since in all such cases cold is harmful—and leaves, and in winter with greasy wool as protection. Do not apply plasters or bandages, employ a reduced diet, and protect (sc. the wound) from cold, excessive weight, pressure, force, and rigid positioning, being aware that all these are detrimental. If properly treated, patients will survive, but be left with an unsightly lameness: for if the deformity is near the foot, the foot will be drawn up, and if elsewhere there will be a corresponding deformity. The bones themselves are not very likely to break off, since they are not extensively denuded, and a thin scar forms. The largest and most proximal joints cause the greatest danger. The only hope of safety is not to reduce dislocations of this type, except the fingers and the bones of the hand; the surgeon should explain the risks beforehand. Undertake the reduction on the first or second day, if not then, about the tenth—but least of all on the fourth. Reduction is performed with small levers. Treatment is as for the bones of the head, using warmth. It is also advisable to give a dose of hellebore immediately after the operation and later. But for other bones (sc. than of the fingers and hand) you must appreciate that reduction means death, and the larger and more proximal the joints, the more certainly and the sooner. An open dislocation of the foot

πους δὲ ἐκβάς, σπασμός, γάγγραινα· καὶ γὰρ ἢν ἐμβληθέντι ἐπιγένηταί τι τούτων, ἐκβάλλοντι ἐλπίς, εἴ τις ἄρα ἐλπίς· οὐ γὰρ ἀπὸ τῶν χαλώντων οἱ σπασμοί, ἀλλ' ἀπὸ τῶν ἐντεινόντων.

34. Αἱ δὲ ἀποκοπαὶ ἢ ἐν ἄρθρῳ ἢ κατ' ὀστέα, μὴ ἄνω, ἀλλ' ἢ παρὰ ποδὶ ἢ χειρί, ἐγγὺς περιγίνονται, ἢν μὴ αὐτίκα μάλα λειποθυμίῃ ἀπόλλωνται. ἴησις, δ' ὡς κεφαλῆς, θερμή.[24]

35. Ἀποσφακελίξιος μέντοι σαρκῶν, καὶ ἐν τρώμασιν αἱμορρόοις ἀποσφιγχθέν, καὶ ἐν ὀστέων κατήγμασι πιεχθέν, καὶ ἐν δεσμοῖς ἀπομελανθέν. καὶ οἷσι μηροῦ μέρος ἀποπίπτει καὶ βραχίονος, ὀστέα τε καὶ σάρκες ἀποπίπτουσι, πολλοὶ περιγίνονται, ὡς τά γε ἄλλα εὐφορώτερα· οἷσι μὲν οὖν καταγέντων ὀστέων, αἱ μὲν περιρρήξιες ταχεῖαι, αἱ δὲ τῶν ὀστέων ἀποπτώσιες, ᾗ ἂν τὰ ὅρια τῆς ψιλώσιος ᾖ, ταύτῃ ἀπο-
378 πίπτουσι, βραδύτερον δέ. χρὴ δὲ τὰ κατωτέρω | τοῦ τρώματος καὶ τοῦ σώματος τοῦ ὑγιέος προσαφαιρέειν—προθνήσκει γάρ—φυλασσόμενον· ὀδύνῃ ἅμα γὰρ λειποθυμίῃ θνῄσκουσι. μηροῦ ὀστέον ἐκ τοιούτου ἀπελύθη ὀγδοηκοσταῖον, ἡ δὲ κνήμη ἀφῃρέθη εἰκοσταίη· κνήμης δὲ ὀστέα κατὰ μέσην ἑξηκοσταῖα ἀπελύθη. ἐκ τοιούτων ταχὺ καὶ βραδέως, αἱ πιέξιες αἱ ἰητρικαί. τὰ δ' ἄλλα ὅσα ἡσυχαίως, τὰ μὲν ὀστέα οὐκ ἀποπίπτει οὐδὲ σαρκῶν ψιλοῦται, ἀλλ' ἐπιπολαιότερον. προσδέχεσθαι ταῦτα χρή· τὰ γὰρ πλεῖστα φο-

[24] θερμή Littré: θερμῆς M

often entails spasms and gangrene: if these signs appear on reduction, there may be a hope from re-dislocation, if indeed there is hope at all, for spasms are not the result of relaxation of the parts, but of their tension.

34. Amputations at a joint or in the length of bones—not proximally, but toward the foot or hand—usually recover, unless the patient suddenly loses consciousness and dies. Treatment is as for the bones of the head, with warmth.

35. Necrosis of the tissues is caused by constriction in bleeding wounds, compression in cases of fractured bones, and mortification from bandages. Even in cases where part of the thigh and arm fall away, and bones and muscles slough off, many patients survive, and in other respects the loss is well tolerated. Now in the cases of fractured bones, the lines of demarcation soon appear, although the actual bone loss—it is at the border of the denudation that this occurs—comes later. You must carefully remove what is lying beneath both the lesion and the sound part of the body, for these are the first tissues to mortify; such patients die from pain and syncope combined. In one case the femur separated on the eightieth day, when the lower leg had been lost around the twentieth day; bones of the leg separated in the middle around the sixtieth day. In such cases, the compression applied during treatment influences whether the sloughing will be quick or slow. In other milder cases the bones do not fall away, nor are the tissues abraded, with the necrosis remaining more superficial: such cases should be taken on, since they are in fact usually more terrifying than danger-

βερώτερα ἢ κακίω. ἡ ἴησις πρηεῖα, θερμῇ διαίτῃ ἀκριβεῖ· κίνδυνος αἱμοῤῥαγιῶν, ψύχεος· σχήματά τε ὡς μὲν ἀνάρροπα, ἔπειτα ὑποστάσιος πύου εἵνεκα ἐξ ἴσου ἢ ὅσα ξυμφέρει. ἐπὶ τοῖσι τοιούτοισι καὶ ἐπὶ τοῖσι μελασμοῖσιν, αἱμοῤῥαγίαι, δυσεντερίαι, περὶ κρίσιν, λαῦροι μέν, ὀλιγήμεροι δέ. οὐκ ἀπόσιτοι οὐδὲ πάνυ πυρετώδεες, οὐδέ τι κενεαγγητέον.

36. *῞Υβωσις, ἡ μὲν ἔσω ἐπιθάνατος, οὔρων σχέσιος, ἀποναρκώσιος· τὰ δὲ ἔξω, τούτων ἀσινέα τὰ πλεῖστα, πολὺ μᾶλλον ἢ ὅσα σεισθέντα μὴ ἐξέστη. αὐτὰ μὲν ἑωυτοῖσι κρίσιν ποιησάμενα, κεῖνα δὲ ἐπὶ πλέον τῷ σώματι ἐπιδιδόντα, καὶ ἐν ἐπικαίροις ἐόντα.*

Οἷον *πλευραὶ κατεαγεῖσαι μέν, ὀλίγαι πυρετώδεες καὶ αἵματος πτύσιος καὶ σφακελισμοῦ, ἤν τε μία, ἤν τε πλείους μὴ καταγεῖσαι δέ· καὶ ἴησις φαύλη, μὴ κενεαγγοῦντα, ἢν ἀπύρετος εἴη. ἐπίδεσις ὡς νόμος· ἡ δὲ πώρωσις ἐν εἴκοσιν ἡμέραις, χαῦνον γάρ. ἢν δ᾽ ἀμφιφλασθῇ, φυματίαι, καὶ βηχώδεες, καὶ ἔμμοτοι, |*
380 *καὶ πλευρὰς ἐσφακέλισαν· παρὰ γὰρ πλευρὴν ἑκάστην ἀπὸ πάντων τόνοι εἰσίν.*

37. Τὰ δὲ *ἀπὸ καταπτώσιος ἧσσον δύναται ἐξιθύνεσθαι· χαλεπώτερα δὲ τὰ ἄνω φρενῶν ἐξιθύνεσθαι. οἷσι δὲ παισίν, οὐ συναύξεται, ἀλλ᾽ ἢ σκέλη καὶ χεῖρες καὶ κεφαλή· ηὐξημένοισιν ὕβωσις, παραχρῆμα μὲν τῆς νούσου ῥύεται, ἀνὰ χρόνον δ᾽ ἐπισημαίνεται δι᾽ ὧνπερ καὶ τοῖσι νεωτέροισιν, ἧσσον δὲ κακοήθως. εἰσὶ δὲ οἳ εὐφόρως ἤνεγκαν, οἷσιν εὔσαρκον καὶ πιμελῶδες τράπηται· ὀλίγοι δὲ τούτων περὶ ἑξήκοντα*

ous. Management should be gentle, with warmth and a strict regimen: the dangers are hemorrhages and cold. The limb should be held in a more elevated position, but later level or however best favors the drainage of pus. These cases, as well as those with mortification, are followed by hemorrhages, and around the crisis dysenteries, which are copious but last for only a few days. Patients do not lose their appetite, become febrile, or need to be fasted.

36. Spinal curvature: inward curvature is lethal, involving the retention of urine and the loss of sensation. Outward curvature, in most cases, causes little harm, much less than spinal concussion without displacement: the curvatures bring about their own crisis, whereas the concussions affect the body more seriously and involve its vital parts.

Similarly, fractured ribs—whether one or more—rarely cause fever, hemoptysis, or necrosis unless the fractured rib pierces the inward parts. Treatment is easy and does not require fasting as long as there is no fever. Bandaging should be according to normal practice, and consolidation follows in twenty days due to the bone's sponginess. But if there is severe contusion, it will be followed by tubercles and coughs, wounds that require the application of bandages, and necrosis in the ribs, since (sc. nervous) cords lead to each rib from all the parts.

37. Curvatures resulting from a fall are less treatable, and those above the diaphragm are more difficult to straighten. In children there is a cessation of growth, except of the legs, arms, and head. Curvature in adults relieves a disease at the moment, but with time the same signs as in children recur, although less severely. Some patients bear up well, tending toward much flesh and ro-

382 ἔτεα ἐβίωσαν. ἀτὰρ καὶ ἐς | τὰ πλάγια διαστρέμματα γίνεται· συναίτια δὲ καὶ τὰ σχήματα ἐν οἷσιν ἂν κατακέωνται· καὶ ἔχει προγνώσιας.

Πολλοὶ δὲ καὶ αἷμα ἔπτυσαν καὶ ἔμπυοι ἐγένοντο. ἡ δὲ μελέτη, ἴησις, ἐπίδεσις ὡς νόμος· διαίτης τὰ πρῶτα ἀτρεκέως, ἔπειτα ἁπαλύνειν· ἡσυχίῃ, σιγῇ· σχήματα, κοιλίῃ, ἀφροδίσια. ἀτὰρ οἷς ἄναιμα, ἐπωδυνώτερα[25] τῶν καταγνυμένων καὶ φιλυποστροφώτερα χρόνοισιν· οἷσι δὲ καταλείπεται μυξῶδες, ὑπομιμνήσκει ἐν πόνοις. ἴησις· καῦσις, τοῖσι μὲν ἀπ᾽ ὀστέου, μέχρι τοῦ ὀστέου, μὴ αὐτὸ δέ· ἢν δὲ μεταξύ, μὴ πέρην, μηδὲ ἐπιπολῆς· σφακελισμός. καὶ τὰ ἔμμοτα πειρᾶσθαι· εἰρήσεται ἅπαντα τὰ ἐσιόντα. ὁρατά, λόγοις δὲ μή· βρώματα, πόματα, θάλπος, ψῦχος, σχῆμα· ὅτι καὶ φάρμακα, τὰ μὲν ξηρά, τὰ δὲ ὑγρά, τὰ δὲ πυρρά, τὰ δὲ μέλανα, τὰ δὲ λευκά, τὰ δὲ στρυφνά, ἐπὶ ἕλκη, οὕτω καὶ δίαιται.

38. Νόμος ἐμβολῆς καὶ διορθώσιος· ὄνος,[26] μοχλός, σφηνίσκος, ἶπος·[27] ὄνος μὲν ἀνάγειν, μοχλὸς δὲ παράγειν. τὰ δὲ ἐμβλητέα ἢ διορθωτέα διαναγκάσαι δεῖ ἐκτείνοντα, ἐν ᾧ ἂν ἕκαστα σχήματι μέλλῃ ὑπεραιωρηθήσεσθαι· τὸ δ᾽ ἐκβάν, ὑπὲρ τούτου ὅθεν ἐξέβη. τοῦτο δέ, ἢ χερσὶν ἢ κρεμασμῷ ἢ ὄνοις ἢ περί τι. χερσὶ μὲν οὖν ὀρθῶς κατὰ μέρεα· καρπὸν δὲ καὶ

[25] ἄναιμα, ἐπωδυνώτερα Littré: ἂν αἷμα ἀνωδύνωι τέρα- M
[26] ὄνος Littré: νόμος M
[27] ἶπος Littré: ὕπος M

bustness, but few of these live to sixty years. There are also cases of lateral distortion (sc. of the spine), and the positions in which such patients lie are accessory causes; these also have prognostic significance.

Many of these patients also cough up blood and suppurate internally: management, treatment and bandaging follow the normal rules. The first phase of the regimen should be strict, but then restorative: employ quiet, silence, correct positioning, and attention to the bowels and to sexual activity. In cases where there is no sign of blood, however, the injuries will be more painful than in cases of fractures of the ribs, and more likely to relapse with time. If a collection of mucus is left behind, patients are likely to suffer a return of their pains. Treatment: where bone is involved, cauterize down toward the bone, but not the bone itself; where the disorder is between the bones, neither through the bones nor remain just on the surface. Necrosis: also try linen pads—complete details about this will be given in my text. Things must be actually seen, not just talked about: foods, drinks, warmth, cold, positioning. As for medications, these include the dry, moist, flame-colored, dark, white, and astringent that are directed against lesions. The same with regimens.

38. Practice of reduction and adjustment: windlass, lever, wedge, press; the windlass for stretching, the lever for bringing into place. Parts that need to be reduced or adjusted must be separated through extension, until each part is in sufficiently elevated position, with the part which has been dislocated above the one from which it has separated. This is done with the hands, or by suspension, or with a windlass, or around some object. Use of the hands must be adjusted to the different parts. The wrist and el-

ἀγκῶνα ἀπόχρη διαναγκάζειν, καρπὸν μὲν ἐς ἰθὺ ἀγκῶνος, ἀγκῶνα δὲ ἐγγώνιον πρὸς βραχίονα ἔχοντα, οἷον παρὰ τῷ βραχίονι τὸ ὑπὸ τὴν χεῖρα ὑποτεινόμενον. ἐν οἷσι δὲ δακτύλου, ποδός, χειρός, καρποῦ, ὑβώματος τὸ ἔξω, διαναγκάσαι δεῖ καὶ καταναγκάσαι, τὰ μὲν ἄλλα ὑπὸ χειρῶν αἱ διαναγκάσεις ἱκαναί, καταναγκάσαι δὲ τὰ ὑπερέχοντα ἐς ἕδρην πτέρνῃ ἢ θέναρι ἐπί τινος· ὥστε κατὰ μὲν τὸ ἐξέχον ὑποκέεσθαι ὄγκον σύμμετρον μαλθακόν· κατὰ δὲ τὸ ἕτερον μήστωρα ἅμα[28] χρὴ ὠθέειν ὀπίσω καὶ κάτω, ἤν τε ἔσω ἤν τε ἔξω ἐκπεπτώκῃ· τὰ δὲ ἐκ πλαγίων, τὰ μὲν ἀπωθέειν, τὰ δὲ ἀντωθέειν ὀπίσω ἀμφότερα κατὰ τὸ ἕτερον. τὰ
384 δὲ ὑβώματα, τὰ μὲν | ἔσω, οὔτε πταρμῷ οὔτε βηχί, οὔτε φύσης ἐνέσει, οὔτε σικύῃ· δεῖ δέ τι, ἡ κατάστασις· ἡ δὲ ἀπάτη, ὅτι οἷόν τέ ποτε κατεαγέντων τῶν σπονδύλων καὶ τὰ λορδώματα διὰ τὴν ὀδύνην δοκέει ἔσω ὀλισθηκέναι· ταῦτα δὲ ταχυφυᾶ καὶ ῥᾴδια. τὰ δὲ ἔξω, κατάτασις, τὰ μὲν ἄνω ἐπὶ πόδας, τὰ δὲ κάτω τἀναντία· κατανάγκασις δὲ σὺν κατατάσει, ἢ ἕδρῃ ἢ ποδὶ ἢ σανίδι. τὰ δ᾽ ἔνθα ἢ ἔνθα, εἴ τις κατάτασις, καὶ ἔτι τὰ σχήματα ἐν τῇ διαίτῃ.

Τὰ ἄρμενα πάντα εἶναι πλατέα, προσηνέα, ἰσχυρά, εἰ δέῃ· μὴ δεῖ ῥάκεσι προκατειλίχθαι. ἐσκευάσθαι

[28] μήστωρα ἅμα Kühlewein: μήστωρ δ᾽ ἂν M

---

[3] This is Adams' interpretation: Withington, "The reduced bone shall be (kept) as far as possible from the place where it was dislocated."

bow can be sufficiently separated by hand power, by holding the wrist in a straight line with the elbow, but the elbow at right angles to the upper arm—positioned like the forearm is when held in a sling. For dislocations of the toes, fingers or wrist, or a kyphosis of the spine, double extension and downward pressure on the projecting bone are required. In other cases separation by hand power is sufficient, but projecting parts must be forced back into their normal position with a heel or a palm of the hand over something, being careful that a suitable soft pad is placed under the projection. On the other side, a skilled assistant must simultaneously press backward and downward, if the dislocation is toward either the inside or the outside. If the dislocation is lateral, one part is pressed away and the other side back against it, so that the two parts are approximated to one another. In cases of spinal curvature, those toward the inside are not (sc. correctable) by applying sneezing, coughing, inflation (sc. of air into the intestine), or a cupping instrument; in fact, the remedy is extension.[3] This misunderstanding arises—just as when vertebrae are fractured—because the lordotic position assumed due to pain appears to be the result of a dislocation toward the inside. Such conditions heal quickly and are not serious. For curvatures toward the outside: extension directed, if the disorder is high up, toward the feet, if it is low down, in the opposite direction: while simultaneously causing extension, apply pressure to the spot by sitting on it, or using a foot or a plank. If there is lateral curvature toward one side or the other, combine some degree of extension with positioning, according to the regimen.

All surgical apparatus should be wide, adjustable, and sufficiently strong and they should not need first to be

πρὶν ἢ ἐν τῇσιν ἀνάγκῃσι πάντα συμμεμετρημένως τὰ μήκεα καὶ ὕψεα καὶ εὔρεα. διάτασις, οἷον μηροῦ· τὸ παρὰ σφυρὸν δεδέσθαι καὶ ἄνω τοῦ γούνατος, ταῦτα μὲν ἐς τὸ αὐτὸ τείνοντα· παρὰ δὲ ἰξὺν καὶ περὶ μασχάλας, καὶ κατὰ περίναιον καὶ μηροῦ τὰ μεταξύ, τῆς ἀρχῆς τὸ μὲν ἐπὶ τὸ στῆθος, τὸ δὲ ἐπὶ νῶτον τείνοντα, ταῦτα δ' ἐς τὸ ὑπὸ ἅπαντα τείνοντα, προσδεθέντα ἢ πρὸς ὑπεροειδέα ἢ πρὸς ὄνον. ἐπὶ μὲν οὖν κλίνης ποιέοντι, τοῦτο μὲν τῶν ποδῶν πρὸς οὐδὸν χρὴ
386 *ἐρεῖσαι, πρὸς δὲ τὸ ἕτερον, ξύλον | ἰσχυρὸν πλάγιον παραβεβλῆσθαι, τὰ δὲ ὕπερθεν ὑπεροειδέα πρὸς ταῦτα ἀντιστηρίζοντα διατείνειν, ἢ πλήμνας κατορύξαντα, ἢ κλίμακα διαθέντα, ἀμφοτέρωθεν ὠθέειν. τὸ δὲ κοινόν, σανὶς ἑξάπηχυς, εὖρος δίπηχυς, πάχος σπιθαμῆς, ἔχουσα ὄνους δύο ταπεινοὺς ἔνθεν καὶ ἔνθεν, ἔχουσα δὲ καὶ μέσον στυλίσκους ξυμμέτρους, ἐξ ὧν ὡς κλιμακτὴρ ἐπέσται ἐς τὴν ὑπόστασιν τῷ ξύλῳ, ὥσπερ τῷ κατ' ὦμον· καταγλύφους δὲ ὥσπερ ληνοὺς λείας ἔχειν, τετραδακτύλους εὖρος καὶ βάθος, καὶ διαλιπεῖν τοσοῦτον ὅσον αὐτῇ τῇ μοχλεύσει ἐς διόρθωσιν· ἐν μέσῳ δὲ τετράγωνον καταγλυφὴν ὥστε στυλίσκον ἐνεῖναι, ὃς παρὰ περίναιον ἐὼν περιρρέπειν τε κωλύσει ἐών τε ὑποχάλαρος ὑπομοχλεύσει. χρὴ δὲ τῆς σανίδος, ἢ ἐν τῷ τοίχῳ τὸ ἄκρον, καταγεγλυμμένον τι ἐχούσης, τοῦ ξύλου ὦσαι τὸ ἄκρον, ἐπὶ*

---

[4] This passage is condensed to the point of inconstruability. The translation is based on the description of the process in *Joints* 47 (pp. 295–97, above); see Withington, p. 442, n. 1.

wrapped in rags. Prepare them before they are brought to bear by properly coordinating all their lengths, heights, and breadths. In double extension, as applied for example to the thigh, attach the bands at the ankle and above the knee exerting their force in the same direction. (sc. For counterextension) the bands are attached at the waist and under the axillae, and another one at the perineum drawn between the thighs: draw one end to the chest and the other to the back, and then bring them all together and tie them together to a piece of wood shaped like a pestle or to a windlass. If operating on a patient in bed, its legs at one end should be pressed up against the threshold, while a strong piece of wood should be laid across at the other end. Then, use either of these as fulcra to draw on the pestle-shaped piece of wood from above; or, burying wheel naves in the ground or setting a ladder horizontally, press from both directions. Routinely the plank used is six cubits long, two cubits wide, and a span thick, and has two windlasses set low down at each end; in the middle there are suitable props on which something like the rung of a ladder can act as a fulcrum for the wood, the way it is employed for the shoulder joint. The plank should have grooves in the form of smooth troughs four fingers broad and deep, leaving enough space between them to carry out the adjustment by levering. In the center make a square indention for a prop to fit into, which when set against the perineum will prevent the patient from slipping, or if left slightly loose will serve as a lever. As to the use of the plank, the maneuver consists of, by pushing on one end, which is set in either a hollowed-out space in a post, or in the wall,[4] to exert force on the opposite end, δὲ θάτερα καταναγκάζειν, ὑποτιθέντα μαλθακά σύμμετρα.

39. Οἷσιν ὀστέον ἀπὸ ὑπερῴης ἀπῆλθε, μέση ἵζει ἡ ῥὶς τούτοις. οἱ δὲ θλώμενοι κεφαλὰς ἄνευ ἕλκεος, ἢ πεσόντος ἢ κατάξαντος ἢ πιέσαντος, τούτων ἐνίοισι τὰ δριμέα ἔρχεται ἀπὸ κεφαλῆς κατὰ τὰς φάρυγγας, καὶ ἀπὸ τρώματος ἐν τῇ κεφαλῇ ἐς τὸ ἧπαρ καὶ ἐς τὸν μηρόν.

40. Σημεῖα παραλλαγμάτων καὶ ἐκπτωμάτων· καὶ
388 ᾗ καὶ ὅπως | καὶ ὅσον διαφέρει ταῦτα πρὸς ἄλληλα· καὶ οἷσι κοτύλη παρέαγε, καὶ οἷσι νευρίον ἀπεσπάσθη, καὶ οἷσιν ἐπίφυσις ἀπέαγε, καὶ οἷσι καὶ ὡς, καὶ ἓν ἢ δύο, ὧν δύο ἐστίν· ἐπὶ τούτοισι κίνδυνοι, ἐλπίδες οἷσι κακαί, καὶ ὅτε κακώσιες θανάτου, ὑγιείης, ἀσφαλείης. καὶ ἃ ἐμβλητέα ἢ χειριστέα καὶ ὅτε, καὶ ἃ οὒ ἢ ὅτε οὔ· ἐπὶ τούτοισιν ἐλπίδες, κίνδυνοι· οἷα καὶ ὅτε χειριστέα, καὶ τὰ ἐκ γενεῆς ἔξαρθρα, τὰ αὐξανόμενα, τὰ ηὐξημένα, καὶ ὅ τι θᾶσσον, καὶ ὅ τι βραδύτερον, καὶ ὅ τι χωλόν, καὶ ὡς καὶ οὔ· καὶ διότι καὶ ὅ τι μινυθήσει, καὶ ᾗ καὶ ὡς καὶ οἷσιν ἧσσον καὶ ὅτι τὰ καταγέντα θᾶσσον καὶ βραδύτερον φυόμενα, ᾗ αἱ διαστροφαὶ καὶ ἐπιπωρώσιες γίνονται, καὶ ἀκὴ τούτων. οἷσιν ἕλκεα αὐτίκα ἢ ὕστερον γίνονται· οἷσι καὶ ὀστέα καταγεῖσι μείω, οἷσιν οὔ· οἷσι καταγέντα ἐξέσχεν, καὶ ᾗ ἐξίσχει μᾶλλον· οἷσιν ἐκβάντα ἢ ἄρθρα
390 ἐξίσχια· ἀπατῶνται | καὶ δι᾽ ἅ, ἐν οἷσιν ὁρῶσιν, ἐν οἷσιν διανοεῦνται, ἀμφὶ τὰ παθήματα, ἀμφὶ τὰ θεραπεύματα.

under which some appropriate soft material has been placed.

39. In cases where a piece of bone comes away from the roof of the mouth, the nose collapses in the middle. Patients with contusions in the head (but no wound) which are the result of falling, or a fracture, or a compression, sometimes have a flow of acrid material from the head into the throat, and from the wound in the head to both the liver and a thigh.

40. Signs of subluxations and dislocations, and how they differ in location, kind, and degree; and also in which cases the socket is fractured, a small ligament is torn, or the epiphysis is detached. And when two bones are present together, if and how one or both of them are broken; in these cases the dangers and bad outcomes that are to be expected, and when there is a threat of death, or a sure hope of recovery. Which cases can be adjusted, or at least attempted, and when, and which cases cannot, and when; in these what are the hopes and dangers. In what cases and when treatment should be undertaken for injuries present from birth, or in adolescents or in adults; which heal promptly and which more slowly, and which bring lameness and which do not and how. Why and in which cases there will be atrophy, of which part and how, and in which cases it will be less; and that broken bones will consolidate more quickly or slowly; where distortions and ankyloses arise, and the cures of these. Cases where lesions arise immediately or later; where fractured bones are shortened, and where they are not; in which fractures bones will project outside, and where they project most. The confusion between dislocations and prominent joints, and the causes of this in observers observing and considering maladies and treatments.

41. Νόμοισι τοῖσι νομίμοισι περὶ ἐπιδέσιος· παρασκευή, πάρεξις, κατάτασις, διόρθωσις, ἀνάτριψις, ἐπίδεσις, ἀνάληψις, θέσις, σχῆμα, χρόνοι, δίαιται. τὰ χαυνότατα τάχιστα φύεται, τὰ δὲ ἐναντία, ἐναντίως· διαστροφαί, ᾗ κυρτοί· ἄσαρκοι, ἄνευροι. τὸ ἐμπεσὸν ὡς προσωτάτω † ἢ τὸ ἐκπεσὸν †[29] ἔσται τοῦ χωρίου ᾗ ἐξέπεσεν. νεύρων, τὰ μὲν ἐν κινήσει καὶ ἐν πλάδῳ, ἐπιδοτικά· τὰ δὲ μή, ἧσσον· ἄριστον ᾗ ἂν ἐκπέσῃ, εἰ[30] ἐμπέσοι τάχιστα· πυρεταίνοντι μὴ ἐμβάλλειν, μηδὲ τεταρταῖα, πεμπταῖα, ἥκιστα ἀγκῶνα. καὶ τὰ ναρκώδεα πάντα, ὡς τάχιστα ἄριστα, <ἢ>[31] τὴν φλεγμονὴν παρέντα. τὰ ἀποσπώμενα, ἢ νεῦρα ἢ χόνδρια ἢ ἐπιφύσιες, ἢ διϊστάμενα κατὰ συμφύσεις, ἀδύνατα ὁμοιωθῆναι· διαπωροῦται ταχέως τοῖσι πλείστοισιν· ἡ δὲ χρῆσις σώζεται. ἐκβάντων, τὰ ἔσχατα, ῥᾷον· τὰ ῥᾷστα ἐκπεσόντα ἥκιστα φλεγμαίνει· τὰ δὲ ἥκιστα θερμαίνοντα, καὶ μὴ ἐπιθεραπευθέντα, μάλιστα αὖτις ἐκπίπτει. κατατείνειν ἐν σχήματι τοιούτῳ, ἐν ᾧ μάλιστα ὑπεραιωρηθήσεται, σκεπτόμενον ἐς τὴν φύσιν καὶ τὸν τόπον ᾗ ἐξέβη. διόρθωσις· ὀπίσω ἐς ὀρθὸν
392 καὶ ἐς πλάγιον παρωθεῖν· τὰ[32] δὲ ταχέως | ἀντισπάσαντα ἀντισπάσαι ταχέως ᾗ δεῖ[33] ἐκ περιαγωγῆς· τὰ πλειστάκις ἐκπίπτοντα ῥᾷον ἐμπίπτει· αἴτιον νεῦσις ἢ

[29] ἢ τὸ ἐκπεσόν M: del. Kühlewein
[30] εἰ Foes in note: ἢ M
[31] Add. Petrequin
[32] π. τ. Littré: παρωσθέντα M
[33] ᾗ δεῖ Kühlewein: ἡ δεῖ M: ἢ δὴ Aldina

41. Regular rules of bandaging: preparation, presentation, extension, adjustment, friction, bandaging, suspension, positioning, posture, times, regimens. The spongiest bones consolidate fastest, the opposite kind the opposite. Distortions where there is kyphosis; atrophy of flesh and ligaments. Force used in reduction will be applied at as great a distance as possible (sc. from the dislocation).[5] Ligaments in mobile and moist parts tend to be slack, while those elsewhere less so. It is best, wherever a dislocation occurs, if it is reduced as quickly as possible; in a person with a fever, do not make a reduction on the fourth or fifth day, and least of all of the elbow. All cases with loss of sensation are best reduced as soon as possible, or wait until the inflammation has subsided. Parts pulled apart—whether ligaments, cartilages or epiphyses, or separations at symphyses—cannot return to normal, and in most cases there is soon an ankylosis, although function is preserved. Of dislocations, those in the extremities are the easiest; and those which dislocate easily have the least inflammation. Where there is the least warmth and no after-treatment, the chance of redislocation is greatest. Extend the bones in such a position that the dislocated bone is raised over the (sc. socket), taking into account its conformation and the place from which the separation has occurred. Adjustment: apply force backward either straight or laterally; when there has been a rapid twist, make a rapid twist back in the required direction by applying circumduction. Joints that have been dislocated many times are easier to reduce, due to the disposition of the ligaments and bones:

[5] It is unclear whether "easily put out" or "easily treated" is meant here.

νεύρων ἢ ὀστέων. νεύρων μὲν μῆκος ἢ ἐπίδοσις·[34] ὀστέων δέ, κοτύλης ὁμαλότης, κεφαλῆς φαλακρότης· τὸ ἔθος τρίβον ποιέει· αἰτίη καὶ σχέσις καὶ ἕξις καὶ ἡλικίη. τὸ ὑπόμυξον ἀφλέγμαντον.

42. Οἷσιν ἕλκεα ἐγένετο, ἢ αὐτίκα ἢ ὀστέων ἐξισχόν-
των, ἢ ἔπειτα, ἢ κνησμῶν ἢ τρηχυσμῶν, ταῦτα μὲν
ἢν αἰσθῇ, εὐθέως λύσας, πισσηρὴν ἐπὶ τὸ ἕλκος ἐπι-
τιθείς, ἐπιδεῖν ὡς ἐπὶ τὸ ἕλκος πρῶτον τὴν ἀρχὴν
βαλλόμενος, καὶ τἆλλα ὡς οὐ ταύτῃ τοῦ σίνεος ἐόν-
τος· οὕτω γὰρ αὐτό τε ἰσχνότατον καὶ ἐκπυήσει[35] τά-
χιστα καὶ περιρρήξεται, καὶ καθαρθέντα τάχιστα
φύσεται. νάρθηκας δὲ μήτε κατ᾽ αὐτὸ τοῦτο προσά-
γειν μήτε πιέζειν· καὶ ὧν ὀστέα μὴ μεγάλα ἄπεισιν
ὡσαύτως, ὧν δὲ μεγάλα, δῆλον· πολλὴ γὰρ ἐμπύησις
καὶ ταῦτ᾽ οὐκ ἔτι οὕτως, ἀλλ᾽ ἀνέψυκται τῶν ὑποστα-
σίων εἵνεκα. τὰ δὲ τοιαῦτα ὁπόσα ἐξέσχε, καὶ εἴτε
ἐμβληθῇ εἴτε μή, ἐπίδεσις μὲν οὐκ ἐπιτήδειον, διάτα-
σις δέ. σφαῖραι ποιηθεῖσαι οἷαι πέδαι, ἡ μὲν παρὰ
σφυρόν, ἡ δὲ παρὰ γόνυ, ἐς κνήμην πλατεῖαι, προσηνέες, ἰσχυραί, κρίκους ἔχουσαι· ῥάβδοι δὲ σύμμετροι
καὶ μῆκος καὶ πάχος κρανίης, ὥστε διατείνειν· ἱμάν-
τια δὲ ἐξ ἄκρων ἀμφοτέρωθεν ἔχοντα εἰς τοὺς κρίκους
394 ἐνδεδέσθαι, ὡς τὰ | ἄκρα ἐς τὰς σφαίρας ἐνστηριζό-
μενα διαναγκάζῃ. ἴησις δέ, πισσηρὴν θερμήν· σχή-
ματα καὶ ποδὸς θέσις καὶ ἰσχίου· δίαιτα ἀτρεκής.

[34] μ. ἢ ἐ. Foes: μήκιος ἢ ἐπιδόσιος M
[35] ἐκπυήσει Littré: ἐκπέσει καὶ M

ligaments are favorable in their length and flexibility, and bones in the evenness of the socket or the roundness of the head. Use gives rise to a friction joint in accordance with the patient's constitution, condition, and age. A small amount of mucus does not engender inflammation.

42. In cases where lesions arise at once because of the projection of bones outside, or later due to either irritation or roughness, if you become aware of this, immediately remove the dressing and apply a pitch cerate to the wound. Then apply a bandage beginning first from on top of the lesion itself, but elsewhere as if no wound was present. This will cause the part to swell least, so that the pus will be generated and expelled as soon as possible, and on its being cleaned away, the lesion will heal up most rapidly. Splints should not be attached over the wound, and no pressure should be applied, as long as no larger pieces of bone break off. Where they do, it is clear (sc. what to do), since much suppuration occurs making the previous treatment unsuitable, rather the wound is left open on account of the depositions. Bones that project outside like this, whether they are expelled or not, are not suitable for bandaging, but need stretching. Rounds are made like shackles, one at the ankle and another at the knee, flat on the side facing the calf, soft and strong, and provided with rings; rods of cornel wood of appropriate length and thickness to keep the limb stretched are attached on both sides by straps at their ends to the rings so that they will provide traction. Treatment: heated pitch cerate; positioning and placement of the foot and hip bone; strict diet. Reduce the

ἐμβάλλειν τὰ ὀστέα τὰ ὑπερίσχοντα αὐθήμερα ἢ δευτεραῖα· τεταρταῖα ἢ πεμπταῖα, μή, ἀλλ' ἐπὴν ἰσχνὰ ᾖ. ἡ δὲ ἐμβολὴ τοῖσι μοχλικοῖσιν· ἢ τὸ ἐμβαλλόμενον τοῦ ὀστέου, ἢν μὴ ἔχῃ ἀποστήριξιν, ἀποπρῖσαι τῶν κωλυόντων· ἀτὰρ καὶ ὡς τὰ ψιλωθέντα ἀποπεσεῖται, καὶ βραχύτερα τὰ μέλεα.

43. Τὰ δὲ ἄρθρα, τὰ μὲν πλέον, τὰ δὲ μεῖον ὀλισθάνει· τὰ μεῖον ἐμβάλλειν ῥᾷον · τὰ δὲ ἐαθέντα μείζους τὰς κακώσιας καὶ ὀστέων καὶ νεύρων καὶ ἄρθρων καὶ σαρκῶν καὶ σχημάτων. μηρὸς καὶ βραχίων ὁμοιότατα ἐκπίπτουσιν.

projecting bones on the same day or the next one, not on the fourth or fifth day, but only later when the edema has gone down. The reduction is performed with small levers: if the fragment of bone requiring reduction does not offer any hold, saw off the piece that is in the way. Furthermore, the limb will become shorter by the length of denuded bone that falls away.

43. Joints are dislocated to greater or less degrees: if less, they are easier to reduce, if more they cause more injury to the bones, ligaments, joints, muscles, and postures. The thigh and upper arm are very similar in their dislocations.

INDEX